Concurrent Design of Products, Manufacturing Processes and Systems

AUTOMATION AND PRODUCTION SYSTEMS
Methodologies and Applications

A series edited by **Hamid R. Parsaei,** University of Louisville, Kentucky, USA

About the series editor

Hamid R. Parsaei, PhD, PE, is a professor of industrial engineering and director of the Manufacturing Research Group at the University of Louisville in Kentucky. He received his MS and PhD in industrial engineering from Western Michigan University and the University of Texas at Arlington, respectively. His teaching and research interests include aspects of engineering economy, robotics, concurrent engineering, computer integrated manufacturing systems and productivity analysis.

Volume 1 GROUP TECHNOLOGY AND CELLULAR MANUFACTURING: METHODOLOGIES AND APPLICATIONS
Edited by **Ali K. Kamrani** and **Rasaratnam Logendran**

Volume 2 QUANTITATIVE ANALYSIS: AN INTRODUCTION
Roy M. Chiulli

Volume 3 CONCURRENT DESIGN OF PRODUCTS, MANUFACTURING PROCESSES AND SYSTEMS
Edited by **Ben Wang**

This book is part of a series. The publisher will accept continuation orders which may be cancelled at any time and which provide for automatic billing and shipping of each title in the series upon publication. Please write for details.

Concurrent Design of Products, Manufacturing Processes and Systems

Edited by

Ben Wang

*Florida A&M University–
Florida State University
Tallahassee, USA*

Gordon and Breach Science Publishers

Australia • Canada • China • France • Germany • India • Japan • Luxembourg • Malaysia • The Netherlands • Russia • Singapore • Switzerland

Amsteldijk 166
1st Floor
1079 LH Amsterdam
The Netherlands

British Library Cataloguing in Publication Data

Concurrent design of products, manufacturing processes and
systems. – (Automation & production systems. Methodologies
& applications ; v. 3 – ISSN 1028-141X)
1. Design, Industrial – Data processing 2. Product management
3. Manufacturing processes – Automation 4. Computer-aided
design
I. Wang, Hsu-Pin
670.4´27

ISBN 90-5699-628-2

CONTENTS

SERIES PREFACE

Automation and Production Systems: Methodologies and Applications provides a scientific and practical base for researchers, practitioners and students involved in the design and implementation of advanced manufacturing systems. It features a balance between state-of-the-art research and usefully reported applications. Among the topics that will be presented as part of this series are: strategic management and application of advanced manufacturing systems; cellular manufacturing systems; group technology; design simulation and virtual manufacturing; advanced materials handling systems; quantitative analysis; intelligent feature-base process planning; advanced genetic algorithms; computer-aided process planning; simultaneous engineering; economic evaluation of advanced technologies; and concurrent design of products.

This series encompasses three categories of books: reference books that include material of an archival nature; graduate- and postgraduate-level texts, reference and tutorial books that could also be utilized by professional engineers; and monographs concerning newly developing trends in applications for manufacturing and automation engineering. The volumes are written for graduate students in industrial and manufacturing engineering, as well as researchers and practitioners interested in product and process design, manufacturing systems design integration and control, and engineering management and economics.

Hamid R. Parsaei

PREFACE

The economic success of any business organization depends on its ability to identify customer needs and to profitably develop and deliver products that meet those needs. While new product and process development has always been challenging, a number of trends have emerged in recent years increasing the stakes for manufacturing and service companies and their new development efforts. For instance, the globalization of markets has led to intense competition among a growing number of world-class competitors. Customers around the world have become increasingly sophisticated, knowledgeable and insistent on high quality in products and services at low price. Technological change continues to accelerate, and the growing breadth and depth of scientific and engineering knowledge is leading to powerful and cost-effective solutions to old and new problems. The sum effect of these trends is to make excellence in new product and process development an increasingly important requirement for corporate survival and prosperity.

Responsiveness is definitely a winning characteristic of any successful business entity as we approach the twenty-first century. Two aspects of this responsiveness are:

1. meeting market demand;
2. meeting market opportunity.

The entire manufacturing industry focused its attention and energy on reducing manufacturing cycle times in the 1980s to meet market demands. Flexible manufacturing, cellular manufacturing and just-in-time production, to name a few, are examples of the industry's response to the call for short manufacturing cycles. As a result, many companies were able to meet market demands without increasing their inventories.

A long product and process development cycle prohibits a company's ability to meet market opportunity and seriously hampers profitability. In the traditional product design paradigm, it would take as long as seven years to develop a new car. From the time a company *perceived* a market opportunity to the time it rolled out a product to *address* that opportunity, the market had already changed its preference.

After years of searching and researching, the industry began to recognize that a new business paradigm would be necessary in order to be competitive and successful in the global marketplace. Such a paradigm would tear down the barriers to effective communications and drastically increase a company's responsiveness to market opportunities. That new business paradigm is Concurrent Design of Products, Manufacturing Processes and Systems (CDPPS). Such a CDPPS paradigm will permit effective and efficient predictions of complete product development, manufacturing process design and customer satisfaction.

This book contains fourteen refereed chapters authored by renowned experts in the areas of product design and manufacturing processes, and systems modeling and simulation. Although these chapters are arranged in the sequence of product, process and system design, in a true concurrent design environment the developments of products, processes and systems occur in the same time frame. Important subjects covered include quality function deployment, rapid prototyping, process design and control, quality engineering, system modeling and simulation. The latest developments in concurrent design of products, manufacturing processes and systems are represented.

I would like to express my appreciation to series editor Dr. Hamid R. Parsaei for the opportunity to edit this book, and to editor Frank Cerra for his constant support. I am indebted to Deborah S. Doolittle and her talent for making the complicated task of coordinating fourteen authors simple, so the project always stayed on schedule.

CHAPTER 1

Concurrent Product/Process Development

SABAH U. RANDHAWA[a] and SHEIKH BURHANUDDIN[b]

[a]*Department of Industrial and Manufacturing Engineering, Oregon State University, Corvallis, Oregon 97331, USA*
[b]*Precision Interconnect, 16640 SW 72nd Avenue, Portland, Oregon 97224, USA*

1. INTRODUCTION

During the past decade, global manufacturing competition has increased significantly. Consequently, the manufacturing industry in the United States has been undergoing some fundamental changes, including a move to low-cost, high-quality systems and a shift in focus from large business customers to diffused commodity market for all size and type of customers. The result has been a shift from technology-oriented industry to customer-oriented industry.

Two of the more important elements in today's changing environment are increased product sophistication and variation. The half life of many products have decreased to the point that 50 percent of their sales occur within the first three years. To remain competitive, manufacturers must minimize total costs while being quick to develop and market new products. This involves integrating many diverse functional areas of an organization into a process of creating a better design when viewed across the entire product life cycle.

Today's products are complex, composed of integrated mechanical and electronic components. Industrial products can contain hundreds of parts requiring a variety of processing and assembly operations. A new automobile can take more than 5 years from initial specifications to production and a surface combat ship up to 10 years (Whitney *et al.*, 1988). The focus on customer-driven market, coupled with increased competition, requires fast updating of designs, flexibility in manufacturing systems, and responsiveness in production schedules. Design and fabrication of such complex products require knowledge in traditional and state-of-the art manufacturing processes, electronics, control, and materials as well as experience with concurrent engineering tools. Economic survival in this environment requires complete integration of all engineering and production functions, particularly design and manufacturing.

The focus of this chapter is concurrent product-process design and evaluation. When planning manufacturing systems there are numerous tradeoffs and decisions to be made. Well conceived manufacturing strategies provide the framework for making these decisions.

2. MANUFACTURING STRATEGIES

According to Berrington and Oblich (1995), there are three basic business strategies for competitive advantage: low cost, more value, and focus on a specific segment of industry achieving either a low cost or value competitive advantage (this is a hybrid of the first two strategies). The choice of a specific strategy depends on numerous factors, including markets, customers, competitors, technology, and the current state and potential capabilities of the business.

To decide among alternative strategies requires a planning process that:

- identifies the "to-be" state,
- assesses the "as-is" state,
- develops goals, such that if the goals are attained, a to-be state is attained,
- develops implementation alternatives to achieve the goals, and
- incorporates flexibility in the process to adapt to changes in to-be state.

Assessing the as-is state is an important aspect of planning. It identifies cost, volume, quality, and productivity issues that can make better products more efficiently. However, even more important is a vision for the to-be state. This shifts the focus from the current conditions to future desired conditions. If decision makers focus on current conditions, they jump immediately to fixing what they perceive to be broken rather than on creating what they want. This process forces identification and prioritization of issues crucial to the organization's continuing performance and generates the changes required to move the organization in the desired direction.

Once specific goals are defined and an alternative is selected to achieve the to-be state, the implementation plan should address the complete system development cycle beginning with a feasibility study and ending with post-implementation review. The steps involved in this planning process include:

- Define objectives that help attain the to-be state and are achievable.
- Perform requirements analysis to identify specific needs and identify promising alternatives for implementation.

- Establish functional specifications of the desired product.
- Perform a technology assessment to address technical feasibility of available technologies.
- Perform a cost-benefit analysis for implementing a technology. An important element of this analysis is the recognition and incorporation of factors, human and organizational, that are difficult to quantify in monetary or other easily measurable units.
- Address acquisition and implementation issues, including the physical, information, and human components of the system.
- Address post-implementation issues including quality and productivity monitoring and control, and cost-benefit tracking.

3. MANUFACTURING SYSTEMS

The fundamental purpose of every manufacturing system is to achieve the transformation of the input material into a specified part or assembly configuration. All other functions, including information flow, material handling, tooling, loading and unloading operations, quality control, and support systems such as part scheduling and maintenance, must be considered to be in support of the manufacturing process.

The need for integrated manufacturing systems has long been recognized. However, few truly integrated systems have been installed. Reasons for lack of integrated systems include (White, 1988):

- It is difficult to understand the complexity and interactions inherent in an integrated system.
- Traditionally, manufacturing systems have been divided into components (for example, functional units and cost centers). Managers are responsible for the performance of individual units. Thus, the focus is on units rather than the system as a whole. To overcome such organization barriers is not easy.
- Even though the value of integrated systems is widely recognized, the commonly accepted notion is that integration is expensive, risky, and complex. Managers prefer the insurance offered by existing systems compared to risk involved in a new organizational culture.
- Resources – monetary, human, and information – are often constrained, limiting investment in leading-edge, but not necessarily proven technologies.

An integrated manufacturing system contains numerous components. These include: product design, manufacturing processes, material handling systems, human and knowledge resources, and information and communication networks. Integration requires a sufficient understanding of the components to allow the development of reliable models for designing, optimizing, and controlling unit manufacturing costs.

The traditional product development process, still widely used in industry, involves disjoint design, manufacturing, and evaluation activities (Figure 1). A major drawback with this approach is that without any detailed assessment of how easily a product can be produced, identifying production problems and eliminating design changes resulting from limitations in available technology is not possible. The cost of

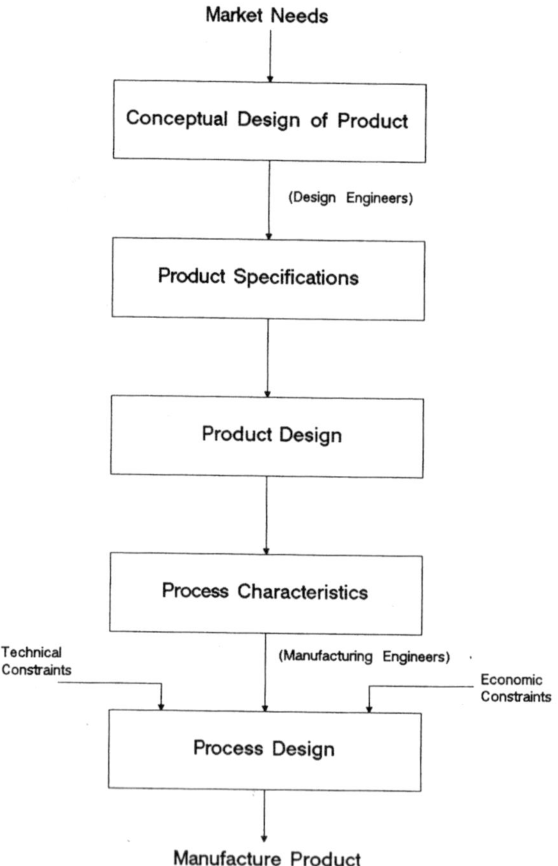

Fig. 1 Traditional product development cycle.

change increases as the product progresses from conception to design to production. The cost of an engineering change during final production could be 10,000 times the cost of change during design (Anderson, 1990). In the later stages of a product's life cycle, change costs are high because of time delays and larger number of personnel and manufacturing components involved. The design phase determines 80 percent of the cost of the product (Anderson, 1990) and deserves much more attention than it is currently receiving. Potential manufacturing and economic problems may be avoided if attention is paid to manufacturing and economic issues during the design process.

Concurrent engineering brings to the design phase consideration of many factors affecting total cost and performance throughout the life-cycle of a product. The complete design-manufacturing cycle is complex, involving market analysis, conceptual design, product design and development, material selection, process planning, production, information and process control, quality and process monitoring, and costing. Concurrent engineering involves integrating the diverse functions in an organization into a process of creating a better product when viewed across the entire product life cycle.

Concurrent engineering is not merely a procedure for automating design practices using computers; it is a process for evaluating product design while considering many life-cycle factors. The three basic components of concurrent engineering are (Terpenny and Deisenroth, 1992): design strategies for the design-manufacturing problem, decision aids to help in the selection of alternative design strategies, and supporting information and knowledge to accomplish the first two functions. A concurrent engineering system requires integration of these three components. Specific tools such as computer-aided design and manufacturing (CAD/CAM) must be integrated within this system. With the right combination of hardware and software, design, manufacturing and quality control functions can work in parallel to reduce lead times and tolerance-related problems.

Recently, much has been written about the importance of concurrent engineering and the "total design" concept (Anderson, 1990; Ettlie and Stoll, 1990; Hartley, 1992; Miller, 1993; Pugh, 1990). However, due to the many life cycle factors and the large amount of knowledge required by each, there is no all encompassing concurrent engineering system in use. Developers of concurrent engineering systems have used a variety of approaches to model components of the total system. El-Gizawy *et al.* (1993) have developed a concurrent engineering with quality approach for manufacturing industry using a parameter design technique for optimization during the product development cycle. Orady and Shareef (1993) have described a framework for simultaneous engineering which involves marketing, finance, design, manufacturing and quality control. Colton and Ouellette (1993) developed a form verification system for the conceptual design of complex mechanical systems. Several companies including Hewlett Packard, Sun Microsystems, Northern Telecom, and Raychem have reported successful implementation of concurrent engineering concepts in specific environments (Shina, 1994). However, much work still needs to be done to attain a fully integrated concurrent engineering environment.

4. DESIGN-MANUFACTURING INTEGRATION

The concurrent engineering process (Figure 2) is complex, involving numerous considerations such as market analysis, concept design, material and process selection, simulations to test design strategies and manufacturing feasibility, information accessibility and integration, costing, manufacturing, and process and quality control. Concurrent engineering is a methodology that uses multi-disciplined teams to focus on the complete design and manufacturing sequence. The essence of concurrent engineering is the integration of product design and production planning into one common activity. The objective is to avoid part features that are unnecessarily expensive to produce and make optimal choices of materials and processes. The result is improved quality of early design decisions and a significant impact on life-cycle cost of the product.

A large amount of information and knowledge is required to support the concurrent engineering process shown in Figure 2. This includes information on geometric and design features, tool and work materials, and process technology. Process, quality and reliability information must be continuously monitored in the manufacturing phase to ensure economic production and quality products. Decision

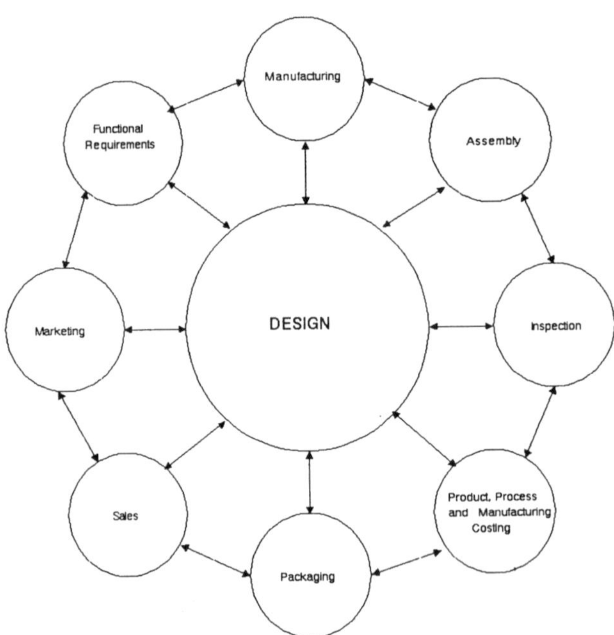

Fig. 2 Concurrent engineering design process.

aids that can be integrated in the evaluation phase include classical optimization models, simulation, multiattribute decision models, and expert systems. The objective is to have an information base that is accessible to all involved in the concurrent engineering process.

To maximize the benefits of concurrent engineering, increased computer-aided design and manufacturing (CAD/CAM) must be used. The use of CAD allows product designers and manufacturing engineers to see the real product, whether in the concept state or in the final design state. Once the basic design is complete, the manufacturing process can be simulated on the screen. By testing the manufacturing programs on the screen, errors can be identified and refinements made before the data are post-processed into the appropriate language for the CNC (computer numerical control) machines, the machines on the shop floor do not have to be tied up for testing, and waste (products not meeting specifications) can be reduced.

5. DESIGN-MANUFACTURING EVALUATION FRAMEWORK

A design-manufacturing evaluation framework is shown in Figure 3. The system consists of four primary components: needs assessment, product design, design-manufacturing evaluation, and product manufacturing. The process starts with a problem definition phase that defines the need requirements for a product. These requirements are then translated into product specifications. The specifications define the characteristics of the product that satisfy its functional requirements, and the conditions and constraints to be met during the transformation of suitable resources to the physical product.

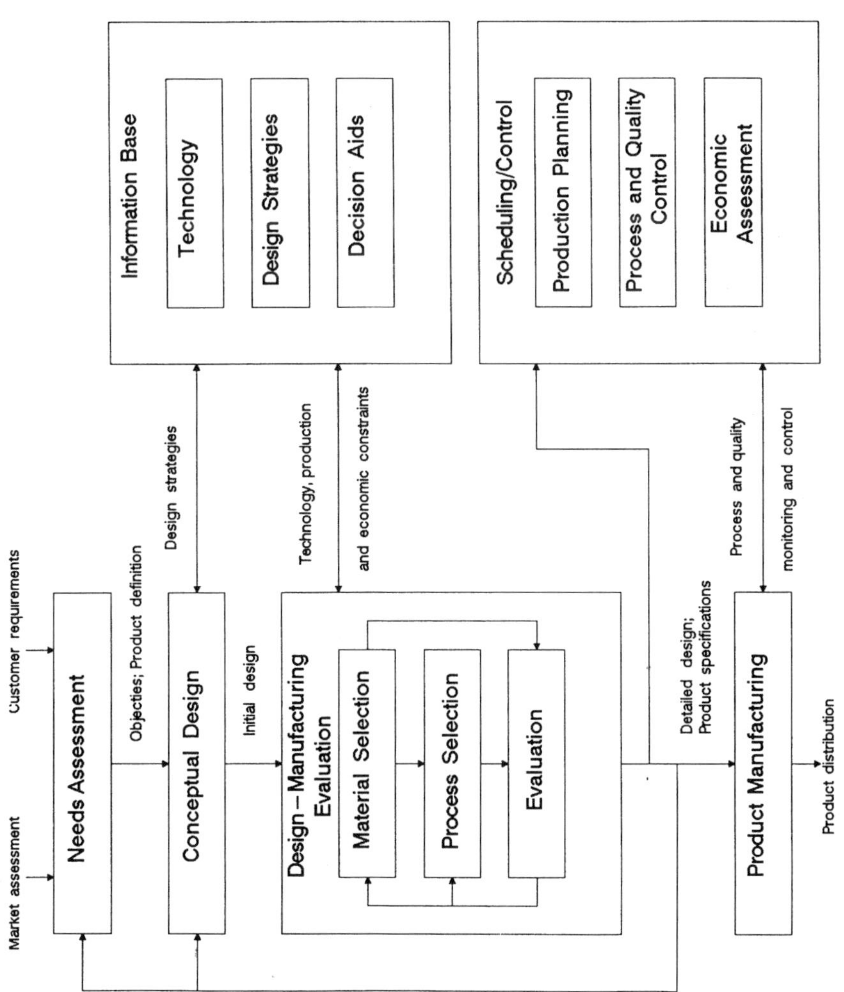

Fig. 3 Concurrent engineering design process.

In the evaluation phase, the design-manufacturing combination is evaluated against economic and technical criteria. Sensitivity analysis on design and production parameters help to identify the best production environment; in some instances it may indicate reconfiguration of product requirements and specifications. The design and product parameters identified at this stage are used in the physical manufacturing process.

5.1. Needs Analysis

The introduction of a new product is both rewarding and risky. It has been reported that approximately one in three new products fail and only one out of seven new product ideas reaches the commercialization phase. The failure rate is as high as 46 percent for new products in a new category (Association of National Advertisers, 1984). Substantial resources are spent on research and development, engineering, and test marketing of products that may or may not be successful. This means that successful products must not only return profit on investment but also pay for the resources spent on unsuccessful products.

Both internal and external pressures stimulate the development of new products. Reduced product life cycle, technical developments, inventions, competition, and customer requests are some of the external stimuli for new product development. Internal pressures for new product development include survival and growth of the company.

Some of the factors which contribute to product failure include poor product quality, insignificant point of difference with existing products in the market, target market being too small, and lack of understanding of customers' desires. Addressing these problems early during product development can reduce the risk of product failure. Techniques that improve chances of introducing a product successfully may be classified as modeling or non-modeling techniques.

- Process/Product Design Modeling: This concept involves evaluating materials, processes, and design alternatives on the basis of an evaluation criterion such as unit cost.
- Non-modeling Methods: Non-modeling methods used for assessing user needs and designing products include Concurrent Engineering Product Development Teams, Focus Groups, Brainstorming, Delphi-based Questionnaires and Surveys, and Quality Function Deployment (QFD).

Some of the non-modeling methods are discussed below. A product/process design modeling framework is discussed in a subsequent section.

5.1.1. Concurrent Engineering Product Development Teams

Product complexity has grown rapidly during the past few decades. A modern automobile containing electrical, electronic and mechanical components may have up to 50,000 parts. The design of a product may be accomplished by an individual designer or a group. Individual designers are effective in designing relatively simple products. Group effort is required for designing complex products requiring thousands of man-hours of design work. The group may be comprised of individuals who are all designers and have a similar role to play in the design process. This type of group is called an "engineering design team". In contrast to this is a concurrent design team where members have different background and play different roles. These teams are typically composed of personnel from design, manufacturing, marketing, purchasing and quality assurance as warranted by the nature of the product and the practices of the company. Concurrent design teams are very effective in recognizing and solving potential manufacturing, assembly, purchasing and quality problems early in the design process.

5.1.2. Focus Groups

A focus group is a carefully selected group of individuals who could be potential users of the product. Focus groups are a way of deriving knowledge about the product and the benchmarks from consumers. Concepts on new products may be presented to the group for evaluation. Exploratory focus groups generate many insights important to consumers. Focus groups explore usage patterns, buying processes, and opinions and attitudes helpful to the marketing effort. Focus groups are very useful in converting customer desires and needs into engineering requirements which in turn guide the design of the product. Focus groups at several stages of product development may be used to continuously steer the product development process.

5.1.3. Quality Function Deployment (QFD)

Quality function deployment (QFD) methodology helps convert customer requirements into technical description or engineering requirements. The design and functionality should not be driven by the perception of the engineer but rather by what the customer wants. By involving the customer, the QFD methodology helps improve customer satisfaction and efficiency of bringing new products to market. Customers involved in developing QFD may be external customers (end users, other companies) or internal customers (for example, manufacturing is a customer of design).

A typical QFD chart is shown in Figure 4. On the left are shown customer requirements which can be obtained using focus groups. Customer requirements are converted to engineering requirements, as shown at the top of Figure 4. The relationship between customer requirements and engineering requirements is shown through numerical values. The values in Figure 4 follow the scheme where 9 represents a strong relation, 3 represents a medium relation, 1 represents a weak relation, and blank is used for no relation. All engineering requirements must have units and target values. These are shown below the customer requirements in Figure 4. Engineering requirements which show strong relation to customer requirements must be given preference for inclusion in product design.

QFD chart also evaluates benchmarks and existing designs against customer requirements. In Figure 4, each benchmark is rated on a scale of 1 to 5 for each customer requirement, where 1 indicates that the design does not meet the requirement at all and 5 indicates complete satisfaction. Benchmarks are also scored against engineering requirements as shown at the bottom of the QFD chart.

The objective of using these techniques is to establish the true needs of the customers. Additional data generated during the needs analysis phase may include:

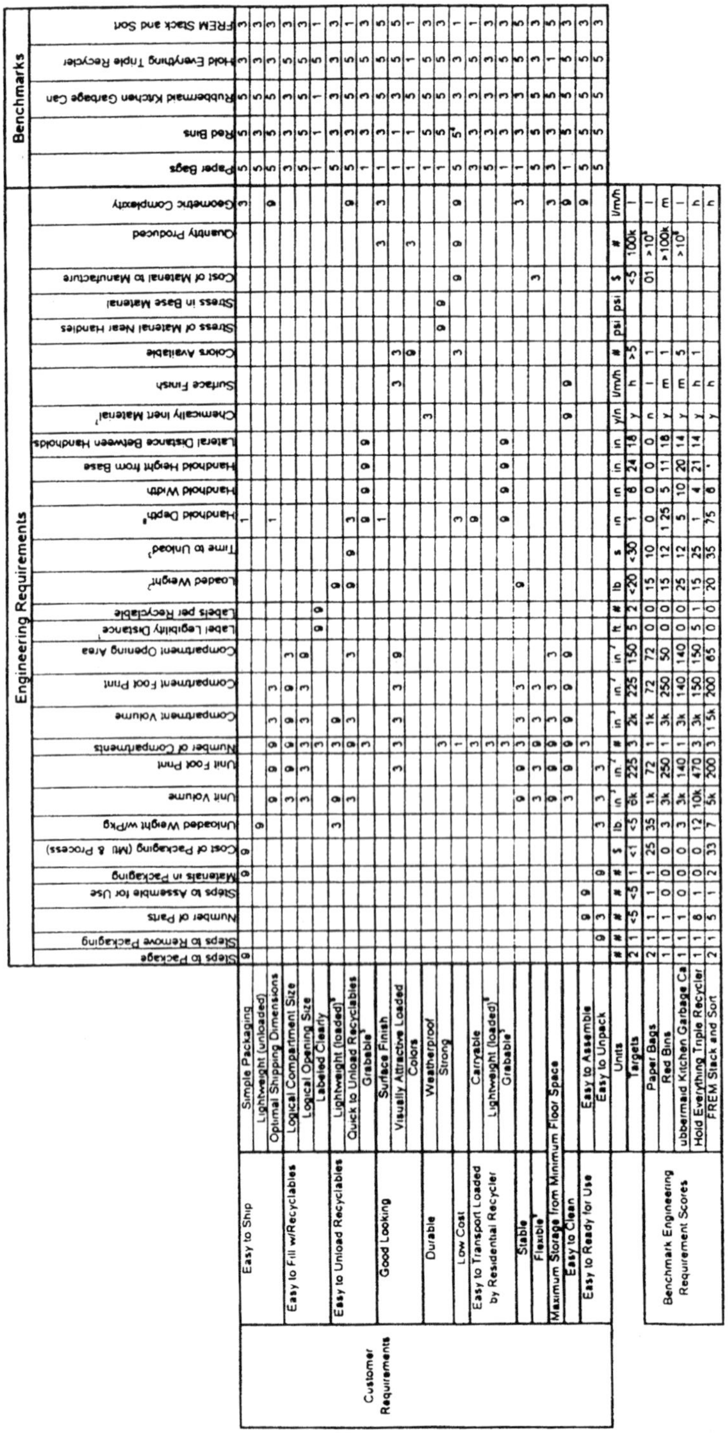

			Engineering Requirements																													Benchmarks				
			Steps to Package	Steps to Remove Packaging	Number of Parts	Steps to Assemble for Use	Materials in Packaging	Cost of Packaging (Mtl & Process)	Unloaded Weight w/Pkg	Unit Volume	Unit Foot Print	Number of Compartments	Compartment Volume	Compartment Foot Print	Compartment Opening Area	Label Legibility Distance[1]	Labels per Recyclable	Loaded Weight[2]	Time to Unload[3]	Handhold Depth[8]	Handhold Width	Handhold Height from Base	Lateral Distance Between Handholds	Chemically Inert Material[7]	Surface Finish	Colors Available	Stress of Material Near Handles	Stress in Base Material	Cost of Material to Manufacture	Quantity Produced	Geometric Complexity	Paper Bags	Red Bins	Rubbermaid Kitchen Garbage Can	Hold Everything Triple Recycler	FREM Stack and Sort
Customer Requirements	Easy to Ship	Simple Packaging	9				9	9												1											3	5	5	5	3	3
		Lightweight (unloaded)							9																							5	3	5	3	3
		Optimal Shipping Dimensions								9	9	9	3	3						1											9	5	5	5	3	3
	Easy to Fill w/Recyclables	Logical Compartment Size								3	9	9	9	9	3																	3	3	3	5	3
		Logical Opening Size								3	3	3	3	3	9																	5	5	5	5	3
		Labeled Clearly										3				9	9															1	1	1	5	1
	Easy to Unload Recyclables	Lightweight (loaded)[8]							3	9		3	9					9														5	3	3	3	3
		Quick to Unload Recyclables								3		9	3		3			9	9	3											9	5	3	5	5	1
		Grabable[3]										3								9	9	9	9									1	3	3	5	3
	Good Looking	Surface Finish																		1										3	3	1	3	5	5	5
		Visually Attractive Loaded									3	3	3	3	9										3	3						1	1	3	5	5
		Colors																								9				3		1	1	5	1	1
	Durable	Weatherproof																						3								1	5	5	5	3
		Strong										3															9	9				1	5	5	5	3
	Low Cost											1								3						3			9	9	9	5	5[4]	3	3	1
	Easy to Transport Loaded by Residential Recycler	Carryable										3								9												3	3	3	5	1
		Lightweight (loaded)[8]										3																				5	3	3	3	3
		Grabable[3]										3								9	9	9	9									1	3	3	5	3
	Stable									9	9	3	3	3				9													3	1	3	3	5	5
	Flexible[7]									3	3	9	3	3															3			5	5	5	3	3
	Maximum Storage from Minimum Floor Space									9	9	9	3	3	3																3	3	3	5	1	5
	Easy to Clean									3	9	9	9	9	9									9	9						9	1	5	5	5	3
	Easy to Ready for Use	Easy to Assemble			9	9						3																			9	5	5	5	5	3
		Easy to Unpack		9	3		9		3	3	3																					5	5	5	5	3
		Units	#	#	#	#	#	$	lb	in³	in²	#	in³	in²	in²	ft	#	lb	s	in	in	in	in	y/n	l/m/h	#	psi	psi	$	#	l/m/h					
		Targets	2	1	<5	<5	1	<1	<5	6k	225	3	2k	225	150	5	2	<20	<30	1	8	24	18	y	h	>5			<5	100k	l					
	Benchmark Engineering Requirement Scores	Paper Bags	2	1	1	1	1	25	35	1k	72	1	1k	72	72	0	0	15	10	0	0	0	0	n	l	1			01	$>10^{5}$	l					
		Red Bins	1	1	1	0	0	0	3	3k	250	1	3k	250	50	0	0	15	12	1 25	5	11	18	y	m	1				>100k	m					
		ubbermaid Kitchen Garbage Ca	1	1	1	0	0	0	3	3k	140	1	3k	140	140	0	0	25	12	5	10	20	14	y	m	5				$>10^{6}$	l					
		Hold Everything Triple Recycler	1	1	8	1	0	0	12	10k	470	3	3k	150	150	5	1	15	25	1	4	21	14	y	h	1					h					
		FREM Stack and Sort	2	1	5	1	2	33	7	5k	200	3	1 5k	200	65	0	0	20	35	75	8	*		y	h						h					

Fig. 4 QFD chart.

market and organization constraints; political, social, environmental and legal considerations, and performance, quality and aesthetic requirements of the product.

5.2. Product Design

The next step in a product life cycle is to convert the statement of needs to product design specifications. The team concept approach discussed above can also be used during the design phase. The design process is an step-wise refinement process that includes design and evaluation of individual components and combination of components into larger components and ultimately into the final product. Approaches that can aid in product design and evaluation include Designing for Manufacturing and Assembly (DFM/DFA) and Simulation.

DFM/DFA guidelines are systematic statements of good design practice derived from years of design and manufacturing experience. These guidelines are general directives that stimulate creativity and show the way to good design for manufacturing and assembly. The objective of these guidelines is to identify product features and concepts that are inherently easy to manufacture and assemble such that the product will have higher quality and lower cost. Design for assembly is an important subset of the total manufacturing process. Depending on the type of the product, as much as 50 percent of the manufacturing costs may be tied up in the assembly process.

Several publications (Anderson, 1990; Boothroyd, 1992; Stoll, 1988) have documented various forms of the DFM/DFA guidelines. An example set of guidelines is shown in Table 1.

Due to the general nature of the above guidelines, many companies have found it advantageous to develop a distinct set of rules and guidelines that apply more specifically to their particular business. Rules are those directives which must not be violated. They must have firm, clear wording and be measurable. Guidelines are recommendations and use words like minimize, maximize, avoid, etc. Rules and guidelines can be grouped on the basis of standardization, compatibility with automation, assembly strategy, repairability, and other such criteria.

Checklists are the most organized way to present rules and guidelines, and document adherence to the formal procedure of product development in a company. A typical format for rules is as follows (Anderson, 1990):

	Followed	*Violated*	*# Violated*
Rule 1:	______	______	______
Rule 2:	______	______	______

The count space is used to keep track of the number of violations.

A similar format for guidelines is as follows:

	# per product
Guideline 1:	________
Guideline 2:	________

TABLE 1
Example DFM/DFA guidelines.

1. Minimize total number of parts. Combine several parts into one part wherever possible, unless:
   (a) The parts move relative to each other.
   (b) The parts must be made of different materials.
   (c) The parts must be able to separate for service.
   (d) The parts need subsequent adjustment.
2. Standardization
   (a) Use standard parts.
   (b) Minimize the number of part types.
   (c) Standardize design features.
   (d) Standardize type and size of materials.
3. Design parts for multiuse.
4. Reduce the difficulty of assembly by:
   (a) Making assembly locations accessible and visible.
   (b) Providing for easy use of assembly tools.
   (c) Providing for easy alignment and positioning during assembly.
   (d) Reducing the depth of insertion.
   (e) Avoiding requiring two hands for assembly.
5. Avoid the need to reposition the partially completed assembly.
6. Use pyramid assembly about one axis of reference, assembling from above.
7. Use the following principles to facilitate insertion:
   (a) Reduce resistance to insertion.
   (b) Provide chamfers.
   (c) Provide generous clearances.
8. Design parts with maximum symmetry.
9. Provide features that will prevent jamming of parts that tend to nest or stack when stored in bulk.
10. Avoid parts that stick together, are slippery, delicate, flexible, very small or large, or are hazardous to the handler.

The purpose of this format is to keep track of the number of design features that are difficult to assemble and manufacture, for example, the number of screws in an assembly and the corresponding guideline would be: "Minimize number of screws".

Product design is usually developed using a CAD (Computer Aided Design) system. The computer model of the product design resulting from the use of a CAD system facilitates modification, evaluation and simulation of the design without the need for describing the model every time a change or manipulation is required. A number of commercial CAD systems provide graphic manipulation facilities in two and three dimensions, interface with graphic input/output facilities, and model analysis functions (such as finite modeling analysis and stress analysis).

Product design teams are increasingly using simulation to validate a product design's performance during the design process. Some CAD systems have the capability to simulate product manufacturing once its design is completed. The simulation results in the generation of a tool path that allows the designer to confirm that the desired part geometry will be created by the machining process.

5.3. Product/Process Evaluation Framework

A key concept in the concurrent engineering approach is to consider process capabilities and material considerations at the design phase. The product/process evaluation process, as shown in Figure 3, starts with the selection of work and tool materials for the part shape to be produced. Next, the characteristics of the manufacturing processes for creating the part shape are specified. The manufacturing process is then simulated to generate machining data for cost analysis. Finally, a sensitivity analysis is performed to select optimal manufacturing environment for part processing.

The design-manufacturing framework in Figure 3 is discussed below using discrete parts manufacturing, the primary focus of flexible manufacturing. In these processes, shape change is achieved by metal removal until the desired geometry is obtained.

Selecting a manufacturing environment involves specification of tool and work materials, manufacturing processes, and machining parameters such as cutting speeds and feed rates. These factors determine production rates and costs. The objective of the design-manufacturing evaluation phase is to identify the best combination of production parameters for manufacturing.

Design and production parameters affect part processing costs. Machining parameters used in part processing need to satisfy conflicting requirements. For example, as cutting speed is increased, it reduces processing time. However, tool wear increases resulting in more frequent tool changing and higher tool cost. An economic model used for evaluation consists of five basic cost components (Lindberg, 1990): machining cost, setup cost, tool cost, tool changing cost, and part handling cost.

$$\textit{Machine cost} = C_D \sum_{i=1}^{k} t_i$$

$$\textit{Setup cost} = \frac{C_s}{L}$$

$$\textit{Tool cost} = \sum_{i=1}^{k} \left(C_{ti} \frac{t_i}{T_i} \right)$$

$$\textit{Tool changing cost} = C_D \sum_{i=1}^{k} \left(t_i \frac{t_c}{T_i} \right)$$

$$\textit{Part loading/unloading cost} = C_D t_p$$

where

C_D direct labor cost, \$/m in
C_{ti} intial tool cost for tool type i, \$/tool type
C_s setup cost per lot
T_i tool life per tool type i
t_i cutting time per piece per tool type i, min
t_c tool changing or reindexing time, min
t_p part loading and unloading time, min
k number of different tools required for part processing
L parts per lot.

The rate of metal removal and power requirements for machining affect machining costs and are dependent upon cutting speed, depth of cut and feed rate. Minimum cost and maximum production rate are two major criteria used for determining the range of cutting speeds (Lindberg, 1990).

Tool life affects tool cost and tool changing cost. Several models have been proposed for estimating tool life; including the Taylor's formula (Schey, 1987):

$$VT^n = C$$

where T is the tool life, V is the cutting speed, and n and c are constants whose values depend on work and tool materials.

An important component of the framework in Figure 3 is support databases. For machining and metal removal processes, two sets of databases are required, one which describes features of tool and work materials, and another which describes the features of the processes used in machining and metal removal.

There are generally two material types that need to be specified in a machining operation: work material and tool material. The processing time, feed rates and other operation characteristics depend on both these materials. Commonly used work materials include low carbon steel, medium carbon steel, high carbon steel, grey cast iron, and aluminum. A number of cutting tool materials are available, but high speed steel and sintered carbide do the bulk of metal cutting.

Material properties appear in tabulated form in handbooks published by technical societies such as the Society of Manufacturing Engineers, The American Society of Metals, The American Society of Mechanical Engineers, and the Society of Automotive Engineers. Appropriate properties can be extracted from these sources for inclusion in the databases.

The process database provides information on specific process parameters, such as turning, cut-off, milling, drilling, shaping, and grinding. The parameters stored in the database would be different for each operation, being a function of the specific operation. For example, the primary specifications in turning are the depth of cut, cutting speed, and feed rate; for drilling, the primary considerations are feed rate, tool peripheral speed, and drill diameter. Also included in the process database are values of constants n and c required to compute tool life for different tool and work materials.

Successful implementation of a design-manufacturing framework requires a user interface for front-end (data input) and back-end (output analysis) system interaction with the user. For example, a user would select work and tool materials and provide information about process and tool setup interactively. This user specified information along with parameters extracted from the database are used in conjunction with the part drawing to simulate part processing on the screen. If processing times and costs are acceptable, the CAD/CAM system then generates numerical control (NC) code for the physical manufacturing process.

Sensitivity analysis concentrates on the simplification of the product specifications to promote ease of manufacture, improve quality, and reduce manufacturing costs. For example, a change in work or tool materials or a change in one or more design and production specifications may result in substantial cost reductions, and improvements in tooling, fixturing and material handling required to support the process. The availability of a CAD/CAM package would allow quick changes in the shape of the part in addition to changes in tool and work materials, and production parameters.

6. COMPARISON OF ALTERNATIVES

Selection and investment in manufacturing technology is a decision with long-term implications for an organization. The correct decision can result in significant productivity and profitability improvements. On the other hand, suboptimal decisions can be the cause of productivity problems resulting from system inefficiencies, cost and quality problems, and worker dissatisfaction.

The result of the design-manufacturing evaluation process described in the previous section is the identification of a set of feasible technology alternatives for manufacturing a specific product (or range of products). Once a set of alternatives have been developed, a multicriteria decision model can be used to compare the alternatives. The structure of the multicriteria model enables the consideration of factors that have different measures, different amount of importance to the decision, and which may represent conflicting objectives.

The multicriteria decision model is a five step process. The activities are: (1) to identify a set of attributes for evaluating alternatives, (2) to develop attribute weights which reflect the relative importance of the attributes in the decision environment, (3) to assess or measure each alternative with respect to each attribute, (4) to aggregate the attribute weights and measures into an overall merit rating for each alternative, and (5) to perform sensitivity analysis.

6.1. Attribute Identification

Attributes are the basis of evaluating and comparing alternatives; hence, these must be carefully identified. Attributes are identified through examination of relevant literature and case studies, administration of questionnaires, and use of delphi-based discussions and surveys with experts and potential users of the system. Identifying an appropriate set of attributes is a multi-step iterative process. A number of these attributes would be identified during the needs assessment phase. There is no "best" set of attributes. The attributes selected for the comparison process depend on the characteristics of the specific environment.

A review of manufacturing literature reveals a wide range of attributes. An example set of attributes are given in Table 2.

6.2. Attribute Weights

Attribute weights determine the relative importance of attributes in the decision environment. A number of importance weighting methods are available. In the rating method, attributes are ranked in order of importance. Each attribute is then assigned a weight with respect to the least important attribute. Attributes may have the same weight if they are considered equal in importance. The resulting weights may be normalized to sum to one.

The importance weights for attributes may also be obtained using a pairwise comparison process, where each attribute is compared with every other attribute.

TABLE 2
Attributes for comparing manufacturing alternatives.

- System Effectiveness
- System Efficiency
- Quality of Output
- Productivity
- Profitability
- Quality of work life
- Flexibility of the system to changes and growth
- Safety and ergonomic considerations
- Manufacturing facility – size and space requirements

Weights are then developed from the sums of the "more important" ratings given to each attribute.

6.3. Attribute Measurement

Once a set of attributes have been identified, the attributes are assessed or measured for each alternative. Data collection, statistical estimates, optimization and simulation models, and economic analysis are some of the methods used to assess quantitative attributes. Qualitative attributes represent subjective factors for which it is generally difficult to define a natural measurement scale. Descriptive classes or interval scales (for example, 0 to 10) can be established to enable a numerical value to be assigned to represent how a site scores with respect to a particular attribute. It may also be desirable for the aggregation process (described below) to translate quantitative measures to an equivalent, dimensionless value scale.

6.4. Aggregation of Attribute Weights and Measures

An overall comparison score for each alternative is computed by combining attribute weights and attribute values using an aggregation model. A commonly used model is the additive-weighted model:

$$A_j = \sum_{i=1}^{n} w_i x_{ij}$$

where A_j = overall score for alternative j
w_i = weight for attribute i
x_{ij} = value of attribute i for alternative j
n = number of attributes

The alternatives can be compared and ranked with a single aggregate measure, A_j.

6.5. Sensitivity Analysis

In the multicriteria procedure described above, attribute weights are subjective assessments. Also, some of the attribute values are based on best guess of experienced personnel or on limited analysis of minimal data. It is, therefore, extremely important to determine the degree of sensitivity of the results to the values used.

Sensitivity analysis consists of varying attribute weights and values over some range of interest and observing the effect on the final rankings of attributes. If the final ordering of alternatives changes greatly with slight variations in some of these parameters, this may provide the motivation and justification for expenditure of more time and money to obtain more accurate estimates. On the other hand, if the results do not change over wide fluctuations in the values of the parameters, no further effort is needed or justified, and the results will help reassure the decision maker of the thoroughness of the study and the validity of the results.

To illustrate the multicriteria decision model, consider two design alternatives, I and II, which are candidates for the final implementation decision. The design team has identified three attributes on which the alternatives are to be compared. These are: production costs, flexibility of the production system to accommodate market changes, and ease of production. The attribute weights are established using a pairwise comparison process, as shown in Table 3. Each column attribute is compared to each row attribute. The value below the principal diagonal is the complement of the value above the diagonal. Thus "0" is the complement of "2", and "1" is the complement of itself. In the table below, production and cost, and production and flexibility are considered equally important, while cost is considered to be more important than flexibility. The importance weights are obtained by normalizing the sum of the scores corresponding to each column attribute.

Table 4 shows the attributes, their respective weights, and the attribute measures for each of the alternative. For consistency, all attribute measures use a 0–10 scale, where 0 and 10 represent the worst and the best values that can be attained in the decision environment. The aggregate scores for the two alternatives are obtained using the additive-weighted model. As can be seen from Table 4, Alternative I is the preferred alternative. Sensitivity analysis can now be used to explore the effect of changing attribute weights and attribute values on the ranking of alternatives.

TABLE 3
Attribute weights using pairwise comparison.

	Cost	Flexibility	Production	Sum	Weight
Cost	–	2	1	3	0.500
Flexibility	0	–	1	1	0.167
Production	1	1	–	2	0.333

TABLE 4
Comparison ratings for alternative designs.

Attributes	Weights	Alternative I	Alternative II
Cost	0.500	6	9
Flexibility	0.167	8	7
Production	0.333	9	3
Aggregate Score		7.33	6.67

7. CONCLUSIONS

The primary objective in concurrent engineering is to consider manufacturing issues early to shorten product development time and time-to-market, and reduce the costs resulting from segregation of design and production. Some basic principles underlie the concurrent engineering design effort. These include:

- Product design must be based on an accurate assessment of user or customer requirements.
- The design effort generally requires multi-disciplinary teams representing individuals with different expertise.
- Developing concurrent engineering systems requires a broad and varied set of information and analysis tools.
- The design process must integrate product characteristics and manufacturing operations, since a product's status cannot be assessed independently of available technology. Furthermore, the system should allow evaluation of the effect of various machining parameters, part geometry, and types of tool and work materials on the production cost and product quality.

References

D.M. Anderson, *Design for Manufacturability*, CIM Press, Lafayette, CA (1990).

Association of National Advertisers. *Prescription for New Product Success*, Association of National Advertisers, Inc (1984).

C.L. Berrington and R.L. Oblich, Translating Business Reengineering Into Bottom-Line Results. *Industrial Engineering*, **27**(1), 24–27 (1995).

G. Boothroyd, *Assembly Automation and Product Design*, Marcel Dekker, Inc (1992).

J.S. Colton and M.P. Ouellette, A Form Verification System for the Conceptual Design of Complex Mechanical Systems. *ASME Winter Annual Meeting*, DE-Vol **66**, 55–66 (1993).

A.S. El-Gizawy, *et al.*, Concurrent Engineering with Quality Approach for Manufacturing Industry. *ASME National Design Engineering Conference*, DE-Vol **52**, 143–149 (1993).

J.E. Ettlie and H.W. Stoll, *Managing the Design-Manufacturing-Process*, McGraw-Hill, Inc., New York, NY (1990).

J.R. Hartley, *Concurrent Engineering*, Productivity Press, Cambridge, MA (1992).

R.A. Lindberg, *Processes and Materials of Manufacture*, Allyn and Bacon, Needham Heights, MA (1990).

L.C.G. Miller, *Concurrent Engineering Design*, Society of Manufacturing Engineers, Dearborn, MI (1993).

E.A. Orady and I. Shareef, Expert System Design Philosophy for Application of Simultaneous Engineering in Industry. *ASME National Design Engineering Conference*, DE-Vol **52**, 151–157 (1993).

S. Pugh, *Total Design*, Addison-Wesley Publishing Company, Wokingham, England (1990).

J.A. Schey, *Introduction to Manufacturing Processes*, McGraw-Hill, Inc., New York, NY (1987).

S.G. Shina, *Successful Implementation of Concurrent Engineering Products and Processes*, Van Nostrand Reinhold Company, New York, NY (1994).

H.W. Stoll, Design for Manufacture. *Manufacturing Engineering*, **100**(1), 67–73 (1988).

J.P. Terpenny and M.P. Deisenroth, A Concurrent Engineering Framework: Three Basic Components. *Proceedings of the Second International FAIM Conference*, Falls Church, VA, June 30–July 2, 237–247 (1992).

J.A. White, Material Handling in Integrated Manufacturing Systems, in *Design and Analysis of Integrated Manufacturing Systems*, (ed. W.D. Compton), National Academy Press, Washington, D.C., 46–59 (1988).

D.E. Whitney, *et al.*, The Strategic Approach to Product Design, in *Design and Analysis of Integrated Manufacturing Systems*, (ed. W.D. Compton), National Academy Press, Washington, D.C., 200–223 (1988).

CHAPTER 2

The Standards: STEP ISO-10303

THU-HUA LIU

Department of Industrial Design, School of Management, Chang Gung University, Tao-Yuan, Taiwan 33333, ROC

1. INTRODUCTION

In the product design process, the product is in an abstract form, for which the physical artifact does not exist. Therefore, until it is manufactured, there needs to be some model(s) of the designed product for those involved to analyze, evaluate, manipulate and refine. One of the most basic roles of models is they can be used by the designer to record and manipulate ideas and to provide a basis for the evaluation of the design (McMahon and Browne, 1994). In a multidisciplinary environment, the models have a major role in the communication of the design between the team of participants involved in the development, manufacturing, and subsequent use of the product.

However, to practice Concurrent Engineering (CE), the models must also provide the *full* communication capability between not only multidisciplinary engineers, but also computer systems and engineering software (Evans, 1988). To achieve this, two issues need to be addressed for the contemporary approach to CE, which uses standardized product data stored in the computer system and does not rely purely on people-to-people communication (Wu *et al.*, 1992):

(1) In an integrated CE environment a generic, shared definition of a product is required to obtain or derive product characteristics for discipline-specific applications.
(2) The abstractions of product characteristics must be consistent across different applications. Such consistency relies on the existence of a shared standard product definition and the ability to extract product characteristics from the shared definition.

The Initial Graphics Exchange Specification (IGES) is the first significant work in CAD data exchange and the dominant standard for CAD/CAM/CAE applications over the years. IGES, which is supported by the US National Bureau of Standards, was established in 1979 and adopted by the American National Standards Institute (ANSI) in 1981. Although IGES is widely used, dissatisfaction that led to the development of variant standards such as the Standard d'Exchange et de Transfer (SET) in France, Standards for "Verband der Automobilindustrie" (VDA/FS) in German, and the European Specific Programme for Research and Development in Information Technology (ESPRIT) project 322 Computer-Aided Design Interfaces (ESPRIT-CAD*I) in the European community. As an example, one of the ESPRIT projects, Neutral Interfaces for Robotics (NIRO), aimed to develop neutral interfaces and a data format for the exchange of geometrical, technological and programming functions and data between CAD, robot control and planning (Bey *et al.*, 1994). As IGES and alternative standards evolved, the need to develop an international standard to integrate previous efforts and to provide an improved fundamental basis for standards activities became clear.

In the US, the IGES organization established the PDES (Product Data Exchange Specification) in 1984 to extend the capabilities of IGES. In 1987, PDES was added to the name of the organization to reflect the new activity. In 1990, the definition for PDES was revised to Product Data Exchange using STEP (Standard for the Exchange of Product Model Data) to underscore that PDES is the US organizational activity that supports the development and implementation of STEP. The IGES/PDES Organization Steering Committee (IPO SC) has decided to make STEP an American National Standard (ANS) (Warthen, 1992), which means that once each STEP document reaches Draft International Standard (DIS) status, the US will undertake a combined review for both the US ballot for the ISO (International Organization of Standards) International Standard and for the ANSI ANS status.

The STEP international standard of ISO has been under development since the mid-1980s. The Initial Release of the STEP standard (ISO 10303) is complete and was accepted as Draft International Standards by the ISO TC 148/SC4 community in February 1993 (IPO, 1993). The following description is excerpted from the ISO document (STEP Part-1, 1992).

> The Standard for the Exchange of Product Model Data (STEP) is a neutral mechanism capable of completely representing product data throughout the life cycle of a product. (Neutral in this context indicates independence from and particular computer-Aided (CAx) software systems.) The completeness of this representation makes it suitable not only for neutral file exchange, but also as a basis for implementing and sharing product databases and archiving. There is an undeniable need to transfer product data in computer-readable form from one site to another. These sites may have one of a number of relationships between them (contractor and subcontractor, customer and supplier); the information invariable needs to iterate between the sites, retaining both data completness and functionality, until it is archived. The most cost effective manner to encapsulate such information is in a neutral format, independent of any CAx software system.

For computer-aided CE, the key point is the development of the information modeling techniques and associated standards that define the representation of product data. STEP provides an avenue for computer-based engineering applications to exchange data between CAD/CAM/CAE/CIM systems (Bloom, 1989). The new international standard, STEP, is a promising standard that acts as a central block to integrate CAx systems and support CE (Liu and Fischer, 1993a). Combining the C^{++} programming and X-window based toolkit, the STEP user can create computer-integrated manufacturing applications more easily than currently possible (Czerwinski and Sivayoganathan, 1994; Qiao *et al.*, 1993).

The following section reviews the STEP technologies and architecture. Sections 3–5 explain three important components of STEP: EXPRESS language, Integrated Resources, and Application Protocols, respectively. Finally, the contributions STEP can make to an enterprise and the future of STEP are discussed in the final section.

2. THE STEP TECHNOLOGIES AND ARCHITECTURE

2.1. STEP Design Goals and Methodologies

STEP is an ambitious, but steadily progressing, new international standard. The design goals of STEP are derived from the following two premises: firstly, there exists a fundamental and non-redundant structure of product data, from which all of the product information that is needed to support all engineering functionality throughout the product life cycle can be derived; and secondly, this standard structure of product data is independent from how the data is used by any application and stored in any particular computer system. Weiss (1988) indicates that STEP is intended to be (1) a standard for information, not just data, (2) a complete representation of a product, (3) expandable to new products and processes, (4) compatible with other standards, (5) independent of computing environment, and finally, to have implementation validity.

To fulfill the above design goals and the critical requirements of open and shareable product data, the STEP developers encountered some technical challenges (Carver and Bloom, 1991): (1) STEP must develop a framework for not only the commerce of data but also the exchange of information, (2) the complexity and scope of STEP is far beyond any existing standards, and (3) the enabling technology of STEP must be developed at the same time the standard is evolving. The consensus approach to meeting the above challenges is to adopt the methodology based upon a three-layer architecture and the use of reference models and formal languages (Smith, 1989). The three level strategy can be described as follows:

Application Layer (interfacing with different applications):

The application level uses application reference models to describe product information. A reference model (Integrated Resource) is a standard representation created for use in constructing an object, a part, or a system for applications.

Logical or Conceptual Layer (forming the integration):

The logic layer uses the formulated information description language – EXPRESS – to define the entities (or objects) in each topical reference model (Integrated Resource model). EXPRESS is an object-oriented information modeling language designed for STEP to specify the representation of product information.

Physical Layer (forming the actual communication medium):

The physical layer is concerned with databases and represents a sequential file using the Wirth Syntax Notation to express formal syntax.

Formal data and information modeling methodologies, such as US Air Force developed IDEF0 and IDEF1x methodologies (IDEF0, 1981; IDEF1X, 1985), Nijssen Information Analysis data Modeling (NIAM) technique (Nijssen and Halpin, 1989), and a graphical subset of EXPRESS called EXPRESS-G (STEP Part 11, 1992), are used to model disciplines in the application layer. The data models produced are then synthesized into a consistent EXPRESS information model called conceptual schema at the logic layer. Finally, at the physical layer, the conceptual schema is used to created a database structure and/or is embedded into an ASCII file format for physical exchange.

As a result of STEP enabling technologies, the content of STEP agrees with the three-layer architecture (Owen, 1993). STEP contains two important information models, e.g. Application Protocols and Integrated Resources, which correspond to the application and conceptual layer, respectively. Also, STEP provides Implementation Methods that comprise the mapping to the physical layer.

2.2. STEP Structure

The structure of the standard itself is reflected by the STEP documentation which is partitioned into several groups of "Parts." The overall structure of STEP standards is as follows:

Industrial Automation Systems – Product Information Representation and Exchange:

Parts 1–9	Introductory
Parts 11–19	Product Data Description Methods
Parts 21–29	Implementation Methods
Parts 31–39	Conformance Testing Methodology and Framework
Parts 41–49	Integrated Resources: Generic Resources
Parts 101–199	Integrated Resources: Application Resources

Parts 201–1199 Application Protocols
Parts 1201–2199 Abstract Test Suites

Figure 1 summarizes the STEP series of Parts. Twelve documents (Parts), as shown in Figure 1, that make up the Initial Release of STEP are marked by '*'. Parts 1 to 9 provide an overview of the structure and intent of the STEP family of standards. Parts 11 to 19 define the structured method and the formal languages used in defining and developing the STEP standards. The EXPRESS lauguage is defined in Part 11, which specifies a language specially designed to deal with information produced or consumed during product manufacturing (STEP Part 11, 1992).

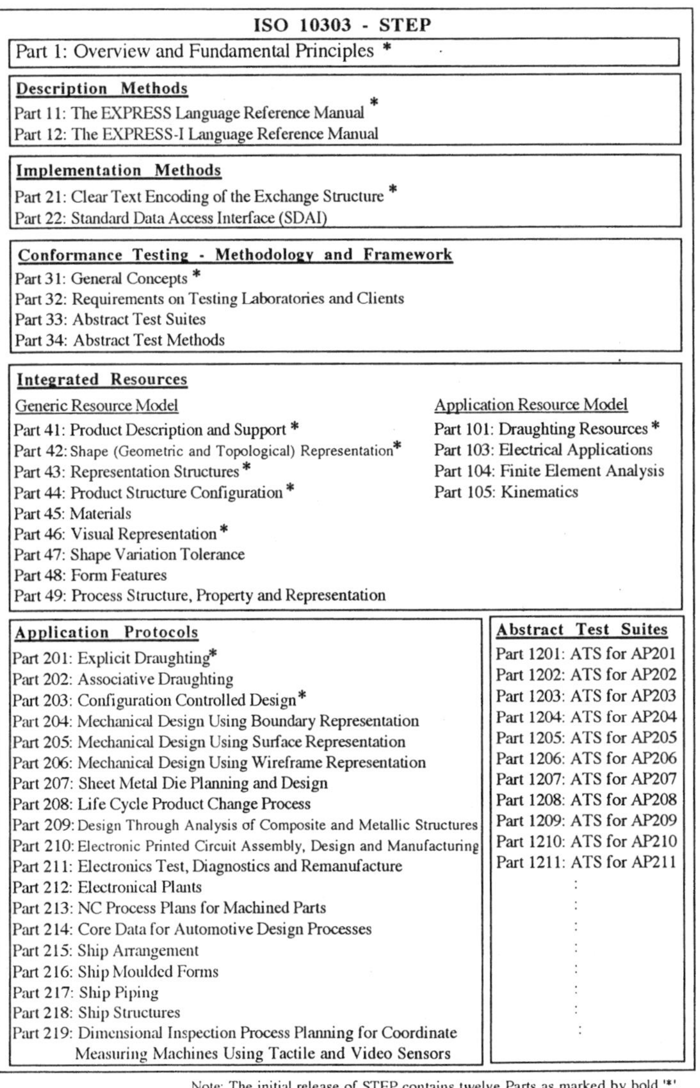

Fig. 1 The part structure of ISO 10303–STEP.

Parts 21 to 29 define the methods and formats used to exchange STEP product data. For the two currently included Parts of this group, Part 21 defines a file format that supports the exchange of data between applications, and Part 22 specifies the SDAI (STEP Data Access Interface) which can be used with EXPRESS schemata to define consistent data access mechanisms. Parts 31 to 39 document the methodology and framework of conformance testing that defines how to achieve validation and conformance of STEP implementations.

Parts 41 to 49 provide the generic resource information models that may be relevant to all application areas, and Parts 101 to 199 provide application resource information models relevant to an identified group of application areas. These integrated resources are definitions of general purpose and content independent information entities required to specify engineering applications and support exchange of product data.

Parts 201 to 1199 are designated for application protocols. Each application protocol is a specification for interfacing with STEP in a standard way which provides the method for defining a subset of STEP entities to be implemented for a particular application (Stark and Mitchell, 1991). The integrated resources are not designed to be used directly as the basis for exchanging product data but as the building blocks for the application protocols through which the data sharing is achieved.

Parts 1021 to 1299 document the group of abstract test suites. Each abstract test suite consists of a set of test cases to be used in conformance testing for its corresponding application protocol. The conformance testing requirements contained in the test suites ensure STEP implements comply with the standard.

The basic structure of STEP consists of three components: an infrastructure, integrated resource information models, and application protocols. The infrastructure has two elements: the Description Methods (DM) and Implementation Methods (IM). STEP is described by a formal language called EXPRESS that has been developed to describe product information for exchange between engineering applications. The implementation methods make EXPRESS-defined data available to application programs. An implementation method is defined by a mapping from the EXPRESS language onto a formal notation (language). Three identified implementation methods are:

(1) physical file exchange, which is the reading and writing of the data corresponding to Application Interpreted Models (AIM) into a sequential file format;
(2) direct access, that allows engineering programs to directly use the data corresponding to AIMs; and
(3) database access, that reads, writes, and modifies a database whose internal schema is determined by the schema of an AIM.

The integrated resource information models (Integrated Resources-IR) contain the generic definition of various data entities that occur in engineering applications. The entities in each resource model are the building blocks of construction for the application information models. As STEP is expected to cover the whole life-cycle of engineering, further resource models will be developed.

Application Protocols (AP) select definitions they need for a particular application from the Integrated Resources (only those entities defined in the IRs). In addition,

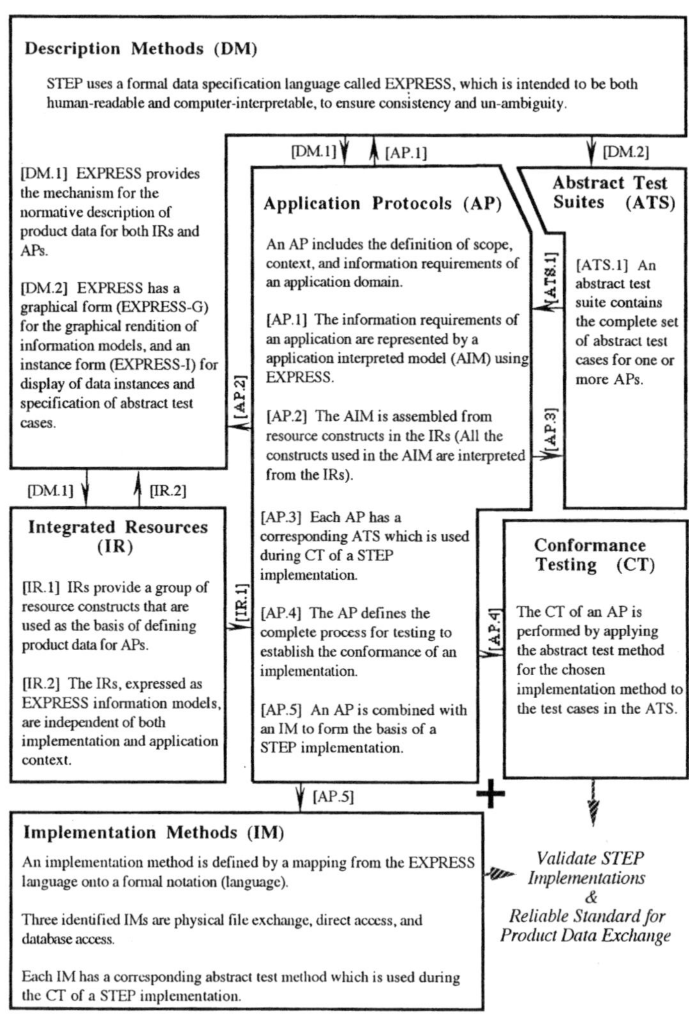

Fig. 2 Inter-relationship of STEP parts structure.

each AP defines its own conformance requirements and has a corresponding Conformance Testing Suite (CTS). Combining the CTS and IMs, the APs are the main result of the STEP development process. The construction procedures of APs enable the standard definitions of application-oriented views of a product.

Figure 2 shows a map indicating the inter-relationships between the STEP parts. STEP separates the implementation method from the information content and contains conformance testing for each AP. This has resulted in a better definition and reliable standard, and achieves efficient physical file exchange. As STEP evolves, more IRs are anticipated and a wider range of APs will be developed based on the interpretations of these resources. According to Owen (1993), the following APs have been purposed:

- Electric/electronic
- Composites – design to manufacturing
- Product life-cycle
- Manufacturing process plans
- Polymer testing
- Sheet metal
- Shipbuilding
- Life-cycle management
- Product operation
- Product procurement
- Near net shape processing
- Process plant functional data and its schematic representation
- Ships electrical systems
- Ships HVAC systems
- Printed circuit assembly manufacturing planning
- Exchange of design and manufacturing product information for cast parts
- Mechanical products definition for process planning using form features
- Constructional steelwork

3. THE EXPRESS LANGUAGE AND OBJECT-ORIENTED CONCEPTS

3.1. Modeling of Things with Object-Orientation

The separation of abstract and concrete ideas is important in modeling objects. Abstraction separates the idea an entity represents from the set of attributes chosen to represent it. For example, a point is an idea, and x, y, z are the names of the attributes commonly used to realize that idea. It is important to be able to deal with either the idea or the details that make up the idea separately and selectively. The term "abstract" is used for the idea the entity represents. The term "concrete" is used for the details of the entity's composition. For example, at the abstract level, "circle" is self-evident. However, this definition is not very useful when the circle is processed by a computer. The attributes named Center, Axis and Radius are elected to realize the concept, where Center is a point (type), Axis is a vector, and Radius is a floating point number.

In object-oriented systems and languages, any real world entity is modeled as an object. Every object has a state and a behavior. The state of an object is a set of attributes of the object, and the behavior of an object is a set of methods. These methods are analogous to functions or procedures in conventional programming languages. The process of invoking a method is called sending a message to the object. Such a message contains arguments just like a procedure or function call in conventional programming languages. Sending a message to an object (invocation of a method) usually modifies the state of that object.

The EXPRESS is an object-oriented information modeling language used to define the logic layer, and is designed to satisfy: (a) modeling the object of interest, (b) defining the constraints that are to be imposed upon interested objects, (c) defining the operations which establish how they are to be used, and (d) modeling in a computer-sensible manner (i.e., processable by computer).

Object-orientation is a general industry term that describes an emerging method of software engineering. Under object-oriented programming, software is assembled from discrete blocks of code called objects or classes. Each object is entirely self-sufficient and discrete from the rest of the code. It has the main advantage that new objects can be added modularly without disturbing existing code. The most fundamental object-oriented concepts that are found in many object-oriented languages and systems are: (1) classification and abstract data type, (2) encapsulation and information hiding, and (3) polymorphism through inheritance (Henderson-sellers, 1991; Ellis and Stroustrup, 1991; Wiener and Pinson, 1988).

The abstract data type or user-defined type is used to support the modularization. Each object class is used as an object type and within the class the object data and functionality are bundled together. Classification is the concept of grouping software ideas into classes of things which represent a generic set of information. Modularization is the idea that each object should be autonomous and self-contained. Encapsulation and information hiding are related to localizing items and to protect the object from interactions with the environment. The inheritance mechanism is one of the ways to achieve the software engineering goal of offering reusability, extendibility, and low maintenance cost. Booch and Vilot (1990) define the polymorphism as a concept of type theory in which a name may denote objects of many different classes related by some common base class. Thus, any object denoted by this name is able to respond to some common set of operations in different ways.

3.2. Computer Processable Language

EXPRESS is a formal language that could be used as part of a methodology aimed at creating an information model. The problem of constructing an information model, capable of describing product data throughout the life of a product, is too large to manage without computer assist. For this reason, the modeling language must be processable by a computer while still being usable by people. The EXPRESS compiler, that translates EXPRESS information models into codes for application programs to read, write and manipulate data defined by that model, is available now. Using a suitable EXPRESS programming environment, the object-orientation style of EXPRESS can provide the solution to achieve the qualities of robustness, extendibility, reusability, and compatibility for information modeling and software engineering.

Software tools that help to implement STEP and develop STEP based applications often contain the following functional units:

(1) EXPRESS Compiler, as mentioned above, to interface with high-level languages;
(2) Browser and Editor for both EXPRESS and EXPRESS-G (graphical editor);
(3) Database Translator for persistent storage repositories;
(4) Tester to test the STEP data for conformance to the standard and to the constraints of defined entities;
(5) Schema Checker helps check the validity of EXPRESS schemata and the semantic consistency of entities; and
(6) IGES and DXF Format Translators as interfaces to the existing standards.

Most of these STEP development tools or software environments are implemented with the following criteria: (1) compliance to STEP Parts 21 and 22, (2) object-orientation, (3) database dependence, and (4) consistency and integrity with STEP technologies.

3.3. EXPRESS Basics

EXPRESS is a conceptual schema language that provides for specification of (1) the objects belonging to a universe of discourse, (2) the information units pertaining to those object, and (3) the constraints on those objects (STEP Part 11, 1992). The text of an EXPRESS information model is similar to that of a PASCAL program with the exception that EXPRESS language does not provide input and output statements, and an EXPRESS information model only specifies data structures and constraints. Main descriptive elements of EXPRESS are as follows:

Schema
Type (character of data that is used to describe entity)
Entity (describes the things of interest to a UoD-universe of discourse)
Algorithm
  Function
  Procedure
Rule (define constraints)

- *Schema*: A schema defines a universe of discourse in which the declared objects have a related meaning and purpose. A schema may contain the following declaration of objects: Constant, Type, Entity, Function, Procedure, and Rule.
- *Type*:
  - base types: REAL, INTEGER, NUMBER, BOOLEAN, LOGICAL, BINARY, STRING
  - aggregation types: ARRAY, LIST, SET, BAG

```
ENTITY Aggregation;
 matrix: ARRAY [1:10] OF                  -- first dimension
 ARRAY [1:20] OF REAL;                    -- second dimension
 bag_of_points:BAG OF Point;              -- zero or more points
 set_of_points:SET [1:?] OF Point;        -- at least one point
 list_of_circles:LIST [0:10] OF Circle;   -- the list can be zero to ten circles
END_ENTITY;
```

  - entity type (user-defined type, an entity may be used as a data type)
  - enumeration type (enumerated item becomes an unique identifier in the current scope, and has a value, from small to large, depending on its position in the list)
  - defined data type (user-defined type created in a TYPE declaration and is a user extension to the base types)
  - select type (a selection list defining a named collection of other types)
- *Entity*: An entity is an object, concept, or idea, defined by a set of attributes. The behavior of the entity is defined by a set of rules and algorithms.

- *Algorithm*: An algorithm (function or procedure) is a sequence of operations on data that produces some desired end state(s). Algorithms are an integral part of a schema and are used to define the behaviors of entities and attributes. A function operates on parameters and produces a single resultant value of a specific type. A procedure operates on parameters to produce the desired end state that may change a set of formal parameters.
- *Rule*: Rules define the constraints on entities and attributes that must be enforced by the information system. Functions, procedures, and rules are used to describe the behavior of things.

3.4. Statements for Constraints

The executable statements for specification of constraints are summarized as follows:

- *Null*
- *Alias*
- *Assignment* (:=)
- *CASE*:

```
a := 3;
x := 34.97;
CASE a OF
    1: x := sin(x);
    2: x := exp(x);
    3: x := sqrt(x); -- executed
    4: x := log(x);
END_CASE;
```

- *COMPOUND*:

```
BEGIN
    A := A + 1;
    IF .......
    END_IF;
END;
```

- *ESCAPE*: Escape causes an immediate transfer to the statement following the end of the block.

```
REPEAT UNTIL (A = 1)
  ...
IF (A < 0) THEN
ESCAPE; -- When executed control passes to the statement after
END_REPEAT
  ...
END_REPEAT;
  ...;
```

- *IF-THEN-ELSE*
- *Procedure call*: PROCEDURE TT (VAR A: INTEGER; B: REAL);

- TT (a, 37) is a valid procedure call, since a type conversion takes place automatically.
- TT (1, 2) is invalid since first parameter is not a variable.
- TT (a) is invalid for the TT procedure needs two parameters.

- *REPEAT statement*
  - Increment control
  - UNTIL-control
  - WHILE-control
- *RETURN statement*: RETURN terminates the execution of a function or procedure.

  RETURN (50); – – for function
  RETURN (work_point); – – for function
  RETURN; – – for procedure
- *SKIP statement*: SKIP causes an immediate transfer to the end of the repeat-block in which it appears.

3.5. Inheritance by Supertype and Subtype

EXPRESS uses supertype and subtype mechanisms to enable the inheritance of entity properties. Inheritance provides explicit support of reusable code and data. Refining an inherited algorithm adds new functionality to existing code. A new entity may be defined as a specilization of an existing entity. The new entity (subtype) inherits desired attributes and constraints of the existing entity (supertype); and the user may specify additional attributes, algorithms, and rules for the new entity. Figure 3 shows the inheritance structure of some entities (objects) declared by an

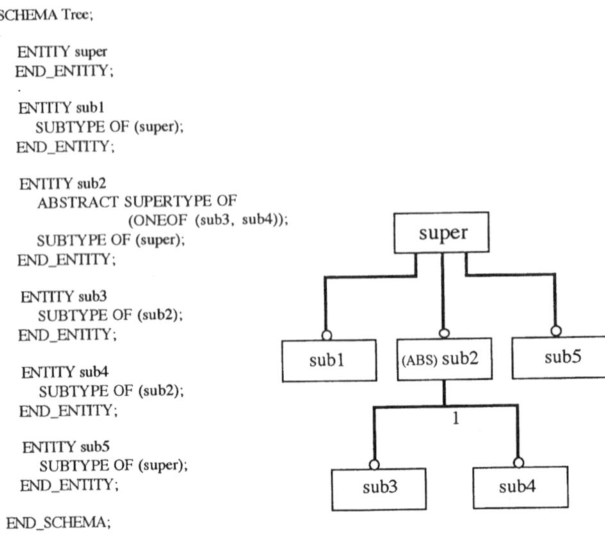

Fig. 3 An example schema and inheritance graph.

example EXPRESS schema. The inheritance graph shown in Figure 3 is illustrated by the graphic symbols of EXPRESS-G, which will be introduced later.

3.6. Some Useful Tools

- *Derived attribute*: A derived attribute is computed in some manner, for example:

```
ENTITY Circle;
    P1, P2, P3: Point;
    DERIVE
    radius: REAL := f1(p1, p2, p3); -- function f1 calculates the radius of the
                                    -- circle
    axis: vector := f2(p1, p2, p3); -- function f2 calculates the axis of the
                                    -- circle
    area: REAL := p1*radius**2;
END_ENTITY;
```

- *Inverse attribute*: An inverse attribute captures the relationship between the declared entity and an attribute of another entity. The following example enforces the constraint on a declared Point such that it can only exist if it is used as the start point of an instanced line.

```
ENTITY Line;
    start,
    end:  Point;
END_ENTITY

ENTITY Point;
    x, y, z : REAL;
INVERSE
    start_line: Line FOR start;
END_ENTITY;
```

- *Unique rule*: The uniqueness constraint enforces that within a given information base the named attributes or combination of attributes shall be different throughout.

```
ENTITY Person;
    name: STRING;
    address: STRING;
    project_ID: STRING;
    SSN_ID: STRING;
UNIQUE
    name, project_ID; -- joint uniqueness constraint
    SSN_ID;           -- simple uniqueness constraint
END_ENTITY;
```

- *Local rule*: All local rules must follow the WHERE keyword.

```
ENTITY Unit_vector;
    a, b: REAL;
    c: OPTIONAL REAL; -- attribute c can have null value
```

WHERE
length_1: $a^{**}2 + b^{**}2 + \mathrm{NVL}(c, 0.0)^{**}2 = 1.0$;
-- NVL is a standard function that handles the OPTIONAL
-- value of c
END_ENTITY;

- *Interface specification*: Two specifications, USE and REFERENCE, are used to effect an interface between two schemata. The example schemata shown in Figure 5 demonstrates the two interface specifications. The USE specification acts only upon entities, treating the foreign entity declaration as local. The REFERENCE specification treats the referenced objects as remaining remote but access is allowed.

3.7. EXPRESS-G

3.7.1. Basic Icons

EXPRESS-G represents an information model by graphic symbols. There are three types of EXPRESS-G symbols – definition, relationship, and composition – which are illustrated in Figure 4. The definition icons are designed to represent various kind of object types and entities. Relationship lines can be used to described supertype/

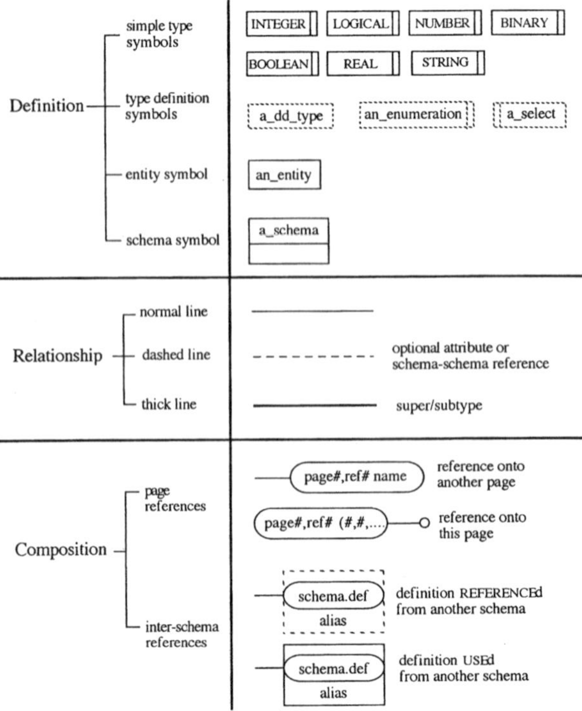

Fig. 4 Basic symbols of EXPRESS-G.

subtype, entity/attribute, schema reference relationships. The composition symbols indicate cross page references and interschema references.

3.7.2. EXPRESS-G Model

The constraint symbol is not supported but entities that are parameters in a rule and attributes of an entity constrained by either a WHERE clause or a UNIQUE clause may be flagged with an asterisk(*). The "to" end of a relationship is marked with an open circle, as shown in Figure 5.

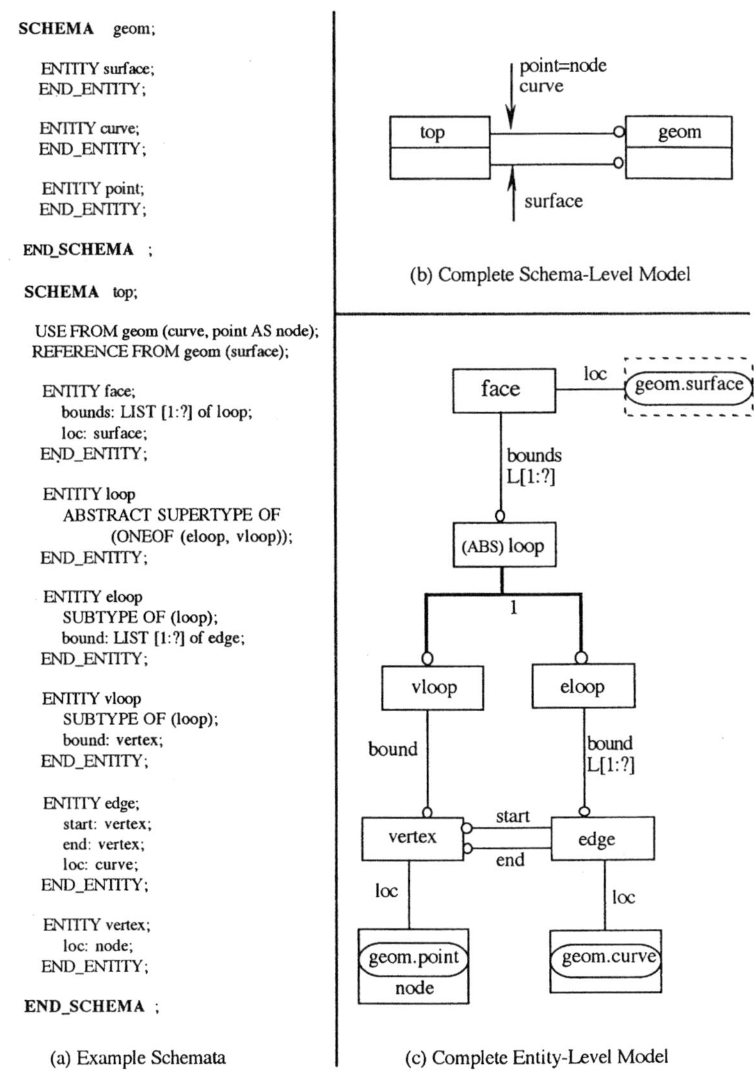

Fig. 5 Example of EXPRESS-G model.

In EXPRESS-G, the text string representing the role name (attribute) of an entity may be placed on the relationship line. An aggregation specification is indicated on the relationship line; for example, in Figure 5(a), the bounds of a face is denoted by "LIST [1:?] OF loop", and its corresponding symbol shown in Figure 5(c) is "L[1:?]" attached to the attribute string. Note that if there is no aggregation specified, then the cardinality is exactly one for a required relation and zero or one for an optional relation.

In EXPRESS, an entity may be part of an inheritance tree, it may have explicit, Derived, and Inverse attributes. A Derived attribute shall be distinguished from an explicit attribute by proceeding the name of the attribute by the characters DER enclosed in parenthesis, i.e. "(DER)". For an Inversed attribute, the attribute shall be proceeded by the symbol "(INV)".

For the super/subtype relationship, the circle end of the relationship line denotes the subtype end of the relationship. When a supertype is ABSTRACT, the symbol "(ABS)" shall precede the name of the entity within the entity symbol box. As shown in Figure 5, the ONEOF relationship can be indicated by a T-branching inheritance line from the supertype to each of its subtypes; the eloop and vloop are in a ONEOF relationship emphasized by the digit "1" being placed at the T junction.

4. THE STEP RESOURCE INFORMATION MODELS

4.1. Integrated Resources

The purpose of STEP is to specify a form for the representation and exchange of computer-interpretable product information throughout the life-cycle of a product. As mentioned previously, STEP separates the representation of product information from the implementation methods. The approach for representation is to provide one definition of product data common to many applications. These common representations, formulated in EXPRESS, are called Integrated Resources. Each Integrated Resource (IR) is represented as a collection of logically related subsets of product data. The definition of each IR is defined as general as possible and is independent of any particular implementation method.

An Integrated Resource is a resource reference model that defines a partial information model for product data. Integrated resources are integrated within a modular structure by removing redundant resource constructs across all IRs. The resource elements and constructs of the IRs are used as the basis of the definition for product data. STEP divides the IRs into two groups. The Generic Resource Model which is completely context independent; and the Application Resource Model which is applicable to a range of applications. The generic resources are independent of all applications and may reference each other. The application resources, which reference and extend the generic resources, correspond to an identified group of similar applications.

As shown in Figure 1, the STEP Integrated Resources contains thirteen parts. The first three generic resource Parts, e.g., Fundamentals of Product Description and Support (Part 41), Geometric and Topological Representation (Part 42), and Representation Structures (Part 43), provide common data elements that will necessarily appear in every application protocol. Part 41 provides three categories of entities (STEP Part 41, 1992): (1) Generic product description resources can be used to describe the framework for all of the integrated resources; (2) Generic management resources enable administrative data to be associated with product data; and (3) Support resources comprise schemata such as document, approval, contract, security classification, person organization, date/time, etc.

Part 42 is a nominal shape information model, which contains three schemata: geometry, topology, and geometric model, used to represent the shape of a component, e.g., atomic part. Part 43 defines the overall structure of representation used for associating groups of representation items into collections distinguishable from each other (STEP Part 43, 1992). This part provides mechanisms to control relationships between such collections and to prevent self-defining cyclic structures of such collections.

The Esprit CADEX project has demonstrated the exchange of files between the major European CAD/CAE vendors based on these three most basic generic resources. Five application protocols have been produced from the CADEX project (Helpenstein, 1993):

- manifold solid boundary representation models (ISO 10303 Part 204)
- surface models (ISO 10303 Part 205),
- wireframe models (ISO 10303 Part 206),
- compound solid geometry models, and
- constructive solid geometry models.

In this section the Geometric and Topological Representation Integrated Resource will be introduced first. For the feature-based applications, STEP provides the Part 48 that specifies the Form Features for the characterization and representation of shape that are of broad industrial interest. At end of this section, the STEP Form Features Generic Resource will be discussed. Note that the ISO-10303 Part 48: Form Features had been suspended. But the structure and information content of this Generic Resource Model are still worth to learn about for feature-based applications.

4.2. Part 42: Geometric and Topological Representation

Part 42 contains all the entities and associated data types and constraint functions used in the explicit representation of the shape of a product model. The nominal shape information model is a conceptual model of information which includes **geometry**, **topology**, and **geometric shapes** (STEP Part 42, 1992). This resource information model is designed to be used to represent the shape of a component (atomic part).

Geometry includes all geometric entities. The entity structure of geometry is shown in Figure 6. Geometry is the supertype of point, vector, axis_placement, coordinate_system, transformation, curve, and surface. The geometry defined in this Part is exclusively the geometry of parameteric curves and surfaces.

The topological entities are required by boundary representation B-rep, but are defined independently from B-rep. The intention is the same topology entities used in B-rep can be used for other purposes as well. For example, in electrical CAE, the topological entities can be used to capture logical connections between system components and their interconnections forming the electronic assembly. Figure 7 shows the classification structure of the topology model schema.

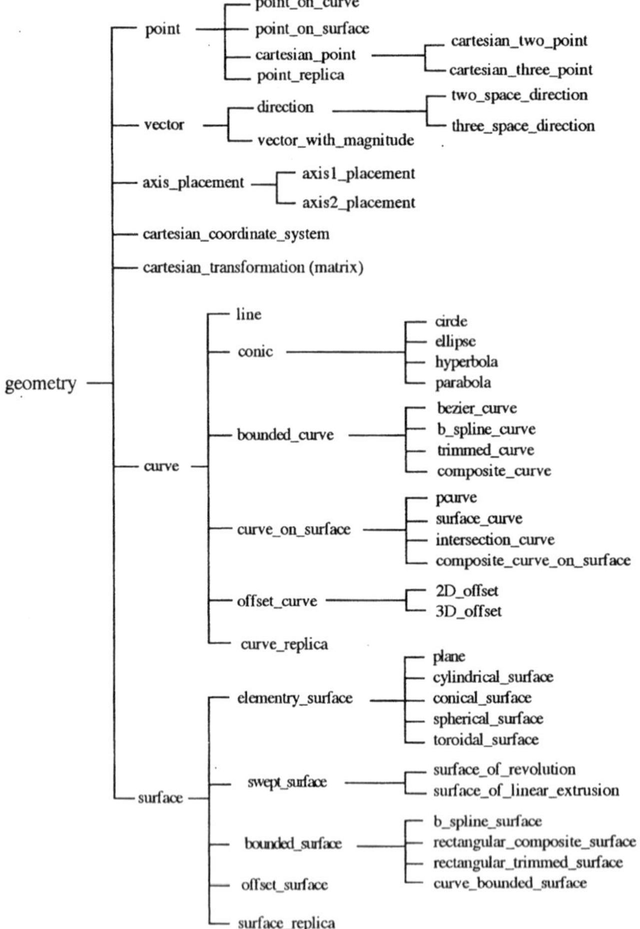

Fig. 6 Entity structure of geometry.

The geometric_shape_model is the supertype of wireframe_model, surface_model, solid_model, and geometric_set. The solid_model is a complete representation of a product's nominal shape. Any point can be classified as being inside, outside, or on the boundary of a solid. The box_domain is an orthogonal box used to limit the domain of a half_space. In a solid model, the domain is either all space (default) or box_domain. Figure 8 shows the structure of shape model entities. The reference relationships between entities illustrated in Figure 6, 7, and 8 are not shown. Readers who are interested in the details of Geometric and Topological Representation should refer to STEP Part 42.

4.3. Part 48: Form Features Representation

A form feature is a stereotypical portion of a shape (e.g., hole, pocket, wall, and rib). Form features are desirable for simplifying the man-machine interface and for

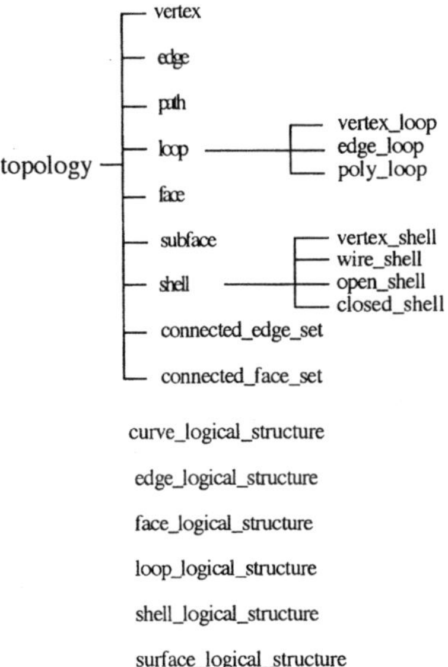

Fig. 7 The classification and structure of topology entities.

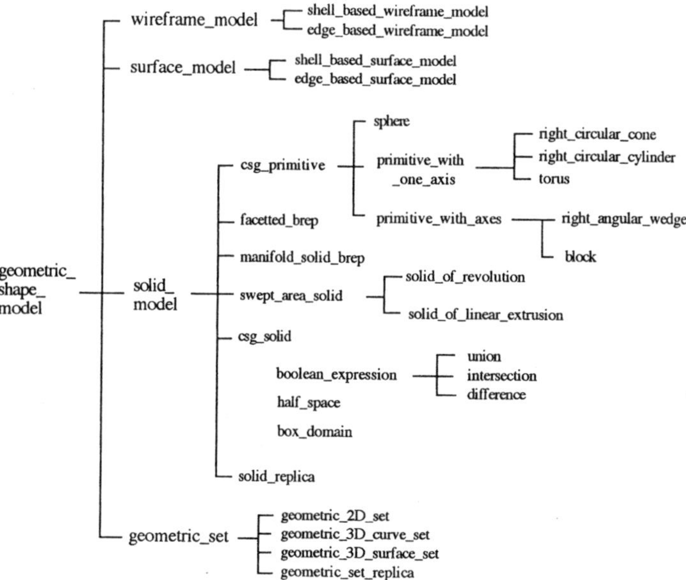

Fig. 8 The structure of geometric shape model entities.

standardizing data structures to permit automated decision systems to interface directly with the part definition. The purpose of the Form Feature Information Model (FFIM) is to develop a form feature representation schema useful for storing and using feature data.

There are two type of form feature representations:

1. Explicit form features are groupings of geometric model elements. The shape elements necessary to define the feature are listed explicitly. For example, an explicit pocket feature might be the identification of five constituent faces.
2. Implict form features model shape information parametrically rather than geometrically. For example, a hole might be described by giving a diameter and center line rather than a cylindrical surface.

A geometric model is an abstract mathematical representation of a shape. B-reps, CSG models, and wireframe models are examples. There is an obvious correspondence between shapes and geometric models – a geometric model is employed to represent a shape. In form feature representation, it is important to recognize the geometric model may differ from the shape of the product with which it corresponds. In particular, the geometric model may be considerably simpler than the product's nominal shape, the difference being made up by implicit features.

A form feature corresponds to a portion of the skin of a shape (i.e., a "dimensionality_2" shape element). As a matter of utility, a form feature should conform to some pattern or stereotype with a name; e.g., bevel gear, hole, bend. The concept of "pre-existing shape" is crucial to implicit feature representations for which an implicit form feature can not exist independently, but only applies to the pre-existing shape (Liu, 1994). Embedded in this concept is the fact that the ordering of implicit feature representations may be necessary. For example, a hole may be installed in a boss, with both being implicitly represented. While the boss representation is understandable without knowledge of the hole, the converse is not true.

Implicit feature representations can be classified according to the effect of the feature upon pre-existing shapes or more-or-less equivalently, upon data associations and interpretations (STEP Part 48, 1991):

1. Passages are subtractions of material from pre-existing shape. The subtracted volume intersects the pre-existing shape's boundary at both ends, increasing the shape's genus by 1.
2. Depressions are subtractions of material whose subtracted volume intersects the pre-existing shape's boundary at one end. The genus of the shape is unchanged.
3. Protrusions are additions of material whose added volume intersects the pre-existing shape's boundary at one end. The genus of the shape is unchanged.
4. Transitions are smoothings or gradualizations of the intersections of elements of the pre-existing shape.
5. Area features are treated as being applied to "dimensionality_2" elements of the pre-existing shape.
6. Deformations involve bending, stretching, etc. of the pre-existing shape.

The structure of implicit form feature entities is shown in Figure 9 (a) and (b). In Figure 9 (a), the form feature can be represented by the implicit form feature, implicit form feature pattern, and replicate form feature. Form features are often identical except for location. The Form Feature Information Model (FFIM) provides two means to "reuse" an implicit representation in such situations – replication and feature patterns. Replication is the modeling of a feature by declaring it to be identical to another, except for location, and specifying its location. Feature patterns

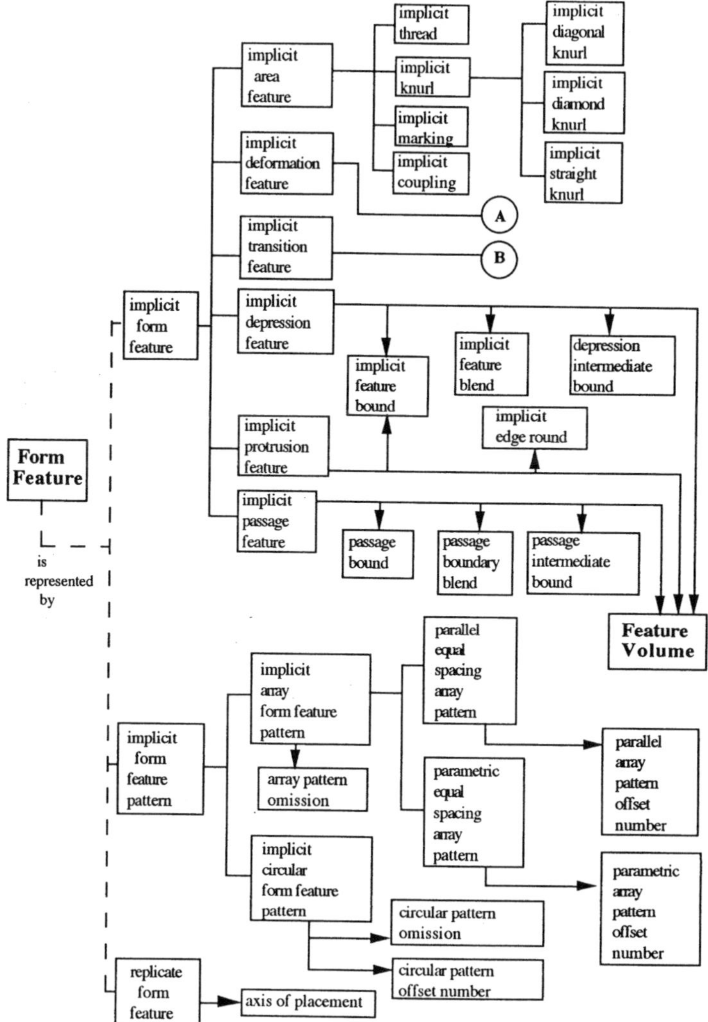

Fig. 9(a) The structure of implicit form feature entities (continued).

are a strong form of replication. An original representation is declared to be repeated in a geometric pattern. The FFIM provides for circular and matrix patterns. In replication and patterns, all information associated with the "original" feature is assumed to apply to the "copies".

In the hierarchical structure of implicit form feature entities (Figure 9 and 10), the arrow means "uses" and the solid line represents the "supertype-subtype" relationship. Also note that the feature volume is used by implicit depression, passage, and protrusion features. These three implicit form features are called volume-associated features which have definite volumetric associations (see Figure 9(a)). That is, their implicit representations are aimed at specifying a volume added to or subtracted from pre-existing shape. Sweeps and rulings are the primary vehicles for defining the

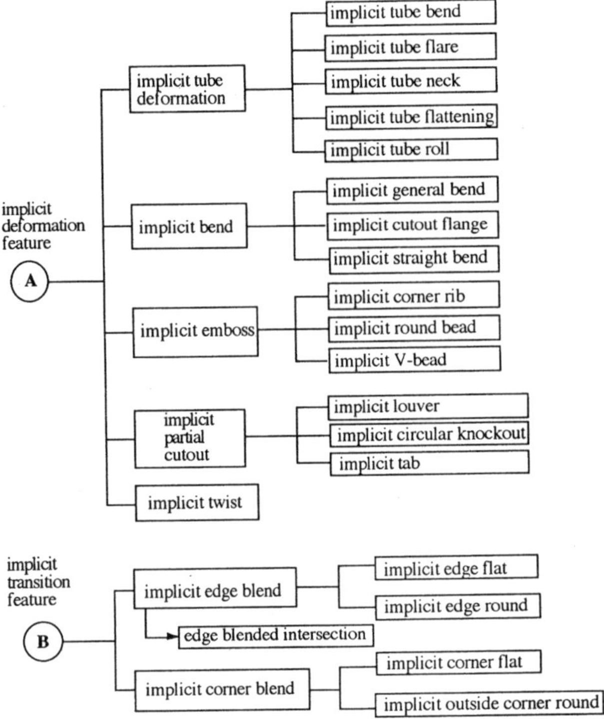

Fig. 9(b) The structure of implicit form feature entities.

added/subtracted volume. The structure of entities regarding feature volume is shown in Figure 10 (a) and (b).

A feature sweep defines a volume by a profile and a sweep path. The feature sweep is a procedural definition of a 2 1/2-dimensional shape consisting of a planar profile, a path along which the profile is to be swept, and shape definitions for either or both ends. The profile is swept normal to the path. The sweep defines a subtracted volume for swept depression and implicit passages, and an added volume for implicit protrusions. The FFIM provides for typical end conditions so the sweeps can be de facto solid sweeps. For example, an axis symmetric feature sweep can have a spherical end condition, which can be used to model sweeps of ball-nosed drill bits. A feature ruling defines a volume using a ruled surface definition. It is a generalization of a feature sweep.

Implicit feature bounds indicate where the added/subtracted volume intersects pre-existing shapes. For example, a through hole modeled as a swept implicit passage might have two such bounds identifying the hole's entry faces(s) and exit face(s). Note that this information may not be necessary, merely convenient. In the example given, the entry and exit faces could, in theory, be deduced from the pre-existing shape and the sweep-defined volume; in practice, this deduction might be quite difficult.

The STEP FFIM gives considerable attention to edge blends (chamfers, fillets, etc.) associated with volumetric features. These blend surfaces occur in three ways:

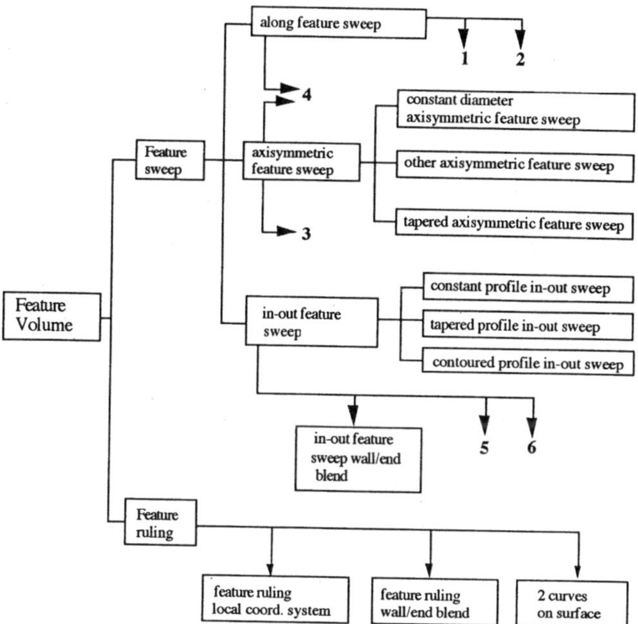

Fig. 10(a) Entities structure of feature volume (continued).

1. Intersections between portions of feature-defining volumes; e.g., between the walls of a sweep-represented pad, and/or between the ruled and one if its end surfaces in a ruling-defined pocket.
2. Intersections between the added/subtracted volumes and pre-existing shape; e.g., between pocket walls and the surface on which the pocket is installed.
3. Instersections between wall surfaces and "ends" of the feature; e.g., between walls and the bottom of a pocket.

Depending on the nature of an implicit form feature, the FFIM uses one of these three ways to locate features:

1. A local coordinate system positions an axis system; the feature's location with respect to that axis system being known.
2. A geometric entity locates the feature; e.g., a bend is located by the bend lone.
3. The feature is located by reference to the geometry of pre-existing shape; e.g., a fillet is located by identifying the surfaces it blends.

5. APPLICATION PROTOCOLS

5.1. The Role of STEP Application Protocol (AP)

Application protocols define the context, use, and kind of product data to support a specific engineering activity or related activities within the product life cycle. An application protocol is a specification for interfacing with STEP in a standard way that provides the method for defining a subset of STEP data that can be

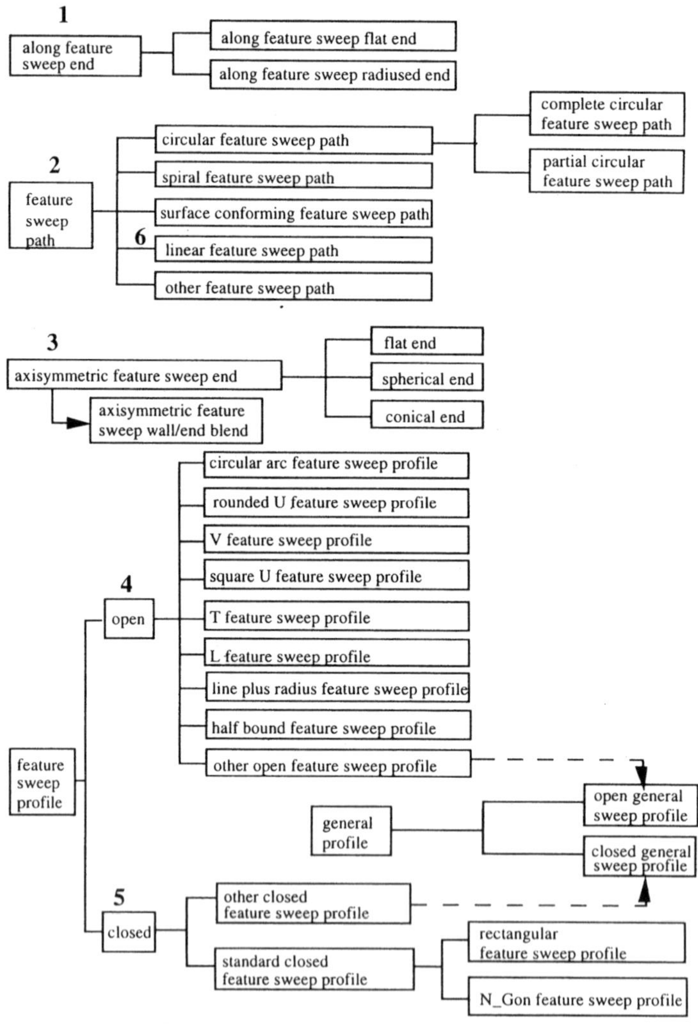

Fig. 10(b) Entities structure of feature volume.

implemented for a particular application. The AP adapts and refines the general purpose constructs in the IRs. Without APs, applications and system vendors would be free to implement non-standardized subsets of the Integrated Resources, repeating the problems of existing standards such as IGES (Owen, 1993).

Understand the development process of STEP AP is important not only for venders who want to implement standard APs, but also for the people who attempt to follow the most up-to-date information techniques to conduct STEP-based applications. Figure 11 shows the important concept of the AP. Based on the function and scope of an application for a product, the required information is extracted from STEP Integrated Resources and reorganized into an application-oriented context-driven information model which provides the basis for the AP.

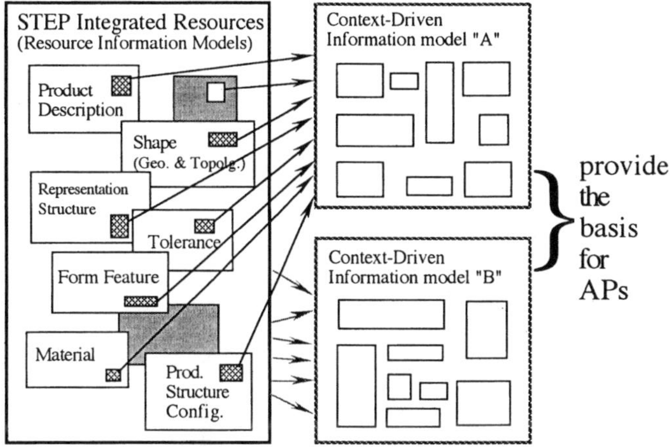

Fig. 11 Application resource models of application protocols.

5.2. The Development Process of an Application Protocol

The process of producing an AP includes five steps (Mitchell, 1991):

- developing the Scope of the AP,
- developing the Application Reference Model (ARM),
- developing the Application Interpreted Model (AIM),
- developing the Abstract Test Suite and validating, and
- finalizing conformance requirements.

Figure 12 illustrates the five steps of an AP development process. The first step of developing an AP is to document the context and requirements of the task the AP is designed to address. The result of this development step is documented as an Application Activity Model (AAM) which captures the data necessary to support the information flow within the application. The AAM can be documented and refined by the following process modeling techniques: IDEF0, NIAM, and EXPRESS-G. After clearly describing the intended use of the AP, the overall requirements become the evaluation criteria for subsequent steps.

The second stage is the development of an Application Reference Model. The ARM is the application information model, derived from the required product data, used when performing operations by the specific application. All data elements in the ARM are organized into entity definitions. The ARM can be developed by IDEF1x, NIAM, and EXPRESS and EXPRESS-G. The result of this information modeling process is then reviewed by experts to ensure self-consistency and tested by sample product data to validate the ARM.

The third phase of developing an AP is to form a specification of standardized STEP structures that achieves the detailed requirements described in the ARM. At this stage, the Application Interpreted Model is produced in which the most appropriate entities selected from the Integrated Resources are organized to represent concepts depicted in the ARM. A quality review is conducted by both

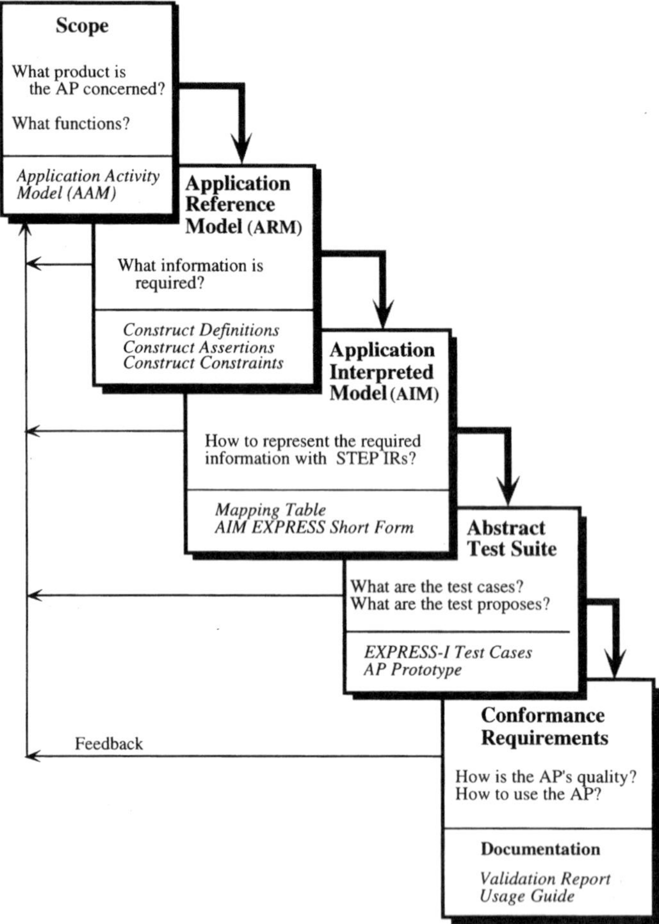

Fig. 12 Process of application protocol development.

application experts and STEP experts to ensure the successful compilation of the EXPRESS model without any loss or change of meaning in the translation from the ARM to the AIM.

In the fourth stage, an Abstract Test Suite is developed for evaluating, if the AIM representation is sufficient for the exchange of product data. The Abstract Test Suite contains a set of test cases, each abstract test case includes a specification of what information is used, how it is used, and the expected outcome of the test (STEP Part 33, 1991). For instance, there are over 120 test cases for the AP-203.

Finally, an evaluation of the conformance requirements and abstract test suite is conducted. This step analyzes the completeness of coverage, correctness, and consistency of the abstract test cases. The development and validation of an AP is an iterative process of progressive refinement. The defects uncovered by the verification and validation testing must be resolved before the final step is complete. Upon completion of the last step, a validation report is summarized to describe the results of the verification and validation activities. After this final step, the AP is ready for implementation.

5.3. Application Reference Model and Application Interpreted Model

The ARM is derived from the AAM. In the ARM, the construct definitions define all the entities and attributes of the information requirements of the AP; the construct assertions enumerate the business rules and the construct constraints explicitly state the constraints of the model. The ARM should undergo peer review by application experts to ensure that it satisfies the stated scope and that the ARM is self-consistent.

ARM ENTITY/attribute	STEP Part	AIM ENTITY/attribute
PART	41	PRODUCT ==>
	203	PART
PART/part_number	41	PRODUCT/id
PART/part_nomenclature	41	PRODUCT/name
PART/standard_part_indicator	41	PRODUCT ==>
	203	PART/standard_part_indicator
PART/part_type	41	PRODUCT_CATEGORY ==>
	41	LOWEST_LEVEL_PRODUCT_CATEGORY/name
	41	LOWEST_LEVEL_PRODUCT_CATEGORY/products
PART/part_class	41	PRODUCT ==>
	203	PART/class
	203 (1)	part_class
SURFACE_BOUNDING _TOPOLOGY/edge	42	TOPOLOGY ==>
	42	EDGE<1>
	42	EDGE/edge_curve -->
	42	curve_or_logical = "CURVE_LOGICAL_STRUCTURE"<2>

RULES:

<1> An EDGE must have an edge curve as a CURVE_LOGICAL_STRUCTURE (no edge can exist without an underlying curve structure).
<2> A CURVE_LOGICAL_STRUCTURE must have a flag attribute of TRUE.

NOTES:

(1) This is an ENUMERATION TYPE created specifically for ISO 10303-203.

LEGEND FOR ARM TO AIM MAPPING

1. Entity names are in upper case letters.
2. Attribute and defined type names are in lower case letters.
3. Selections from SELECT types are in quotes.
4. An AIM attribute that is followed by the '-->' symbol refers to another AIM ENTITY or a SELECT type, specified on the subsequent line.
5. An AIM entity that is precede or followed by the '==>' symbol states that it is a SUPERTYPE of the AIM ENTITY listed above or below it, depending on the direction of the arrow.
6. Rules are enclosed in arrows <n>, where n is the reference number of the rule stated at the end of the mapping table.
7. Notes are enclosed in parenthesis (m), where m is the reference number of the note stated at the end of the mapping table.

Fig. 13 Example of ARM to AIM mapping table.

The AIM is the intepretation of the information defined in ARM. During the AIM interpretation process, the relationship between the constructs in the ARM and in the AIM are documented. In this process, mapping tables are used to represent the correlation of ARM entities and their interpretations in the AIM. This cross reference establishes correspondence between the functionality of the ARM and the constructs in the AIM. Figure 13 illustrates a portion of the mapping table of the Configuration Controlled Design Application Protocol cited from Part 203 (STEP Part-203, 1992). The most appropriate STEP entities for representing concepts or objects depicted in the ARM are selected for use in the AIM specification. The AIM can be represented by the EXPRESS short form which specifies external schema reference, entity and constraint definitions, and type definitions.

The primary objective of STEP is to support product information sharing across a broad range of applications. Support of this sharing by applications will be facilitated by the Application Protocols through the related context among applications. Thus, sharing product data across AP boundaries is an important issue regarding the communication between different application systems. Also, an AP may contain entities not included within existing STEP Integrated Resources. The STEP community has contributed its efforts to resolve these critical issues.

6. CONCLUSIONS

The need to share complex product data across multiple enterprises using different computer systems connected by networks is growing. The ISO-10303 Standards for the Exchange of Product Model Data (STEP) addresses this need by providing information models which unambiguously describe product data for different enterprise applications. The validity of those information models is essential for success in sharing data in an automated multidisciplinary and modern business environment. The validation and conformance testings based on the National PDES Testbed at the US National Institute of Technology (NIST) was founded by the Department of Defense Computer-Aided Acquisition and Logistic Support (CALS) Office in 1990 (McLean, 1990). Validation testing is the process of ensuring STEP is usable and functionally complete, unambiguous, and consistent. On the other hand, conformance testing is the testing of a candidate product's capabilities and behavior; this helps to assure product conformity in implementations.

In addition, the STEP includes the research and development of the necessary information technologies for the envisioned shared-database environment and advanced open data systems. It is a standard for information, not merely for product data. In order to share product data between different applications running on different computer systems within the same or different organizations, STEP is intended to tackle the information technology issues, not only for the standard itsel, but also for supporting automated concurrent engineering (Shaw, 1993). Hence, once the STEP standard and its environment are in place, all types of enterprise information can be more easily shared and, of course, will provide a mechanism for multi-enterprise integration. STEP data exchange and interface standards and their supporting technologies enable and facilitate an automated form of computer-aided concurrent engineering that will initiate a new industrial revolution.

As the evolution of product data standards progresses, product data engineering will eventually become a new engineering discipline that essentially supports the computer-aided concurrent engineering. This new engineering discipline will use the following standard tools to enhance the CE communication for enterprises (Liu and Fischer, 1993b):

(1) Object-oriented concept which is a standard approach for software engineering;
(2) EXPRESS which is a standard information modeling language;
(3) STEP which is a new generation international standard for product definition;
(4) Application protocol which standardizes the interface with the STEP; and
(5) Object-oriented database which is the standard database for data intensive applications (e.g., solid modeling) and data centered architecture systems (e.g., CAD system).

The evolution of these standard technologies is promising. With their maturity, the same concept and methodology can be extended for the enterprise modeling. Just as STEP implies product data modeling, as well as the infrastructure necessary to access and contribute to product information, the enterprise modeling is armed to achieve standardized enterprise integration and to form a foundation of an enterprise integration framework. The enterprise integration framework should include the structure, methodologies, and standards to accomplish the integration of all activities of an enterprise (Caver and Bloom, 1991). The enterprise integration framework will be the next step in the evolution of engineering standards. It will magnify the STEP efforts and provide the standardized organization and formal models for the multi-enterprise concurrent engineering.

The future will likely hold expendable systems where plug-replaceability depends upon an application employing standard data format and protocols to communicate with other applications already employing common methodologies and standard tools (Goldstein, 1994). The ISO has dedicated its efforts to bring together the work of the US PDES project and the European CAD*I project into the unified standard and the IGES/PDES Organization has decided to froze IGES at Version 6.0. This emphasizes the future will lie with STEP.

References

I. Bey, D. Ball, H. Bruhm, T. Clausen, W. Jakob, O. Knudsen, E.G. Schlechtendahl, T. Sφrensen, (eds) *Neutral Interfaces in Design, Simulation, and Programming for Robotics*, ESPRIT Research Reports, Subseries PDT, NIRO Project, Springer-Verlag, Germany (1994).

H.M. Bloom (1989) The Role of the National Institute of Standards and Technology as it Relates to Product Data Driven Engineering, NIST, Center for Manufacturing Engineering, NISTIR #89–4078, Gaithersburg, MD, July 1989.

G. Booch and M. Vilot Object-Oriented Design-Inheritance Relationships, *The C+ + Report* **2** (9), pp. 8–11 (1990).

G.P. Carver and H.M. Bloom, Concurrent Engineering Through Product Data Standards, NISTIR #4573, National Institute of Standards and Technology, Gaithersburg, MD, U.S.A (1991).

A.S. Czerwinski and K. Sivyoganathan, "Development of CIM Applications from PDES/STEP Information Model, "*Concurrent Engineering: Research and Applications* (CERA), **2**(2), pp.133–136 (1994).

M.A. Ellis and B. Stroustrup, *The Annotated C+ + Reference Manual*, Addison-Wesley Publishing Company, New York (1991).

B. Evans, *Simultaneous Engineering*, The American Society of Mechanical Engineers (ASME), New York, NY, Vol. **110**, No. 2 (1988).

D. Goldstein, "An Agent-Based Architecture for Concurrent Engineering,"*Concurrent Engineering: Research and Applications* (CERA), **2**(2), pp.117–124 (1994).

H.J. Helpenstein, (ed.) CAD *Geometry Data Exchange Using STEP*, ESPRIT Research Reports, Subseries PDT, CADEX Project, Springer-Verlag, Germany (1993).

B., Henderson-Sellers, *A Book of Object-Oriented Knowledge: Object-Oriented Analysis, Design, and Implementation*, Prentice Hall, Singapore (1991).

IDEFO "ICAM Architecture, Part II, Volume IV-Function Modeling Manual (IDEFO)," Report Number AFAWL-TR-81-4203 (Available from: Mantech Technology Transfer Center, Wright-Patterson Air Force Base, OH, USA) (1981).

IDEF1*x* "Integrated Information Support System (IISS), Volume V-Common Data Model Subsystem, Part 4-Information Modeling Manual-IDEF1 Extended," Report Number AFAWL-TR-86-4006 (Available from: Mantech Technology Transfer Center, Wright-Patterson Air Force Base, OH, USA) (1985).

IPO (1993) The PRO Exchange, Official Publication of the US Product Data Association, The IGES/PDES Organization and National IGES User Group, National Computer Graphics Association (NCGA), May/June 1993.

T.-H. Liu, "An Approach for Standard Manufacturing Form Feature Applications," *Mingchi Institute of Technology Journal*, Taiwan, Vol. **26**, pp. 1–24 (1994).

T.-H., Liu and G.W. Fischer, "Developing Feature-Based Manufacturing Applications Using PDES/STEP," *International Journal of Concurrent Engineering: Research and Applications (CERA)*, **1**(1), 1993, pp. 34–45 (1993a).

T.-H. Liu and G.W. Fischer, "An Approach for PDES/STEP Compatible Concurrent Engineering Applications," North Atlantic Treaty Organization, Advanced Study Institute (NATO/ASI) Series F, *Concurrent Engineering Tools and Technologies for Mechanical System Design*, (ed. Edward J. Haug), Springer-Verlag, New York, pp. 433–464 (1993b).

T.-H. Liu and G.W. Fischer, "Assembly Evaluation Method for PDES/STEP Based Mechanical Systems," *Journal of Design and Manufacturing*, **4**(1), pp. 1–19 (1994).

C.R. McLean National PDES Testbed Strategic Plan 1990, NISTIR #4438, National Institute of Standards and Technology, Gaithersburg, MD, U.S.A. (1990)

C. McMahon and J. Browne, CADCAM: *From Principles to Practice*, Addison-Wesley Publishing Company, Taipei, Taiwan (1994).

M. Mitchell, A Proposed Testing Methodology for STEP Application Protocol Validation, NIST, National PDES Testbed, NISTIR #4684, Gaithersburg, MD, U.S.A (1991).

G.M. Nijssen and T.A. Halpin, *Conceptual Schema and Relational Database Design: A Fact Oriented Approach*, Prentice Hall, New York (1989).

J. Owen, *STEP: An Introduction*, Information Geometers Ltd., Winchester, UK (1993).

L.-H. Qiao, C. Zhang, T.-H. Liu, H.-P.B. Wang and G.W. Fischer "A PDES/STEP Based Product Data Preparation Procedure for Computer-Aided Process Planning," *International Journal of Computers in Industry*, **21**(1), pp. 11–22 (1993).

N. Shaw, Interfacing Technology for Manufacturing and Industry: From Islands of Automation to Continents of Standardization and Beyond, Proceedings of the IFIP-TC5/WG5.10 Working Conference, Darmstadt, Germany, March 15–17, 1993, North-Holland, IFIP Transactions B-10 (1993).

B.M. Smith, Product Data Exchange: The PDES Project-Status and Objectives, NIST, Center for Manufacturing Engineering, NISTIR #89-4165, Gaithersburg, MD, U.S.A. (1989).

C. Stark and M. Mitchell, Development Plan: Application Protocol for Mechanical Parts Production, NIST, National PDES Testbed, NISTIR #4628, Gaithersburg, MD, U.S.A. (1991).

STEP Part 1 Product Data Representation and Exchange – Part 1: Overview and Fundamental Principles (ISO CD 10303-1), National Institute of Standards and Technology, Gaithersburg, MD, U.S.A. (1992).

STEP Part 11 Exchange of Product Model Data – Part 11: Descriptive Methods: The EXPRESS Language Reference Manual (ISO CD 10303-11). National Institute of Standards and Technology, Gaithersburg, MD, U.S.A. (1992).

STEP Part 203 product Data Representation and Exchange – Part 203: Application Protocol: Configuration Controlled Design (ISO CD 10303-203), National Institute of Standards and Technology, Gaithersburg, MD, U.S.A. (1992).

STEP Part 33 Conformance Testing Methodology and Framework: Structure and Development of Abstract Test Suites," ISO TC184/SC4/WG6, Working Draft N15 (ed. J. Owen), April 17, (1991).

STEP Part 41 Exchange of Product Model Data – Part 41: Integrated Resources: Fundamentals of Product Description and Support (ISO CD 10303-41), National Institute of Standards and Technology, Gaithersburg, MD, U.S.A. (1991).

STEP Part 42 Exchange of Product Model Data – Part 42: Integrated Resources: Geometric and Topological Representation (ISO CD 10303-42), National Institute of Standards and Technology, Gaithersburg, MD, U.S.A. (1991).

STEP Part 43 Exchange of Product Model Data – Part 43: Integrated Resources: Representation Structures (ISO CD 10303-43), National Institute of Standards and Technology, Gaithersburg, MD, U.S.A. (1992).

STEP Part 48 Exchange of Product Model Data – Part 48: Integrated Resources: Form Features (ISO CD 10303-48), National Institute of Standards and Technology, Gaithersburg, MD, U.S.A. (1991).

B.D. Warthen, (ed.) Product Data International, Vol. **3**, No. 6, Warthen Communications, p. 7. (1992).

J. Weiss, "STEP Functional Requirement," ISO TC184/SC4, Document N30, May 1988.

R.S. Wiener and L.J. Pinson, *An Introduction to Object-Oriented Programming and C++*, Addison-Wesley Publishing Company, New York, (1988).

J.-K. Wu, T.-H. Liu and G.W. Fischer, "PDES/STEP Based information Model for CAE and CAM Integration," *International Journal of Systems Automation: Research and Applications (SARA)*, **2**(4), pp. 375–394, (1992).

7. ABBREVIATIONS AND ACRONYMS

(CE)	Concurrent Engineering
(IGES)	Initial Graphics Exchange Specification
(ANSI)	American National Standard Institute
(SET)	Standard d'Exchange et de Transfer
(VDA)	Verband der Automobilindustrie
(ESPRIT)	European Specific Programme for Research and Development in Information Technology
(ESPRIT CAD*I)	ESPRIT Computer-Aided Design Interfaces
(STEP)	STandard for the Exchange of Product model data
(PDES)	Product Data Exchange Using STEP
(IPO SC)	IGES/PDES Organization Steering Committee
(ANS)	American National Standard
(DIS)	Draft International Standard
(ISO)	International Organization of Standards
(NIAM)	Nijssen Information Analysis data Modeling
(DM)	Description Method
(IM)	Implementation Method
(AIM)	Application Interpreted Model
(IR)	Integrated Resource
(AP)	Application Protocol
(CTS)	Conformance Testing Suite
(FFIM)	Form Feature Information Model
(ARM)	Application Reference Model
(AAM)	Application Activity Model
(NIST)	National Institute of Technology
(CALS)	Computer-Aided Acquisition and Logistic Support

CHAPTER 3

Quality Function Deployment as a Tool for Integrated Product and Process Design

JULIE K. SPOERRE

Industrial Engineering Department, College of Engineering, Florida A&M University/Florida State University, 2525 Pottsdamer Street, Tallahassee, Florida 32310, USA

1. INTRODUCTION

In the past several decades, a significant transformation has occurred in the way companies conduct their business. While, at one time, it was sufficient to manufacture a particular product as long as the quality and cost to consumer remained competitive with the industry as a whole, this is no longer the case. Due to the fiercely competitive nature of the consumer markets today, every company must ensure that each step taken in the product life cycle is performed with the customer in mind. If this is not accomplished, the company will certainly suffer the consequences. It cannot be overstated that the satisfaction of the consumer must be the top priority of modern companies if they are to enjoy success now and for years to come.

Although many companies believe that change must take place, and quickly, few fully understand exactly what change is needed and in what direction the change should take place. Without a clearly defined objective, many industries go through the motion of implementing modifications, not realizing that they are, in reality, spinning their wheels. Once the first hurdle of determining a common goal is overcome, the next task is to develop or implement the tools necessary to achieve the overall objective. For example, it is widely agreed upon among all sectors of industry that the customer must be given primary importance. This being satisfied, how are the customer's preferences integrated with the existing products and processes?

The answer to this question lies in quality function deployment (QFD), the focus of this chapter. Quality function deployment has been defined as converting the consumers' demands into "quality characteristics" and developing a design quality for the finished product by systematically deploying the relationships between the demands and the characteristics, starting with the quality of each functional component and extending the deployment to the quality of each part and process (Hauser and Clausing, 1988). QFD can also be used to determine the feasibility of introducing a new product into the market and, thus, can be viewed as a strategic planning tool (Maddux, *et al.*, 1991).

Contrary to traditional technology-driven processes, QFD is a customer-driven process (Brown, 1991). This has created the need for an efficient and accurate method for obtaining specific information on what the customer wants. Without access to such information or the willingness to put a great deal of effort into finding this information, QFD cannot be pursued further. Without customer input, it is impossible for a company to form a successful strategy for their role in the marketplace.

There are obviously many advantages to using QFD. First and foremost, by identifying important product features to the customer, the success of the product is greatly increased (Akao, 1990). Based on the desires of the customer, designers and process planners can work together to create an efficient design and process for the manufacture of the product; i.e. QFD promotes an integrated product/process design approach. Another significant benefit of using QFD is that the information attained is analyzed objectively. Since the customer evaluates all competitors on an equal basis, the data is extremely useful in describing the position of the company relative to its competitors. Finally, quality improvement is realized since the most critical process parameters are identified and efforts are focused on controlling these parameters (Gopalakrishnan, *et al.*, 1992).

Dowlatshahi (1994) proposed two broad categories in defining the advantages of concurrent engineering, and subsequently, QFD. The categories and a brief description are provided below:

1. Reduction in product development lead time.
   During product development in concurrent engineering, all activities are carried out on a parallel basis, as opposed to the traditional serial basis. This approach has the ability to reduce the entire product design cycle time and, at the same time, reduce the duplication of effort and costly product redesigns.
   In order for a reduction in product development lead time to occur, an efficient flow information and communication must exist among the concurrent engineering team in the early design stage.

2. Overall cost savings.
   One of the outcomes of concurrent engineering is designing parts for manfacturability. Through a careful review of process planning activities,

total product costs are lowered via reduction in the number of parts to be manufactured, improved machine utilization time, more easily manufactured parts, and fewer rework and scrap items.

The primary tool in QFD is the house of quality, which pictorially defines all the necessary information to market a successful product – customer requirements, engineering characteristics, bench marking against competitors, etc. Although the house of quality, in itself, requires a significant amount of work, the benefits from this systematic method of product development are well worth the effort.

There are four different phases that comprise the entire production cycle in QFD: product planning, part deployment, process planning and production planning. Each phase is defined by similar house structures. The focus of this chapter will be on the product planning house, commonly referred to as the house of quality. An example will be used throughout the text in order to clarify the discussion and provide a practical representation of QFD. In addition, the other three houses will be described briefly to provide a complete picture of the concept of quality function deployment.

2. DEFINITION OF QFD

Several definitions of QFD have been provided by Prasad (1994):

- QFD is a planning tool for translating customers' needs and expectations into products' requirements.
- QFD is a systematic disciplined method that assists users in prioritizing resources, such as when to focus time and where to concentrate on product improvements.
- QFD is a type of conceptual road map that provides the means for inter-functional planning and communications.
- QFD is a mechanism to bring new products and improved existing products to market sooner with lower cost and higher quality.
- QFD is a system that ensures that product and processes will be designed right the first time and everything will fall in place.

By definition, QFD supports an integrated product/process design approach through improved communication among all departments from the beginning of the product's conception, or design improvement stage, to the final stage of product distribution. Through an interdisciplinary approach, the QFD objectives of designing customer-driven products/processes and manufacturing these products at a higher quality and lower cost are accomplished.

3. PRODUCT PLANNING

3.1. House of Quality

There are six major components in the initial house of quality: (1) customer requirements – WHATs, (2) engineering characteristics – HOWs, (3) customer requirements/engineering characteristics relationship matrix, (4) engineering characteristics correlation matrix, (5) customer competitive assessment and (6) engineering

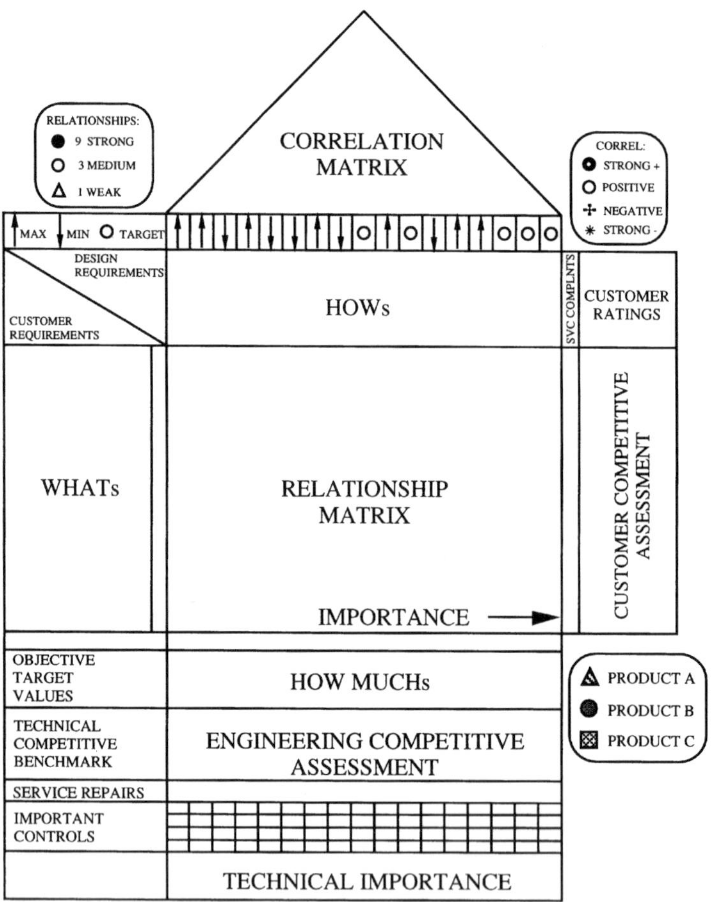

Fig. 1 House of quality components.

characteristics targets – HOW MUCHs. In addition, other components include customer importance ratings, engineering competitive assessment, and technical importance ratings. Figure 1 shows the skeleton of the house of quality with each of the sections labeled.

Each of the components will be explained and an example will be provided. The product that is utilized in this sample case is a full-size upright refrigerator.

3.2. Customer Requirements

The first step in creating the house of quality is to define the customer requirements for the product under study. In Figure 1, the customer requirements are labeled as "WHATs"; i.e., What features and characteristics are required by the end consumer of the product? This task alone takes a considerable amount of time. It is imperative that the information attained at this step is accurate and truly representative of the target consumer population.

It is this step that is the crux of the concept of quality function deployment since, by definition, it is a system for incorporating consumer requirements into the company requirements. Because of its importance, the need to continually redefine the needs of the customer cannot be overstated. Companies must never lose touch with their customers, especially in today's world where customer's attitudes, competitive products and the economic environment are undergoing constant change.

The stage of defining the WHATs has often been referred to as determining the "voice" of the customer. But whose voice? The list below describes some of those voices that may be involved:

- Customers
- Owners
- Those who bought your products
- Those who bought competitor's products
- Those who switched to your competitor
- Those who switched from your competitor to you
- Those who are satisfied with your product
- Those who are not satisfied with your product

In addition to the technical requirements of the product, the QFD team must identify the emotional requirements of the customers. For example, consider the example of a full-size refrigerator. Technical requirements include refrigerator capacity, electrical consumption and temperature consistency within the refrigerator cabinet. On the other hand, emotional requirements may be how well the refrigerator blends with the existing appliances, available colors or comfort features, such as an ice maker or water dispenser.

It is important for the QFD team to consider the three types of quality that customers consider, consciously or subconsciously, when purchasing a product. These quality types are basic features, performance features and excitement features. Basic features are those characteristics of a product that are expected and, in most cases, taken for granted by the consumer. For example, when looking for a refrigerator, the customer never gives a thought to whether or not the refrigerator will work when it is plugged in. These features evoke a low level of satisfaction from the customer, but if absent, a tremendous amount of dissatisfaction occurs. Performance features are those requirements that are spoken by the consumer and obtained through surveys, questionnaires, etc. Some examples of performance features on a refrigerator include adjustable shelves, automatic defrost, water dispenser and ice maker. These features do not prevent the customer from purchasing the product, but yield a high level of satisfaction if they are present. Excitement features, in contrast, are unexpected surprises to the consumer. These features are not asked for or even considered, yet they provide a significant increase in satisfaction to the end user of the product. An alarm that beeps when the refrigerator door is open or an electronic touchpad on the door to adjust the inside temperature are examples of excitement features.

After a sufficient amount of information is collected, the QFD team has the task of organizing and consolidating the customer wants. Figure 2 provides a sample listing of the customer requirements. At this time, an affinity diagram is useful. This tool gathers large amounts of language data and organizes them into groupings

"Easy to clean"	"Makes ice fast"
"Can see food in drawers"	"Easy to reach items in back"
"Coils are easy to clean"	"Easy to clean behind"
"Temperature controls easy to read"	"Door closes easily"
"Temperature controls easy to adjust"	"Handle doesn't pinch"
"Lot of room for vegetables"	"Freezer is large"
"Stays cold"	"Doesn't waste energy"
"Items don't freeze in fridge"	"Shelves are adjustable"
"Good warranty"	"Runs quietly"
"Recyclable"	"etc."
"Built to last"	"etc."
"Holds a lot"	"etc."

Fig. 2 Voice of the customer raw data.

based on the natural relationship between each item (Bossert, 1991). Once the raw data is identified, the items are grouped into categories based on a natural relationship among items. Each of these natural groupings is given a title.

An affinity diagram for the raw data (see Figure 2) is shown in Figure 3. The assigned category headings are "ease of use", "quality", "efficiency", "flexibility" and "size". At this point, the customers' requirements have been organized and consolidated. This information is placed in the WHAT category of the house of quality, as shown in Figure 4.

For each customer attribute (CA), a customer importance rating is assigned. Again, this information is discovered through customer surveys, interviews, company database information or other sources. These importance ratings are vital in defining the most critical technical requirements and in resolving conflicts found in the correlation matrix. This will be discussed in greater detail later.

By developing a product complaint history, a company can identify which features or characteristics of each product give customers the most problems. This can help to pinpoint design flaws and allow the QFD team to focus on the necessary product/process design changes for overall improvement of the product's life cycle. In doing so, the integrated product/process design approach is enhanced.

3.3. Engineering Characteristics

The next logical step in the development of the house of quality is translating the WHATs into the engineering characteristics, or HOWs (see Figure 1); i.e., given the

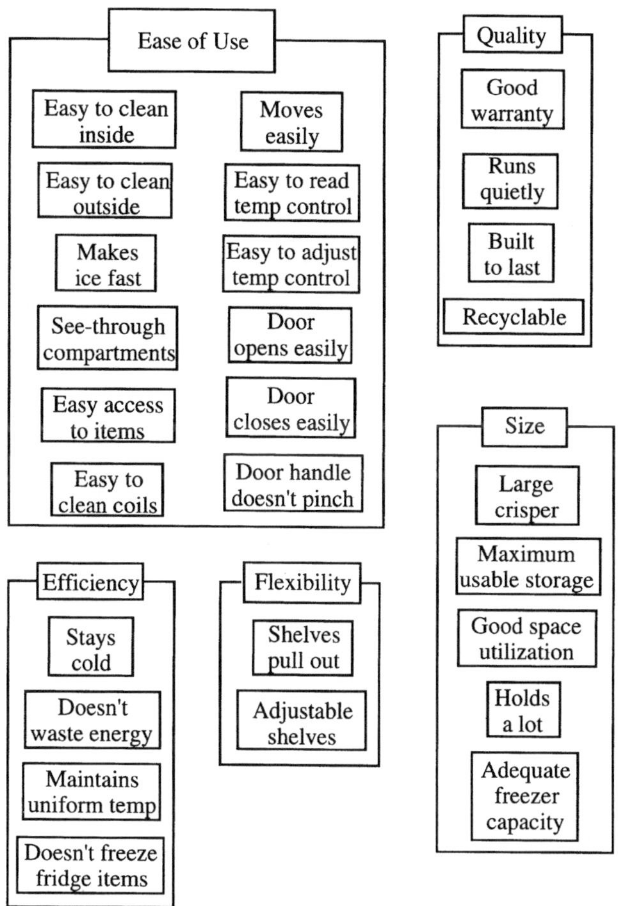

Fig. 3 Affinity diagram for full-size refrigerator.

customer requirements, what are the technical requirements needed to make the desired customer attributes a reality?

There are several major points to consider when establishing the engineering characteristics. Design requirements should meet the following conditions:

- Measurable (quantifiable)
- Meaningful to the manufacturer
- No design constraints (no parts or processes)
- Overall product evaluation criteria

A word of warning is necessary at this stage. Even though accurate customer information has been collected up to this point, the voice of the customer can be easily lost unless the engineering characteristics are developed with considerable thought and study. The QFD team must ensure that the proposed design requirements are fulfilling the true needs of the customer and not just the needs of the

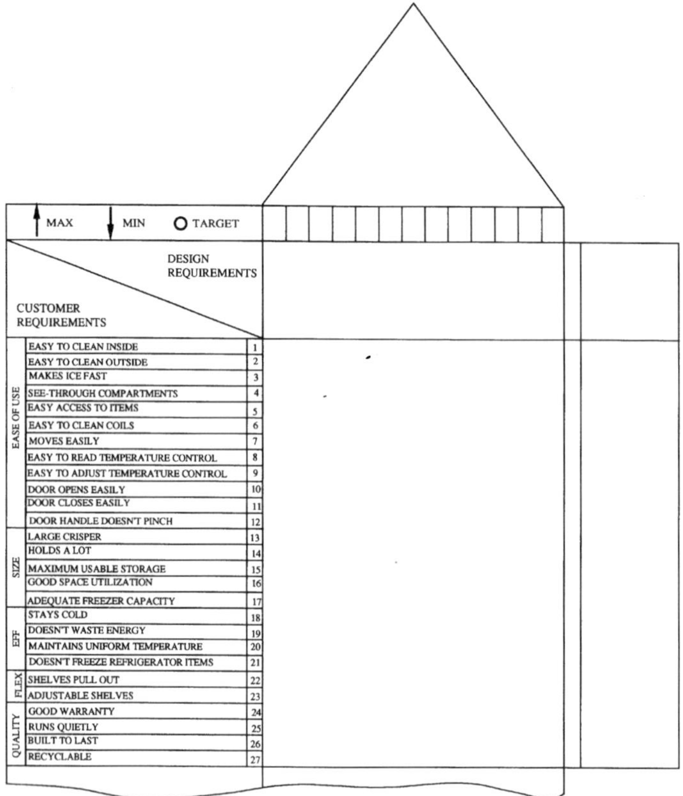

Fig. 4 Customer attributes in product planning phase.

manufacturer. This is why it is so important to have a cross-functional team in which all departments are represented. The generation of the HOWs is a creative process and all members need to have the opportunity to provide input. Each department has invaluable information that is necessary for the continued success of the QFD project.

Due to the critical and creative nature of establishing the HOWs, a brainstorming session is a practical technique. All team members gather to discuss their ideas and thoughts, with the purpose of creating an appropriate and customer-oriented list of engineering characteristics. A sample item list for the previously defined category "efficient" is provided in Figure 5. Once the brainstorming session is complete, a fish-bone diagram will aid in organizing the results of brainstorming. A fishbone diagram is given for the category "efficient" in Figure 6.

A similar brainstorming session is conducted for the remaining headings under the WHATs to determine the engineering characteristics that are required to satisfy the customer attributes. As described in the section on customer attributes, once all the engineering characteristics are defined, an affinity diagram will assist in consolidating and organizing this criteria. In Figure 7, the HOW section of the house of quality is completed for all the items in the WHAT section of the house of quality.

- Amount of heat removed
- Electricity consumed
- SPC
- Operator training
- Degree change in temperature when door is opened/closed
- Temperature gradient in refrigerator
- Temperature gradient in freezer
- Insulating material
- Heat dissipation
- Accurate temperature controls
- Only runs when needed to maintain temperature
- Seal on refrigerator door
- Independent of indoor temperature changes
- Adjusts to changes in number of items stored

Fig. 5 Brainstorming to establish the HOWS.

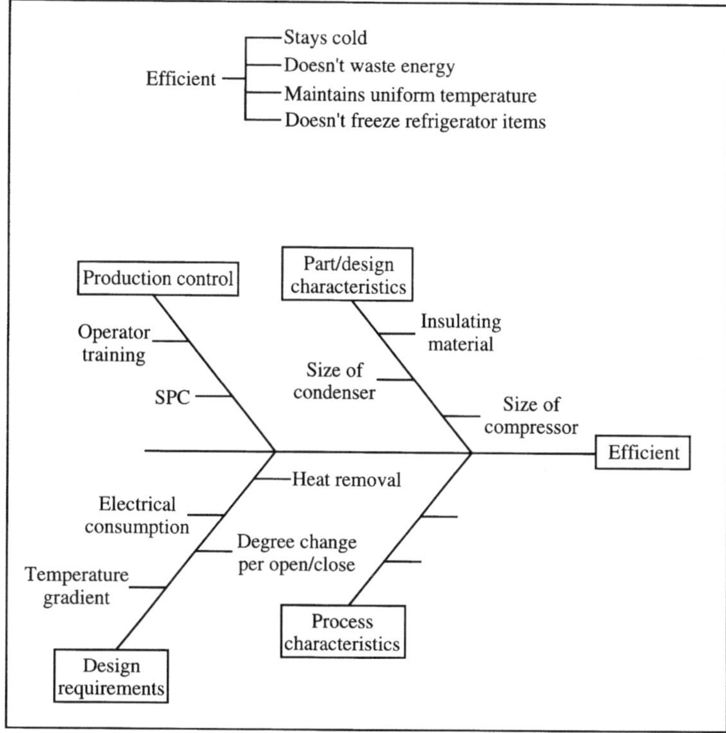

Fig. 6 Establishing the HOWs using a fishbone diagram.

3.4. Relationship Matrix

Referring to Figure 1, the center of the house of quality is labeled "RELATIONSHIP MATRIX". In this matrix, the relationship between the customer attributes

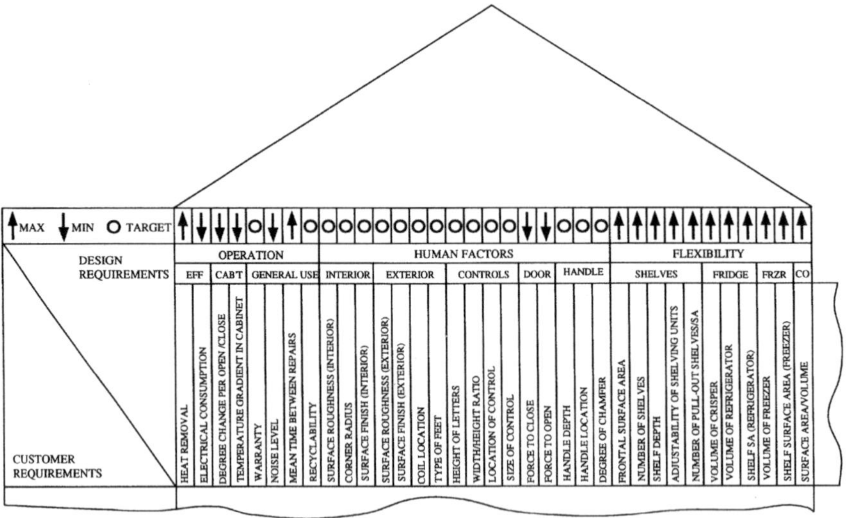

Fig. 7 Engineering characterisitcs in product planning phase.

(WHATs) and the engineering characteristics (HOWs) are defined. Symbols are most commonly used to define the relationships. The notations that will be used in the refrigerator example are provided below:

- • Strong (9)
- o Moderate (3)
- △ Weak (1)

The number in parentheses is the strength given to the relationship between each WHAT/HOW pair when that particular symbol is used. If the design requirement has a significant influence on satisfying the customer want, the intersecting square in the matrix will contain a filled circle which represents nine points. For example, the design requirement "heat removal" has obvious effect on the customer attribute "stays cold"; thus, a value of nine will be assigned to the position in the matrix where these two items intersect. If the design requirement has a lesser influence on the customer want, a value of three is used; a weak influence only contributes a value of one. Figure 8 shows the completed relationship matrix for the refrigerator example.

3.5. Correlation Matrix

The correlation matrix (see Figure 1), located in the roof of the house of quality, is developed to define which of the design requirements are related and the degree of correlation between these requirements. As in the case of the relationship matrix, the correlation matrix uses symbols to define either a positive or negative relationship. The notation that was selected for the refrigerator example is described as:

- • Strong +
- o Positive
- + Negative
- ∗ Strong -

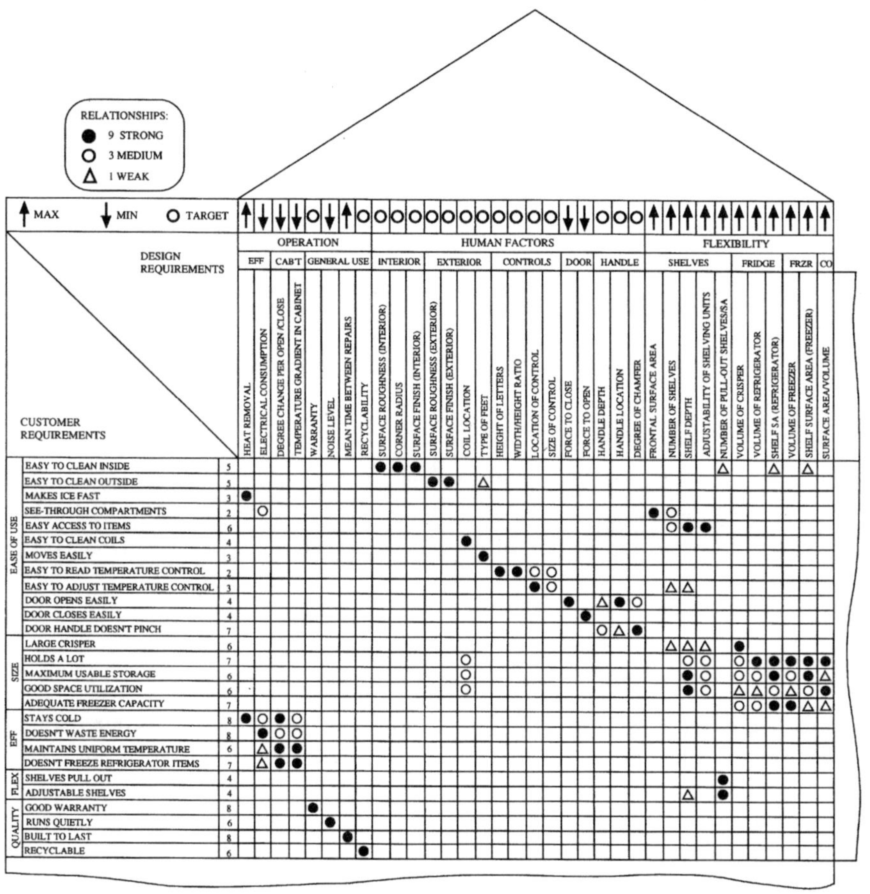

Fig. 8 WHAT vs. HOW relationship matrix in product planning stage.

As an example, the design requirements "heat removal" and "electrical consumption" have a strong negative correlation. The objectives of minimizing the electrical consumption and maximizing the heat removal are conflicting requirements. Once the conflicting objectives are discovered, steps are taken to resolve the issue, keeping in mind the voice of the customer.

The completed correlation matrix is shown in Figure 9 for the full-size refrigerator.

3.6. Customer Importance Rating/Benchmarking

3.6.1. Customer Importance Ratings

Once the customer requirements have been collected, it is necessary to know how important each characteristic is to the consumer. This knowledge serves two main purposes: (1) it provides a logical focus for the QFD team, since the attributes that are regarded as the most important by the customers will be stressed throughout the

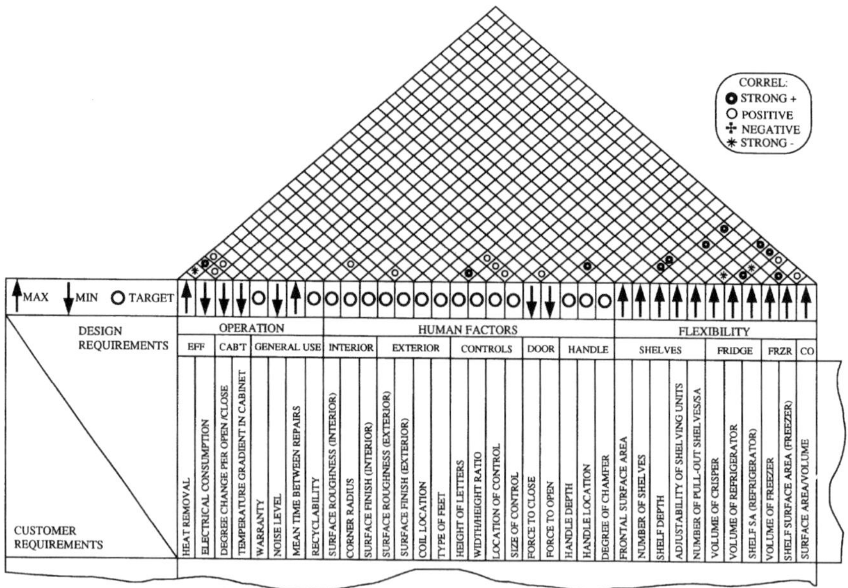

Fig. 9 HOW correlation matrix in product planning phase.

entire quality function deployment process, and (2) when conflicts occur within the correlation matrix, i.e., negative correlations, the customer importance ratings assist in their resolution.

In the refrigerator example, the highest customer importance ratings were "stay cold", "doesn't waste energy", "good warranty", and "built to last", each assigned a rating of eight.

3.6.2. Benchmarking

Customer Competitive Assessment

It is important not only to obtain the relative importance placed on each attribute by the consumer, but to gain information on how well your company compares with the competition in regard to the customer wants. For each customer attribute, the consumer will rate its performance, as well as that of the company's main competitors. Typically, a scale from one to five is used, with five being the best and one being the worst. It is necessary that the same criteria be used for each customer attribute from each of the competitors, i.e., the ratings of one competitor should be relative to the ratings of the other competitors.

For ease in evaluating the ranking of your company relative to that of the competitors, it is beneficial to use symbolic notation instead of numbers. In addition, it is common practice to connect the points related to your company in order to identify where you are positioned relative to other companies. If all the customer attribute ratings for your company are higher than the ratings for your competitors on the customer competitive assessment graph, your company is in a good competitive position for that particular product.

Engineering Competitive Assessment

Once the design requirements are set, they are evaluated relative to the same design requirements of the top competitors. If data on an existing product from the competition is not available, it is sometimes necessary to purchase or rent the competitors' products to conduct a comparative in-house analysis.

In a manner similar to the customer competitive assessment, each engineering characteristic is rated among all competitors and charted on a graph. Once again, the points related to your company are connected in order to provide a quick visual check to assess where you are positioned relative to other companies. For instance, if all the engineering characteristic ratings for your company are higher than the ratings for your competition on the engineering competitive assessment graph, your product is technically superior to that of the competitors.

Figure 10 provides the customer importance ratings for each customer requirement, as well as information in the customer competitive assessment section. The engineering competitive assessment benchmarks are shown in Figure 11.

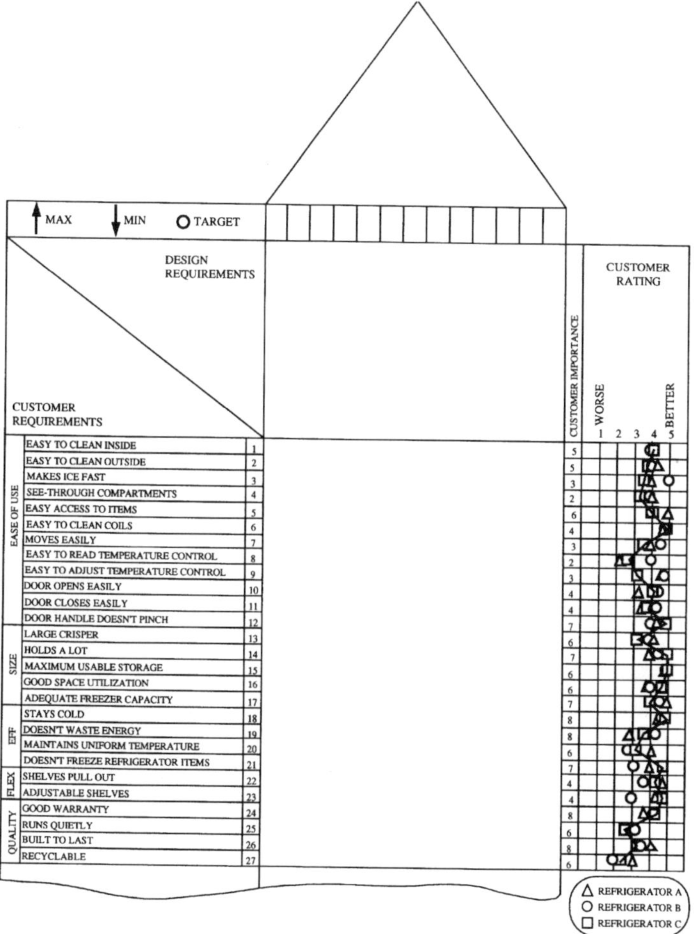

Fig. 10 Customer importance ratings and customer competitive assessment.

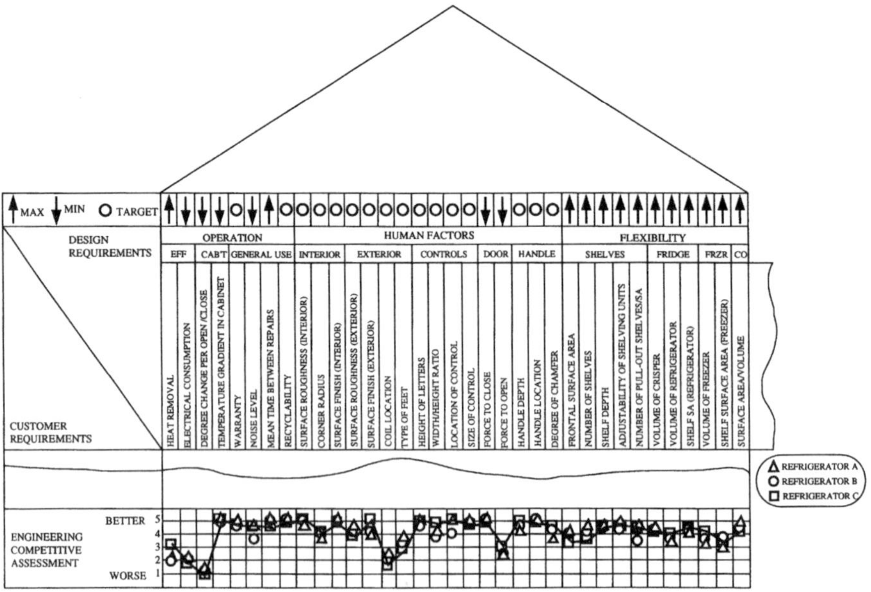

Fig. 11 Engineering competitive assessment.

3.7. Engineering Characteristics Target Values

The engineering characteristics target values are also known as the HOW MUCHs in the house of quality (see Figure 1). The QFD team assigns specific target values for each of the items in the HOW section and these are recorded in the lower portion of the house. Selection of the objective target values serve several purposes in that they provide: (1) an objective means of assuring that the engineering requirements are met, (2) targets for further detailed developments, and (3) a basis for benchmarking.

Figure 12 shows the engineering characteristic target values (objective target values) in the house of quality.

3.8. Technical Importance

After the customer requirements, engineering characteristics, matrices, benchmarking and target values have been determined, the final step in the product planning phase is to assign each engineering characteristic a technical importance value.

At the bottom of Figure 13 are the technical importance ratings of all engineering requirements. The calculations of these values are determined by studying each column of the relationship matrix and multiplying each symbol in that column times the customer importance rating given to the corresponding customer requirement. For example, in column 1, representing "heat removal", the corresponding technical importance is calculated by: $9(3) + 9(8) = 99$ and column 30, or "Volume of crisper", is given a technical importance of: $9(6) + 3(7) + 3(6) + 3(7) = 120$.

The complete house of quality for a full-size refrigerator is shown in Figure 14. Once all sections of the house of quality have been completed, it is important to

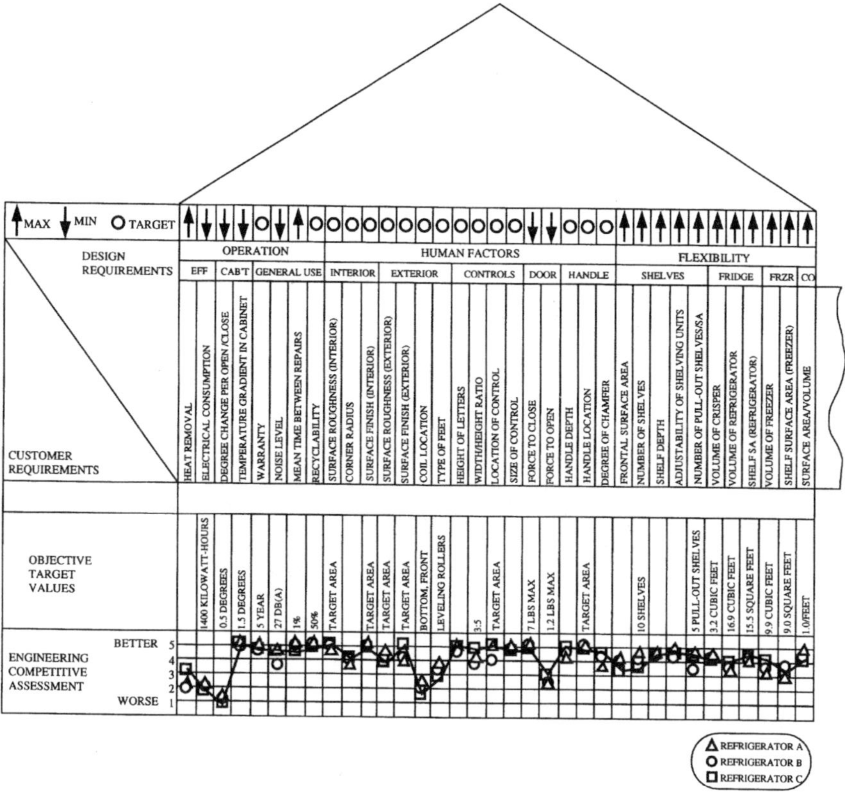

Fig. 12 Engineering characteristic target values.

analyze and diagnose the product planning matrix. In particular, special attention should be given to the following items:

1. Blank row/columns
   A blank row indicates that there is an insufficient number of engineering characteristics. The customer attribute corresponding to that row is not satisfied by any of the HOW items. An additional engineering requirement(s) should be included to satisfy the WHAT. Likewise, a blank column implies that, for the corresponding engineering characteristic, there is no customer requirement that relates to it. This could be a result of an unspoken quality, in which case, the engineering characteristic should remain. Otherwise, there is no need to "overdesign" the product.
2. Conflicts
   There may be conflicts that arise between the customer competitive assessment and the engineering competitive assessment. For each customer attribute rating, the engineering characteristics that have a strong relationship with that customer want should be rated similarly. For example, in the refrigerator house of quality, the customer rating for the attribute "doesn't freeze refrigerator items" is very good. The engineering characteristics that have a strong relationship with this attribute are "degree of change per open/close"

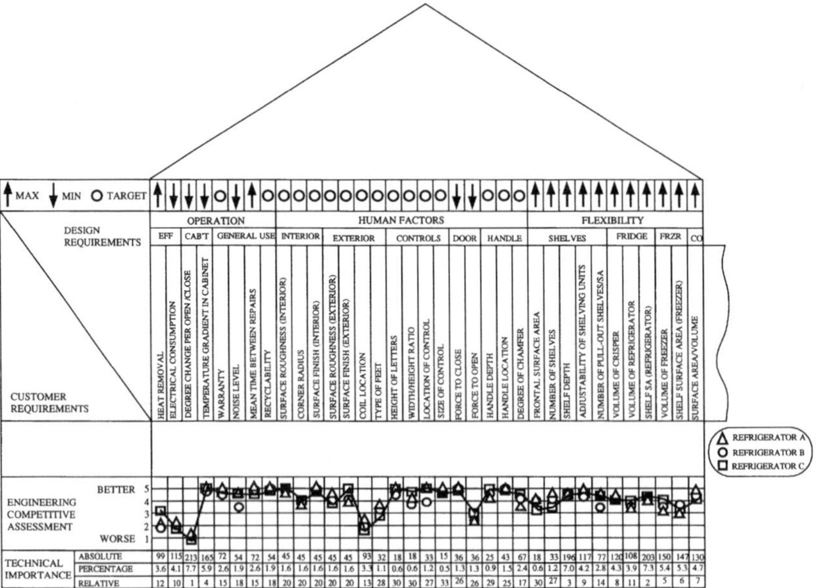

Fig. 13 Technical importance of engineering characteristics.

and "temperature gradient in cabinet"; therefore, the technical assessment of these requirements should also be very good. In this example, the assessment of "degree change per open/close" is low. One of two possibilities exist: (1) the customer demand and competitive assessment is not well understood, or (2) the engineering assessment is not well understood. In either case, further evaluation needs to be conducted.

3. Sales points

   Look for features of your product that are superior to that of the competitors. These strong points can be exploited in advertising schemes. At the same time, the customer assessment highlights features that are clearly inferior and need further work to improve the product and remain competitive.

4. Resolve negative correlations

   Identify research and development areas to minimize the degree of conflict among engineering characteristics. If this is not possible, an adjustment of the target values can help achieve a workable range for the product design. Finally, if the previous two alternatives are not feasible, the technical importance levels of the conflicting items should be noted and utilized in determining which HOW has the greatest overall impact on satisfying the customers' needs.

5. Decide what design requirements should be "deployed" to Phase II

   In conjunction with the technical importance ratings at the bottom of the house of quality, decide which design requirements need further refinement. At this decision-making point, the focus should be on items that are important, new or difficult since these items will result in high rewards if given attention and high risk if not.

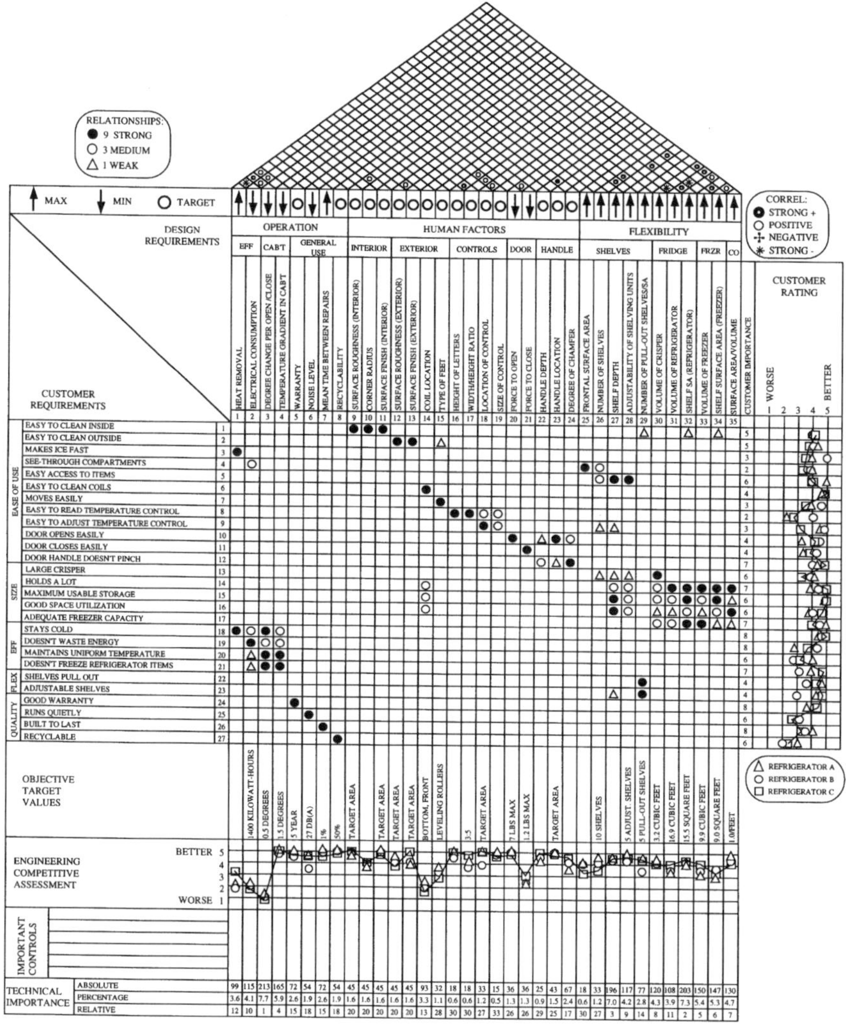

Fig. 14 House of quality for full-size refrigerator.

4. PART DEPLOYMENT

Once the house of quality has been completed in the product planning phase, the most important engineering requirements are carried over to the part deployment phase. In this phase, there are several objectives: selection of the best design concept, determination of critical parts and determination of critical part characteristics.

In surveys, interviews of questionnaires conducted during the product planning phase, there are many basic qualities that will not be mentioned by the consumer. However, it is important that all these attributes be considered in product development to assure completeness of the part deployment phase. As an example, most customers would not mention the insulating material or door seal within the

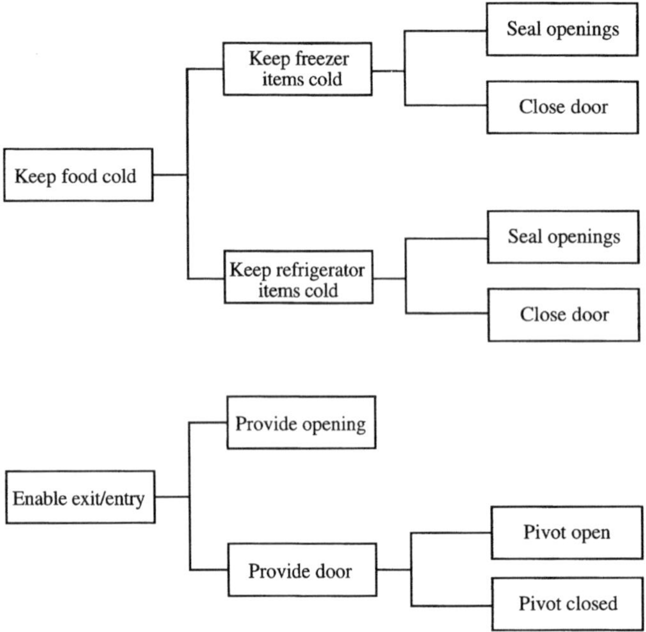

Fig. 15 Function tree for full-size refrigerator.

refrigerator. But such items are critical to the function of the product, therefore, called functional requirements.

The initial step in the part deployment phase is to identify all the functional requirements of the product. In order to organize the results of a brainstorming session to define the functional requirements, a function tree is commonly used. The function tree defines, first of all, the features that will make a product work. Beginning with the global functional requirements, the function tree evolves into more detailed requirements which can be assigned target values. A partial function tree for a full-size refrigerator is provided in Figure 15.

From the function tree, the functional requirements can be obtained and target values assigned to each. These requirements are added to the existing list of engineering requirements that were exported from the product planning phase. The part deployment matrix showing the design requirements is given in Figure 16.

As in the product planning phase, the final step in the part deployment matrix is to identify the critical part characteristic values and the importance rating of each critical part characteristic. Figure 17 provides a representation of all components of the part deployment matrix.

The next step in the part deployment phase is to identify the systems, parts and critical part characteristics of the product. At this step, all components of the product are identified and linked to the design requirements. A partial part deployment matrix is shown in Figure 18 for the systems, parts and critical part characteristics.

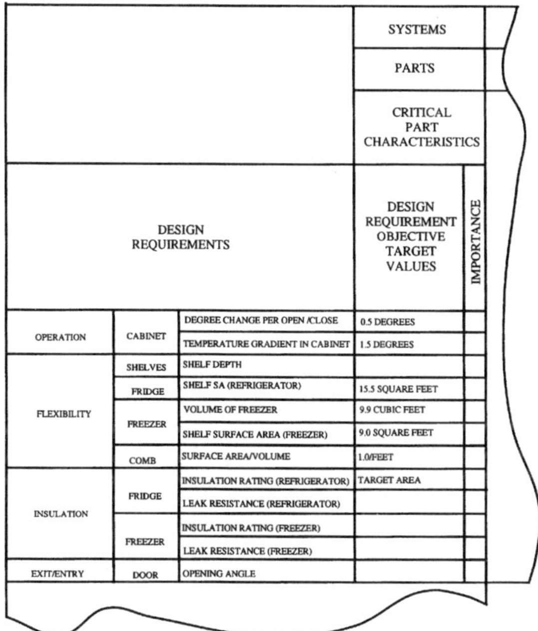

Fig. 16 Design requirements in part deployment matrix.

5. PROCESS PLANNING

The third phase in quality function deployment is the process planning stage. The primary objectives of this phase are to determine the best process/design combination, determine the critical process parameters, establish process parameter target values and determine items for further development.

Initially, critical part characteristics defined in the part deployment stage are selected as key characteristics for the process planning stage. The part characteristics selected should be those that are most sensitive to manufacturing and processing variations. These items require special attention, while others are accommodated by the normal system.

As in the previous stages of quality function deployment, the "upper" portion of the house is completed by defining the relationship between this section and the "middle" region of the house. In this case, the "middle" section represents the critical parts and part characteristics and the "upper" section defines the critical process parameters related to the parts/part characteristics. In Figure 19, some of the critical parts/part characteristics selected from the part deployment matrix are shown, as well as the critical process parameters related to those parts/part characteristics.

At this point, it may be useful to analyze several processing alternatives. There are several techniques for doing this. An explanation of the Pugh concept for process selection is provided in American Supplier Institute QFD Implementation manual (1989) and the process decision program chart (PDPC) is described in Bossert (1991), to provide a few examples.

			SYSTEMS		DOOR SYSTEM						REFRIGERAT'N SYSTEM							REFRIGTOR				FREEZER			
			PARTS		DOOR		GASKET			HD	COND		COMP			EVAP		SHELF		CAB		SHELF		CAB	
			CRITICAL PART CHARACTERISTICS		GASKET HOLE SIZE	SEAL SURFACE LOCATION	SECTION	COMPRESSION	SET	ATTACHMENT	VOLUMETRIC EFFICIENCY	REFRIGERATION EFFECT	COMPRESSION RATIO	SUCTION PRESS	CAPACITY	EVAPORATING TEMPERATURE	COOLING CAPACITY/TEMP Δ	SURFACE AREA	DEPTH	VOLUME	INSULATION RATING	SURFACE AREA	DEPTH	VOLUME	INSULATION RATING
DESIGN REQUIREMENTS			DESIGN REQUIREMENT OBJECTIVE TARGET VALUES	IMPORTANCE																					
OPERATION	CABINET	DEGREE CHANGE PER OPEN /CLOSE	0.5 DEGREES	5																					
		TEMPERATURE GRADIENT IN CABINET	1.5 DEGREES	4																					
FLEXIBILITY	SHELVES	SHELF DEPTH		5																					
	FRIDGE	SHELF SA (REFRIGERATOR)	15.5 SQUARE FEET	5																					
	FREEZER	VOLUME OF FREEZER	9.9 CUBIC FEET	4																					
		SHELF SURFACE AREA (FREEZER)	9.0 SQUARE FEET	3																					
	COMB	SURFACE AREA/VOLUME	1.0/FEET	3																					
INSULATION	FRIDGE	INSULATION RATING (REFRIGERATOR)	TARGET AREA	5																					
		LEAK RESISTANCE (REFRIGERATOR)		4																					
	FREEZER	INSULATION RATING (FREEZER)		5																					
		LEAK RESISTANCE (FREEZER)		4																					
EXIT/ENTRY	DOOR	OPENING ANGLE		3																					

Fig. 17 Critical part characteristics in part deployment matrix.

			SYSTEMS		DOOR SYSTEM					REFRIGERAT'N SYSTEM							REFRIG'R				FREEZER				
			PARTS		DOOR		GASKET			COND		COMP			EVAP		SHELF		CAB		SHELF		CAB		
DESIGN REQUIREMENTS			DESIGN REQUIREMENT OBJECTIVE TARGET VALUES (CRITICAL PART CHARACTERISTICS)	IMPORTANCE	GASKET HOLE SIZE	SEAL SURFACE LOCATION	SECTION	COMPRESSION	SET	VOLUMETRIC EFFICIENCY	REFRIGERATION EFFECT	COMPRESSION RATIO	SUCTION PRESS	CAPACITY	EVAPORATING TEMPERATURE	COOLING CAPACITY/TEMP Δ	SURFACE AREA	DEPTH	VOLUME	INSULATION RATING	SURFACE AREA	DEPTH	VOLUME	INSULATION RATING	
OPERATION	CABINET	DEGREE CHANGE PER OPEN /CLOSE	0.5 DEGREES	5						●	●	△	△	●	△	●				△				△	
		TEMPERATURE GRADIENT IN CABINET	1.5 DEGREES	4			△	△	△		●			●						△				△	
FLEXIBILITY	SHELVES	SHELF DEPTH		5													○	●	○			△			
	FRIDGE	SHELF SA (REFRIGERATOR)	15.5 SQUARE FEET	5													●	○	○			△	△		
	FREEZER	VOLUME OF FREEZER	9.9 CUBIC FEET	4															△		○	○	●		
		SHELF SURFACE AREA (FREEZER)	9.0 SQUARE FEET	3																	●	○	○		
	COMB	SURFACE AREA/VOLUME	1.0/FEET	3													●	○	●		●	○	●		
INSULATION	FRIDGE	INSULATION RATING (REFRIGERATOR)	TARGET AREA	5																●					
		LEAK RESISTANCE (REFRIGERATOR)		4	○	○	●	●	●											○					
	FREEZER	INSULATION RATING (FREEZER)		5																				●	
		LEAK RESISTANCE (FREEZER)		4	○	○	●	●	●															○	
CRITICAL PART CHARACTERISTIC VALUES																									
IMPORTANCE		ABSOLUTE			24	24	76	76	76	45	81	5	5	81	5	45	87	75	61	66	66	40	77	66	1147
		RELATIVE			2	2	7	7	7	4	7	1	1	8	1	4	8	7	5	6	6	3	7	6	

Fig. 18 Complete part deployment matrix.

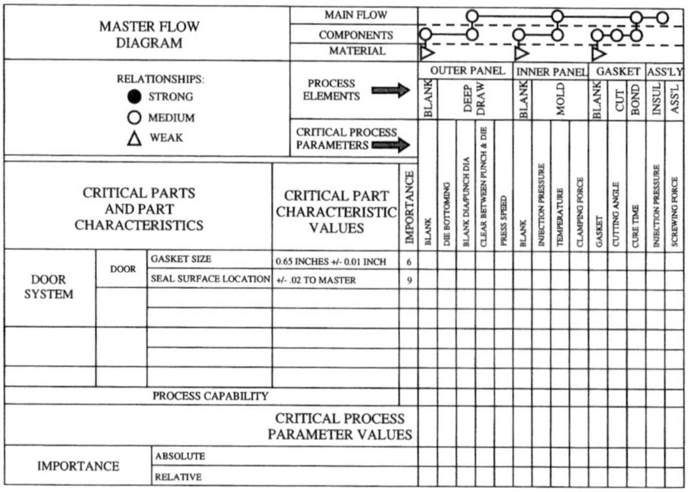

MASTER FLOW DIAGRAM — MAIN FLOW / COMPONENTS / MATERIAL

RELATIONSHIPS: ● STRONG ○ MEDIUM △ WEAK

PROCESS ELEMENTS: OUTER PANEL (BLANK, DEEP DRAW); INNER PANEL (BLANK, MOLD); GASKET (BLANK, CUT, BOND); ASS'LY (INSUL, ASS'L)

CRITICAL PARTS AND PART CHARACTERISTICS			CRITICAL PART CHARACTERISTIC VALUES	IMPORTANCE	BLANK	DIE BOTTOMING	BLANK DIA/PUNCH DIA	CLEAR BETWEEN PUNCH & DIE	PRESS SPEED	BLANK	INJECTION PRESSURE	TEMPERATURE	CLAMPING FORCE	GASKET	CUTTING ANGLE	CURE TIME	INJECTION PRESSURE	SCREWING FORCE
DOOR SYSTEM	DOOR	GASKET SIZE	0.65 INCHES +/- 0.01 INCH	6														
		SEAL SURFACE LOCATION	+/- .02 TO MASTER	9														
PROCESS CAPABILITY																		
CRITICAL PROCESS PARAMETER VALUES																		
IMPORTANCE	ABSOLUTE																	
	RELATIVE																	

Fig. 19 Critical process parameters in process planning matrix.

Once the process elements have been determined, based on the process selection, a master flow diagram will assist in identifying the critical process parameters. This information will then be included in the top portion of the process planning matrix.

Finally, the relationship matrix is completed, target values are defined and importance ratings assigned to the critical process parameters. An example of the process planning matrix for a full-size refrigerator is shown in Figure 20. This figure is only partially complete – the entire system includes the door, refrigerator compartment, freezer compartment and refrigeration system. Process planning on the door, however, provides an adequate representation of this phase in the quality function deployment cycle.

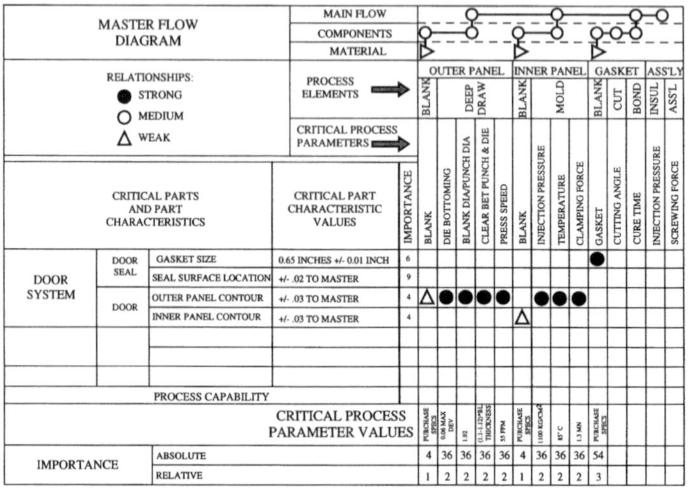

MASTER FLOW DIAGRAM — MAIN FLOW / COMPONENTS / MATERIAL

RELATIONSHIPS: ● STRONG ○ MEDIUM △ WEAK

PROCESS ELEMENTS: OUTER PANEL (BLANK, DEEP DRAW); INNER PANEL (BLANK, MOLD); GASKET (BLANK, CUT, BOND); ASS'LY (INSUL, ASS'L)

CRITICAL PARTS AND PART CHARACTERISTICS			CRITICAL PART CHARACTERISTIC VALUES	IMPORTANCE	BLANK	DIE BOTTOMING	BLANK DIA/PUNCH DIA	CLEAR BET PUNCH & DIE	PRESS SPEED	BLANK	INJECTION PRESSURE	TEMPERATURE	CLAMPING FORCE	GASKET	CUTTING ANGLE	CURE TIME	INJECTION PRESSURE	SCREWING FORCE
DOOR SYSTEM	DOOR SEAL	GASKET SIZE	0.65 INCHES +/- 0.01 INCH	6										●				
		SEAL SURFACE LOCATION	+/- .02 TO MASTER	9														
	DOOR	OUTER PANEL CONTOUR	+/- .03 TO MASTER	4	△	●	●	●	●		●	●	●					
		INNER PANEL CONTOUR	+/- .03 TO MASTER	4						△								
PROCESS CAPABILITY																		
CRITICAL PROCESS PARAMETER VALUES					PURCHASE SPECS	0.06 MAX DEV	1.92	(1.1-1.12)*BL THICKNESS	55 PPM	PURCHASE SPECS	1100 KG/CM2	85° C	1.3 MN	PURCHASE SPECS				
IMPORTANCE	ABSOLUTE				4	36	36	36	36	4	36	36	36	54				
	RELATIVE				1	2	2	2	2	1	2	2	2	3				

Fig. 20 Process planning matrix for full-size refrigerator.

FLOW DIAG.			PROCESS ELEMENTS		CRITICAL PROCESS PARAMETERS	CRITICAL PART CHARACTERISTIC VALUES	PROCESS CAPABILITY	IMPORTANCE	OPERATION EVALUATION					PLANNING REQ'MENTS				OPERATION INFO			
MATERIAL	SUB-ASSEMBLY	MAIN FLOW							DIFFICULTY	FREQUENCY	SEVERITY	ABLE TO DETECT	TOTAL POINTS	QC CHARTS	PREV. MAINT. STD.	MISTAKE PROOF.	ED. & TRAIN. REQ'D	JOB INSTRUCTIONS	LABOR: DIRECT	LABOR: INDIRECT	CYCLE TIME
			OUTER PANEL	BLANK																	
			OUTER PANEL	DEEP DRAW	DIE BOTTOMING	0.06 MX DEV		5	1	1	3	1	15					B1			.012
			OUTER PANEL	DEEP DRAW	BLANK DIA/PUNCH DIA	1.92		4	2	1	3	2	48					A2			
			OUTER PANEL	DEEP DRAW	CLEAR BET PUNCH & DIE	(1.1-1.12)*BT		4	3	2	2	3	144					C3			
			OUTER PANEL	DEEP DRAW	PRESS SPEED	55 FPM		3	1	3	2	3	54					B3			
			INNER PANEL	BLANK																	
			INNER PANEL	MOLD	INJECTION PRESSURE	1100 KG/CM^2		5	3	1	3	1	45					B2			
			INNER PANEL	MOLD	TEMPERATURE	85° C		4	2	2	3	2	96					A1			
			INNER PANEL	MOLD	CLAMPING FORCE	1.3 MN		4	2	1	2	3	48					C1			
			GASKET	BLANK																	
			GASKET	CUT	CUTTING ANGLE																
			GASKET	BOND	CURE TIME																
			ASS'LY	INSUL	INJECTION PRESSURE																
			ASS'LY	ASS'L	SCREWING FORCE																

Fig. 21 Production planning matrix.

6. PRODUCTION PLANNING

The results of the part deployment and process planning stages are the determination of critical part characteristics that must be monitored and the critical process parameters which must be controlled in order to maintain the quality of the manufactured part. The production planning phase ensures that all efforts up until this point have not been in vain, i.e., a quality assurance program is put into place to assure that the voice of the customer is embodied in the end product.

The production planning matrix lists the critical process parameters selected from the process planning stage and includes an operation evaluation (degree of difficulty, frequency, severity and ability to detect) for each item. A total points column, as shown in Figure 21 is determined by the multiplication of every column in the operation evaluation section and the importance rating column. In addition, selection of the necessary planning requirements (quality control charting, preventive maintenance schedules, etc.) for each of the process parameters is included in the production planning matrix, as well as operator information that includes job instruction forms, labor information and cycle times.

7. CONCLUSION

The four phases in QFD – product planning, part deployment, process planning and production planning – have been described in detail in the above text. The product planning phase defines the customer requirements and translates these into the necessary technical requirements. In part deployment, the critical technical requirements are utilized in identifying the parts and critical part characteristics of the product. The critical part characteristics are carried through to the next phase, process planning, where the corresponding process parameters and process parameter values are selected. Finally, in production planning, the critical process parameters from the previous phase are placed into a production planning matrix. This matrix includes information on each operation, planning requirements for each process parameter and operator instructions.

Quality function deployment provides many tools for promoting an integrated environment through interdepartmental cooperation. From the initial planning, all members of the organization are involved, each providing unique insight into the product. In addition to satisfying the customers' needs, significant reductions in product development lead time and an overall increase in cost savings result. As more companies move towards integrated product, process and enterprise design, quality function deployment will become widely accepted and recognized as a tool for attaining high quality, low cost products while keeping the customers' preferences a priority.

References

Y. Akao, *Quality Function Deployment: Integrating Customer Requirements into Product Design,* Productivity Press, Cambridge, MA (1990).

American Supplier Institute, Inc. *Quality Function Deployment Implementation manual*, Version 3.0. (1980).

J.L. Bossert, *Quality Function Deployment: A Practitioner's Approach, ASQC Quality Press, Marcel Dekker, Inc* (1991).

P.G. Brown, QFD: Echoing the Voice of the Customer. *AT&T Technical Journal*, March/April, 18–29 (1991).

K.N. Gopalakrishnan, B.E. McIntyre and J.C. Sprague, Implementing Internal Quality Improvement with the House of Quality. *Quality Progress*, September, 57–60 (1992).

J.R. Hauser and D. Clausing, The House of Quality.*Harvard Business Review*, May/June, 63–73 (1988).

G.A. Maddux, R.W. Amos and A.R. Wyskida, Organizations Can Apply Quality Function Deployment as Strategic Planning Tool. *Industrial Engineering*, September, 33–37 (1991).

B. Prasad, Product Planning Optimization Using Quality Function Deployment, in *Artificial Intelligence in Optimal Design and Manufacturing*, Prentice Hall, Inc., pp. 117–152 (1994).

CHAPTER 4

Computer Aided Tolerance Analysis

CHUAN-JUN SU[a] and TIEN-LUNG SUN[b]

[a]*Department of Industrial Engineering and Engineering Management, Hong Kong University of Science and Technology;* [b]*Knowledge Based Systems Laboratory, Industrial Engineering Department, Texas A&M University*

1. INTRODUCTION

Ever since manufacturing processes introduced imperfections and designers started assigning tolerances to parts, tolerances have significantly affected product cost, quality, and ultimate usability.

Computer-aided approaches to tolerance analysis and synthesis fall into several different categories (Lu *et al.*, 1991). The earliest works focused on conventional tolerances, i.e., manipulation of dimensions and the analysis of tolerance specifications. More recent work has focused on the issues for both the conventional tolerances and the geometric tolerances, i.e., manipulation of dimensions and the analysis of tolerance specifications. More recent work has focused on the issues for both the conventional tolerances, and the geometric tolerances, i.e., the underlying theory of geometric tolerancing, composition, and properties such as validity and sufficiency.

The techniques used for analyzing conventional tolerance come in two distinct flavors (Greenwood *et al.*, 1987). The simpler sort of tolerance analysis is called worst-case analysis in which dimensions have conventional tolerances and the output equals the nominal value of the dimension, and its upper and lower limits. More complex tolerance analysis techniques allow statistical tolerances, and output a statistical distribution for the design function.

The conventional tolerance is ambiguous and inadequate in capturing the desired functionality. Recent works in tolerancing focus on the geometric tolerances, which constrain portions of the part surface to specific regions of space called tolerance zones. Several underlying theories for representation of the geometrical tolerances by computer have been proposed. Based on these theories, research on the synthesis and analysis of geometric tolerance have been suggested. Integration of the tolerance information with the solid model data base also has been studied (Requicha 1983, 1984).

2. CONVENTIONAL TOLERANCES AND THE PARAMETRIC APPROACH

(Hillyard and Braid 1978) present a theory to explain how dimensions and views combine to specify the shape of a mechanical component. They derived the Jacobian criterion from physical principles by using analogies from structural engineering. For the purpose of dimensioning, an object is regarded as an engineering frame structure whose members and joints correspond to the edges and vertices of the object. Adding a dimension is analogous to adding a *stiffener* to the frame. Figure 1 depicts two simple stiffeners among the eleven stiffeners defined in their work: a strut, which is a stiffener specifying a length dimension, and a web, specifying an angle dimension. Information carried by a stiffener includes the *separation*, stored as a unit vector and a magnitude, together with a real number for tolerance. The arrow in Figure 1(a) shows the direction of the separation vector. In Figure 1(b), however, the direction of the separation vector is perpendicular to the paper. In each case, the tolerance is $\pm\tau$.

To determine whether a component is under, over, or exactly defined by a given dimensioning scheme, a dimensioning scheme can be expressed as a matrix form:

$$\boldsymbol{Rd} = \boldsymbol{u} \tag{1}$$

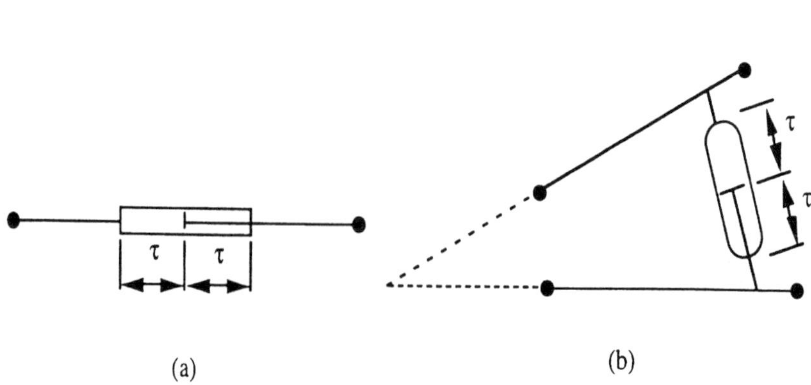

Fig. 1 Examples of simple stiffeners: (a) A strut (b) A web.

where **R** is a rigidity matrix, **d** is the displacement vector, and **u** is the variation vector. If **R** is singular, then there exists some redundancy in the scheme, i.e., over toleranced. If **R** is nonsingular, we can rewrite Eq. (1) as:

$$\boldsymbol{d} = \boldsymbol{R}^{-1}\boldsymbol{u} \tag{2}$$

and use this equation to perform the sensitivity analysis to examine the relationship between the nominal shape and small changes in the dimensions.

The work of Hillyard and Braid, however, only focused on the conventional tolerances and on the polyhedral representation.

(Hoffmann, 1982) described a method to solve the problem of part design analysis and tolerance allocation by reducing the problem to the analysis of systems of linear inequalities. The corresponding inequality for a tolerance specification is written as:

$$L \leq f(x) \leq U \tag{3}$$

where x is the parameter vector of a part, f is the tolerance function, and L and U are the lower and upper bounds of the tolerance zone. Using the first two items of Taylor's expansion for $f(x)$ in equation Eq. (3), a system of linear inequalities can be derived:

$$\boldsymbol{L} \leq \boldsymbol{f}(\boldsymbol{x_0}) + \boldsymbol{T}\Delta\boldsymbol{x} \leq \boldsymbol{U} \tag{4}$$

where **L** and **U** are vectors of tolerance bounds, **f** is a vector of tolerance function, **T** is the matrix of partial derivatives of **f** in $\mathbf{x_0}$, and the parts $\mathbf{x_0}$ are nominal parts of the given tolerance specification. For a consistent, determinate, and not over-dimensioned tolerance specification, the matrix **T** should be invertible. Since the operations and setups of a manufacturing process are inaccurate, different executions of a given process plan will result in parts with different parameter vectors. If S_p denotes the set of parameter vectors for a process plan P, then S_p is a function of operation and setup working dimensions and inaccuracies:

$$S_p = \{\boldsymbol{x} = \boldsymbol{x_0} + \Delta\boldsymbol{x_0} + \Delta\boldsymbol{x} : \Delta\boldsymbol{x} \in \Pi(\boldsymbol{h})\} \tag{5}$$

where $\mathbf{x_0} + \Delta\mathbf{x_0}$ is the part produced if all operations result in their working dimensions, **h** is a vector of *process and setup* inaccuracy parameters and Π is a function of the working dimension inaccuracies caused by operation and setup.
We may assume that $\Pi(\boldsymbol{h})$ has the form:

$$|\,\Delta\boldsymbol{x}\,| \leq P.\boldsymbol{h}, \quad \boldsymbol{h} > 0 \tag{6}$$

The process plan P satisfies a given tolerance specification if $S_p \subseteq M$, where M is the set of different solutions of the tolerance specification. Considering Eq. (3) to Eq. (6), the condition $S_p \subseteq M$ can be expressed as:

$$\boldsymbol{L} - \boldsymbol{f}(\boldsymbol{x_0}) \leq \boldsymbol{T}\Delta\boldsymbol{x_0} + |\boldsymbol{T}|\boldsymbol{P}\boldsymbol{h} \leq \boldsymbol{U} - \boldsymbol{f}(\boldsymbol{x_0}) \tag{7}$$

Using this model, the author proposed solutions for several basic tolerancing problems:

1) *Recognizing the consistency, determinacy and stability of a part design.* This problem corresponds to the recognition of consistency of the system of linear inequalities and the recognition of the boundedness and dimensionality of the set of solutions to the system of linear inequalities.
2) *Calculating resulting tolerances between components of a part.* This problem corresponds to the calculation of the maximal and minimal value of a tolerance function over the set M.
3) *Allocating tolerance.* Given a tolerance specification and an operation sequence plan, working dimensions and uncertainties of all operations and setups have to be found such that the resulting parts meet the specification. If the tolerance specification is consistent and stable, then Eq. (7) is consistent and has an $\mathbf{h} > 0$ solution.

The mathematical model can be applied to the conventional tolerances and some tolerances of geometric tolerancing. However, only 2-D components can be handled by this approach.

(Light and Gossard, 1982) described a procedure by which geometry is determined from a set of dimensions. The geometry vector **x** is defined as a set of characteristic points expressed by their Cartesian coordinates:

$$\boldsymbol{x} = (X_1 Y_1 Z_1 \ldots X_N Y_N Z_N)^T = (x_1 x_2 x_3 \ldots x_{n-2} x_{n-1} x_n)^T \tag{8}$$

where $n = 3N$. Dimensions impose constraints on the permissible location of the characteristic points. The constraints can be analytically expressed by nonlinear equations:

$$F_j(\boldsymbol{x}, \boldsymbol{d}) = 0 \qquad j = 1, 2, \ldots, m \tag{9}$$

where **x** is the geometry vector, **d** is the vector of the dimensional values and m is the number of constraints. The geometry corresponding to an altered dimension is found through the simultaneous solution of the set of constraint equations in Eq. (9). When the Newton-Raphson method is used, changes in the geometry vector for each iteration are found by the following equation:

$$\boldsymbol{J}\Delta \boldsymbol{x} = \boldsymbol{r} \tag{10}$$

where **J** is the Jacobian matrix of F, $\Delta\mathbf{x}$ is the vector of displacements and **r** is the vector of residuals.

$$\Delta \boldsymbol{x} = \{\Delta x_1, \Delta x_2, \cdots, \Delta x_n\}^T$$

$$\mathbf{r} = \{-F_1, -F_2, \cdots, -F_m\}^T$$

The displacement vector is added to the geometry vector to obtain a new geometry vector, and the process is repeated until the residuals, **r**, vanish. A procedure for reducing the number of constraints and the effect of sparse matrix methods in reducing the time required to solve the equations are also proposed by the authors. Only 2-D geometry and conventional tolerances are mentioned in this work.

(Requicha, 1984) reviewed the inverse parameterization approaches described above. He concluded that these approaches are attractive because they control objects through meaningful geometrical parameters such as distances and angles. But it is difficult to determine whether a set of geometrical constraints defines a valid object unambiguously. He then suggests that such parametrizations are used only as an input technique and the parameterization is internally converted to a primary representation, e.g., Brep or CSG, by the system.

2.1. Statistical Approach to Tolerancing

Statistical tolerances specify some statistical distribution for the dimensions. For example, a dimension might be specified as a random variable modeled by a normal distribution density function with some mean and standard deviation. Many works have been done toward the statistical analysis of conventional tolerances (Greenwood *et al.*, 1987).

3. INTEGRATING GEOMETRIC TOLERANCING WITH SOLID MODELERS

As of recent, manual drafting is rapidly being replaced by computerized systems for defining the geometry of mechanical parts and assemblies. Of the many methods that have been studied and used to represent a part on a computer, the solid modeling technique is the most interesting and challenging one because of its ability to represent the 3-dimensional parts unambiguously. Solid models generated by computer aided design (CAD) are expected to become the primary sources of geometric information. One deficiency of current solid modelers is that they are only capable of representing the nominal geometry of parts. In other words, they lack tolerance facilities. Providing means for annotating solid models directly would eliminate the need to deal with a separate drafting system, but would not contribute to the completeness of geometrical specifications or ease of computer interpretations. The lack of tolerancing facilities blocks the use of solid modelers from supporting fully automatic manufacturing planning tasks and from performing some of the spatial reasoning tasks required for assembly planning. Thus, the meaningful integration of geometric tolerance information into solid geometric models is clearly of importance. The integrated representations must be complete, unambiguous, consistent, and meaningful in order to ensure that various application algorithms using the representations will in turn produce other meaningful information (Requicha, 1983, 1984; Jayaraman and Srinivasan, 1986). Such informationally complete representation schemes for mechanical design specifications remain to be an open question.

(Requicha, 1984) points out three main issues which need to be addressed when incorporating tolerancing information in solid modelers:

1) *Representation of tolerance.*

   How are tolerances to be represented in a solid modeler?

What is the geometric meaning of such representation?
What is the criteria for deciding the validity of the representation?

2) *Synthesis and analysis of tolerance specification.*

In the design session, can we use the functional requirements of the part to generate, i.e. to synthesize, the tolerance specification for the part ? Given two toleranced parts, we need to decide whether there exist instances of parts "in spec.," such that they will interfere with each other and therefore will fail to assemble.

3) *Tolerancing allocation.*

How are tolerancing data to be used for planning automatically the manufacture, inspection, and assembly of mechanical components?

In addition to Requicha's criteria, (Etesami, 1987) has outlined two facilities which are provided by ANSI and must also be provided by any proposed tolerancing theory:

1) *Assembly specification facility.* **ANSI** standards use size, position, and orientation tolerances to specify the assembly requirements. These tolerancing constraints construct the assembly or alignment requirements for a mechanical design.
2) *Form specification facility.* **ANSI** standards specify a set of geometric form tolerances which are used to construct the functional requirements for a design.

Below we discuss several proposed approaches toward integrating geometric tolerancing with solid modelers.

3.1. Tolerance Zones Approach

Requicha has discussed the semantics of tolerancing for mechanical parts and the alternative theoretical approaches (Requicha, 1984). He concluded that the *perfect-form semantics* are not suitable for representing tolerances in solid modelers because they ignore the rationale behind the geometric tolerancing methods described by **ANSI (ANSI, 1982)**. For example, assertions that specify allowable limits for parameters are insufficient for expressing position tolerance at **MMC** (maximum material condition). Under the *imperfect-form semantics*, the author suggests using *tolerance zones* as an approach to deal with the imperfect form. A part in spec. if the appropriate portions of its boundaries lie within the tolerance zones. A tolerance specification T is defined as a triple:

$$T = (S, \{F_i\}, \{A_{ij}\}) \tag{11}$$

where S is a representation for the nominal solid, F_i's are called nominal surface features and they form a partition of the boundary of S, and $\{A_{ij}\}$ is a set of assertions about F_j. To construct a tolerance zone, several issues have to be considered (Requicha, 1984):

- How are 3-D zones to be constructed?

- How does one identify surface features of an actual object?
- How are tolerance zones to be defined for complex features such as slots and pockets?

Requicha gives two approaches to construct tolerance zones for complex features. The first one is based on the "parameterization" of the object. In this approach, tolerances are regarded as constraints imposed on compositions of solid description parameters, such as feature shape and position parameters. The parameterization is used to construct a perfect-form tolerance zone within which the imperfect form features of the actual object must lie. The tolerance zone is feature F is a solid (possibly unbounded) defined by a Boolean composition of half-spaces H_i such that:

1) $\partial H \supset F$.
2) For every half-space $H_i \in H, H_i$ must contribute two-dimensional subsets to F. For example, the Solid H shown in Figure 2(b) violates this condition because H_3 contributes a "face" to but not to F.
3) The material side of S and H must agree.

The concepts of symmetric features and the measured axes (or centerplanes) can be constructed based on the extended features. Figure 3 illustrates the center entities and the measure planes associated with a symmetric feature and planar features, respectively. A datum is a measured axis, centerplane, or center associated with a symmetric feature, or a measured plane associated with a planar feature.

To decide whether a feature is "in spec.", Requicha rejected the notion of a measured size because the concept "size" has no obvious meaning for a feature of imperfect shape (Requicha, 1983). Instead, he suggests using mathematical rules based on notions of maximum material condition (MMC) and least material condition (LMC). Given a nominal feature F, its associated extended feature H, and two numbers $D_p > 0$ and $D_n < 0$ called the *positive* and *negative offsets*, the MMC

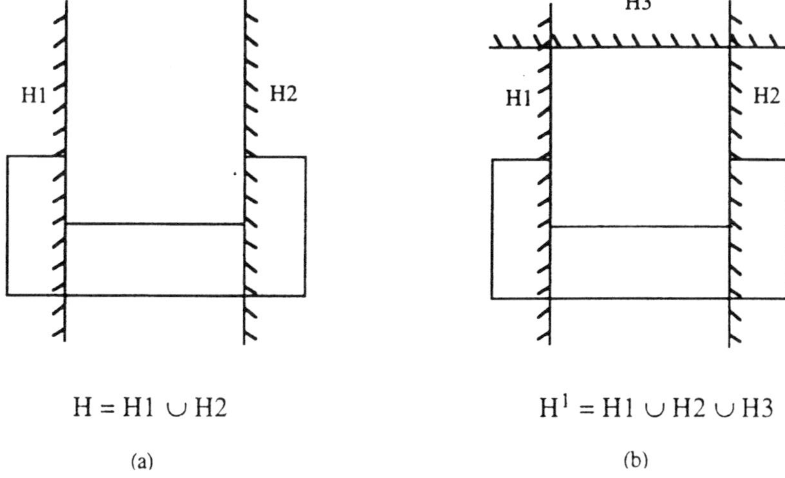

Fig. 2 Examples of simple stiffeners: (a) A strut (b) A web.

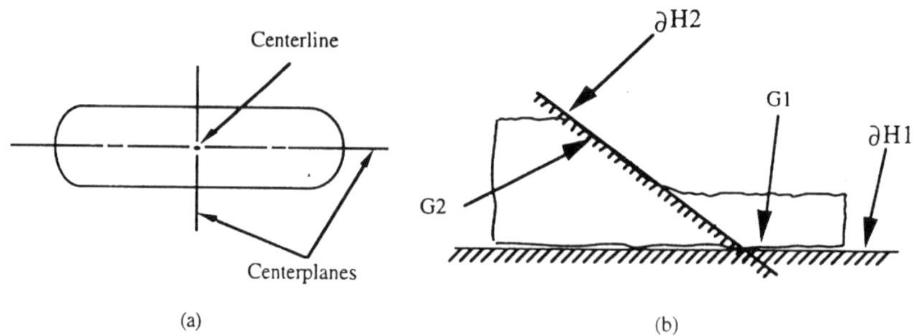

Fig. 3 (a) A symmetric feature and its associated center entities (shown in two-dimensional projection). (b) Two planar features $G1$ and $G2$ and associated measured planes $\partial H1$ and $\partial H2$.

and LMC solids are defined as follows:

$$\text{MMC}\ (\text{D}_p; H) = O(\text{D}_p; H), \tag{13.1}$$

the set difference between a largest and smallest acceptable object, called a *maximal material condition* (MMC) object and *least material condition* (LMC) object.
The advantages of the parameterization approach are (Etesami, 1987):

1) Since the tolerance zone is the set difference between MMC object and LMC object, implementation of the theory in CSG representation system is straight forward. In simple typical cases, tolerance verification is reduced to simple comparisons of real numbers.
2) The parameters could be used to represent features and prescribe a simple method of inspec. verification.

The problem with the parameterization approach is: how can one determine in more complex situations what are the "right" parameter values for MMC and LMC object? (Etesami, 1987) points out another two drawbacks of this approach:

1) This approach lacks abstraction power in dealing with a general class of solids.
2) This approach is insufficient in representing some spatial constraints because of a lack of direct parameters to characterize the situation.

The second approach for constructing tolerance zones is based on the *offset* operation (Requicha, '83). More precisely, a tolerance zone is the regularized set difference between two *offset* objects with specified offset values. The offset object that corresponds to a solid S and offset Δ is defined as follows:

$$O\,(\Delta; S) = \begin{cases} \{p : d(p : d(p, S) \leq \Delta\} & \text{if } \Delta \geq O \\ S_{-}{}^{*}O(-;)C^{*}S) & \text{if } \Delta \leq O \end{cases} \tag{12}$$

where $_{-}{}^{*}$ and c^{*} denote the regularized difference and complement operations, and $d(p, S)$ is the distance between a point p and the set S.

To use the offsetting operation to construct the tolerancing specification, the concepts of extended features, symmetry, and measured entities are defined (Requicha, 1983). An extended feature H associated with a nominal surface

$$\mathrm{LMC}(D_n; H) = O(D_n; H). \tag{13}$$

where O is the offsetting operation defined as Eq. (12).

If G is the actual feature corresponding to F and H, G satisfies a *size tolerance* with single offsets D_p, D_n if and only if there is a congruent instance $H^1 = R(H)$, where R is a rigid motion, such that

$$G \subset \mathrm{MMC}(D_p, H^1) -^* \mathrm{LMC}(D_n, H^1) \tag{14}$$

Similarly, G satisfies a *form tolerance* with value T_f if there is a congruent instance H^1 of H and two numbers D_1 and D_2 such that

$$G \subset O(D_1; H^1) -^* O(D_2, H^1), \tag{15}$$

where

$$D_1 > D_2$$
$$D_1 - D_2 \leq T_f.$$

G satisfies an *unqualified position tolerance* T_p if

$$G \subset O(T_p/2; H^1) -^* O(-T_p/2; H^1) \tag{16}$$

where H^1 is a congruent instance of H that positioned correctly.

Two *qualified position tolerances* are discussed by Requicha, i.e., MMC and RFS. Given an actual feature G corresponding to a nominal feature F with size tolerances characterized by offsets D_p and D_n, an MMC position tolerance $T_p (T_p \geq 0)$ relative to some datum system has the following meaning. Let H^1 be an instance of the extended feature H at the appropriate location and orientation relative to the datum system. G satisfies its position constraints if

$$G \subset O\,(T_p + D_p;\ H^1). \tag{17}$$

A regardless of feature size (RFS) position tolerance is an assertion on a planar or symmetric feature that defines a solid tolerance zone within which a measured entity of the feature must lie. An RFS tolerance zone for an axis is given in Figure 4; the tolerance specification consists of (axis = z, value = 0.1, $Z_{max} = 3$, $Z_{min} = 0$), defines a cylindrical tolerance zone and the line L is an acceptable axis.

The offsetting operation based tolerance zone theory provides a basic schema to integrate the tolerancing information with the solid modeler. However, some part of the theory such as the extended features must be modified in order to adequately satisfy the industrial requirement and to perform tolerance analysis and manufacturing and assembly planning.

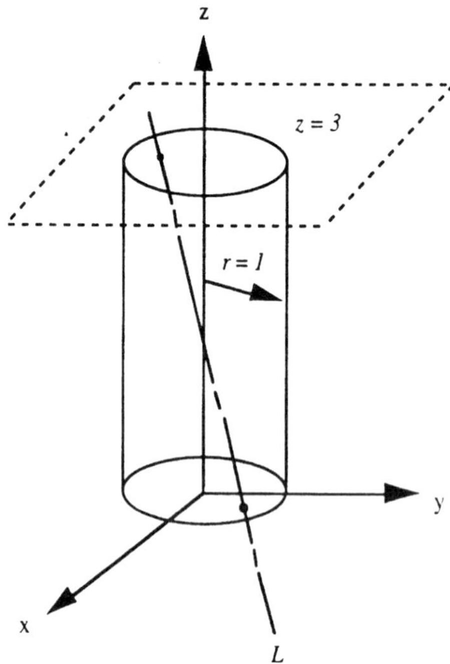

Fig. 4 An RFS tolerance zone for an axis.

(Etesami, 1987) reviewed the offset solid theory and concluded that some of the ANSI specifications can not be well described. The author suggests two enhancements to this theory:

1) In addition to surface features, the curve and point features and appropriate offset solids for them can be defined.
2) The feature of the offset theory can be integrated with the parameterization approach to form a unified theory which can address a larger spectrum of applications.

In summary, the non-parametric approaches are more general and avoid some ambiguity problems. Nonetheless, some manner of parameterization is required to efficiently measure for conformance.

3.2. Representation of Geometric Features and Tolerances in a Solid Model

Based on the "tolerance zone by offsetting" theory, Requicha *et al.*, proposed a scheme to represent surface features in a CSG based solid modeler and to associate tolerances and other attributes with the surface features (Requicha and Chan 1986). The variational information associated with a solid is represented by a graph called VGraph. The lowest-level entities in the VGraph are NFace nodes, which are the nominal faces of an object. Next come VFace nodes, which in most cases are the whole NFace. The surface feature nodes, SFeats, are groups of VFaces. DatSys nodes represent datum systems and contain an ordered set of data. Each datum is represented by pointing to an SFeat and has an associated qualifier, e.g. MMC or RFS. Figure 5 shows the structure of the VGraph.

The main links between the VGraph structure and the nominal representations in a solid modeler are established through NFaces. One can simply attach the VGraph to a modeler's Brep by identifying NFaces with the faces in the Brep. Some problems with this approach are that 1) the validity of the variational representation is not well understood, and 2) the propagation of attributes in the VGraph can be computationally complicated.

3.3. Manufactured Part Modeling Based on Tolerance Zone Theory

Based on Requicha's tolerance zone theory, Etesami presents a theory for modeling slightly imperfect manufactured parts such that relevant manufacturing information is recorded (Etesami '88). In the suggested theory, the manufactured part model is composed of three models: a boundary model for part surfaces, a model for axes and curves, and a datum model. The boundary model is represented by a set of solids referred to as "boundary solids". The boundary solids are constructed as follows:

Let a closed ball with radius R be associated with each point on a subset of the boundary of a solid S and denote this subset as $@S$. A surface constructor for $@S$ is defined as:

$$\#S = \{p : d(p, @S) \leq R\} \tag{18}$$

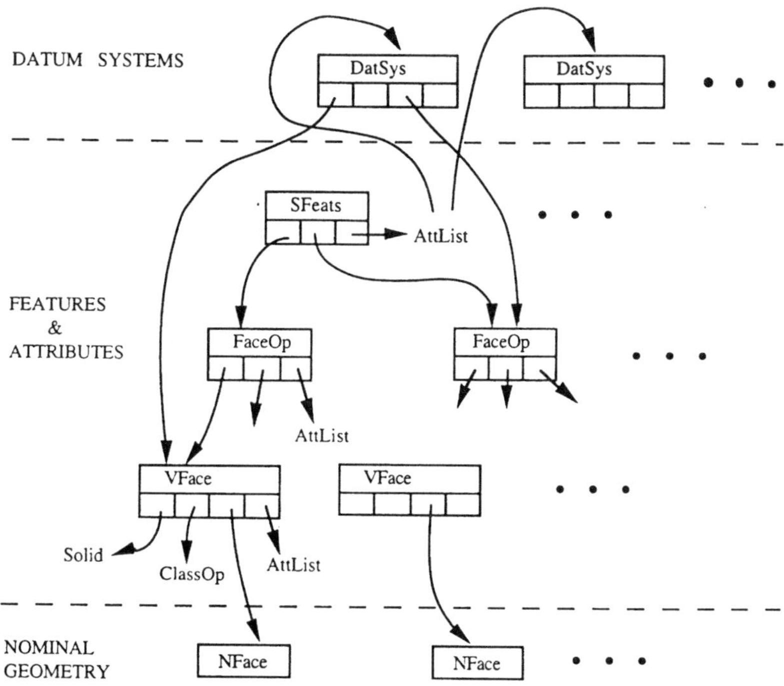

Fig. 5 Logical structure of the VGraph.

where

$$d(p, @S) = \text{Min} \parallel p - q \parallel; \quad \text{for all } q \in @S$$

Figure 6 depicts planar and cylindrical feature constructors and their respective depths, denoted as d in the figure.

The surface of an object represented by a boundary model divides the boundary point-set into two subsets: 1) S_i, the one which is entirely inside the object and 2) S_o, the one which is entirely outside of the object. The boundaries of S_i and S_o define the LMC and MMC boundaries of S respectively.

The ANSI tolerancing standards are concerned with four classes of callouts: forms, orientations, positions and runouts. Two examples are: 1) the "flatness" specification directly corresponds to the depth of a planar constructor and 2) the measure of "position" of a centre axis is the diameter of a cylinder which enclose the axis feature such that the center position and orientation of the cylinder is constrained by the datum specification. The boundary model, once constructed, provides an easy assessment to those imperfect measurements similar to results obtained by ANSI's prescribed tests for in-spec. verification.

With the presented interpretation of tolerances, the effect of tolerance qualifiers is easy to assess. For example, if the feature is a depression feature (e.g. a hole), the associated miimum size of the feature is the functional size of the feature. The difference between this functional size and the allowed MMC size of the feature determines the amount of bonus tolerance which can be allocated to applicable tolerances of position, orientation and form.

3.4. Virtual Boundary Requirements (VBR) Theory

(Jayaraman and Srinivasan 1989) propose a theory to represent the geometric tolerances in a solid modeler from the perspective of two classes of functional requirements:

1) The *assembly requirement*, which is referred to as the specification of the positional relationships among the parts in a assembly.
2) The *material bulk requirement*, which is referred to as a specification of the geometric characteristics of critical material portions of a part.

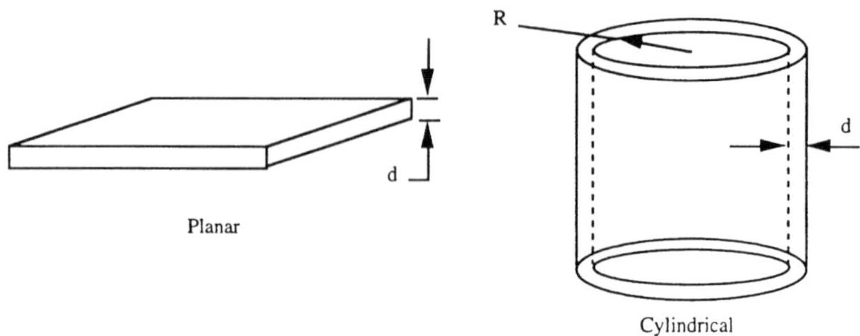

Fig. 6 Planar and cylindrical surface constructor.

The two functional requirements can be captured in a specific form of tolerances called *virtual boundary requirements* (VBRs). In a VBR environment, the functional requirements are expressed through a rigid collection of virtual half-spaces. The VBRs are built through the following processes. First, with desired assembly configuration, the relevant surface features are identified for each piece of the part. Then an appropriate virtual half-space is associated with each pair of features identified. For example, in Figure 7, "base" and "lip" are two surface features identified for assembly requirement, and a virtual half-space *vhs_1* can be associated with the feature pair (base, lip). Similarly, another virtual surface *vhs_2* can be associated with a feature pair (shank, hole).

The VBRs are generalizations of current MMC and LMC tolerances. The authors further demonstrate that Requicha's theory (Requicha '83) is inadequate for dealing with VBRs. Accordingly, a theoretical basis for the rigorous statement and interpretation of VBRs is proposed.

In their companion paper, (Jayaraman and Srinivasan 1986), present a method to convert VBRs to another form of tolerances designated as conditional tolerances (CTs). A theoretical basis for converting VBRs to CTs is proposed and some common and practical VBRs are demonstratedly converted to CTs. For example, in the case of the assembly requirement for a stud and the material bulk requirement for a hole as shown in Figure 8, the parameters maintained for the primitive include

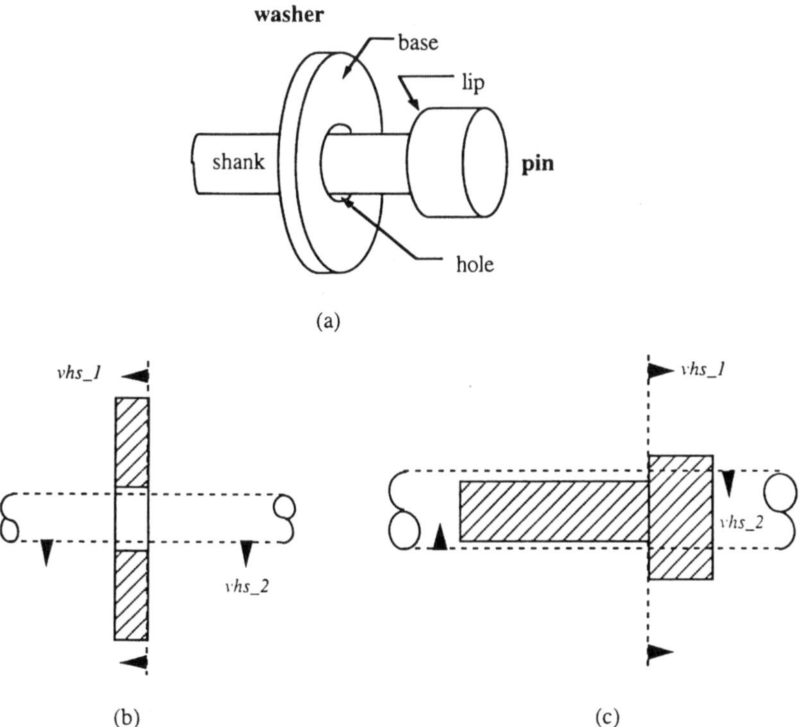

Fig. 7 (a) pin and washer. (b) virtual half-spaces for washer. (c) virtual half-spaces for pin.

locational parameter c_1, orientation parameter c_3, nominal diameter s_{N1}, actual diameter s_1, and length of slab parameter l_1. It is necessary that $s_1 \leq s_{N1} + 2 \mid a_1 \mid$ and the conditional tolerance on c_3 is given by:

$$0 \leq c_3 \leq 2\tan^{-1}\left[\frac{\frac{(s_{N1}+2|a_1|-s_1)}{l_1}}{1+\sqrt{1+\left(\frac{(s_{N1}+2|a_1|+s_1)}{l_1}\frac{(s_{N1}+2|a_1|-s_1)}{l_1}\right)}}\right]$$

However, a general-purpose algorithm for deriving CT zones for VBRs is not known. Deriving tolerance zones from VBRs is often difficult even for simple features.

3.5. Tolerance Synthesis Based on VBR

Based on the VBR theory, Lu *et al.* developed a framework called CASCADE-T for addressing tolerance synthesis and analysis (Lu *et al.*, 1992). The functional requirements described using VBRs are coverted to conditional tolerance relationships. Tolerance primitives are derived to represent the conditional tolerance relations that must hold for parameterized geometric features. Each primitive includes

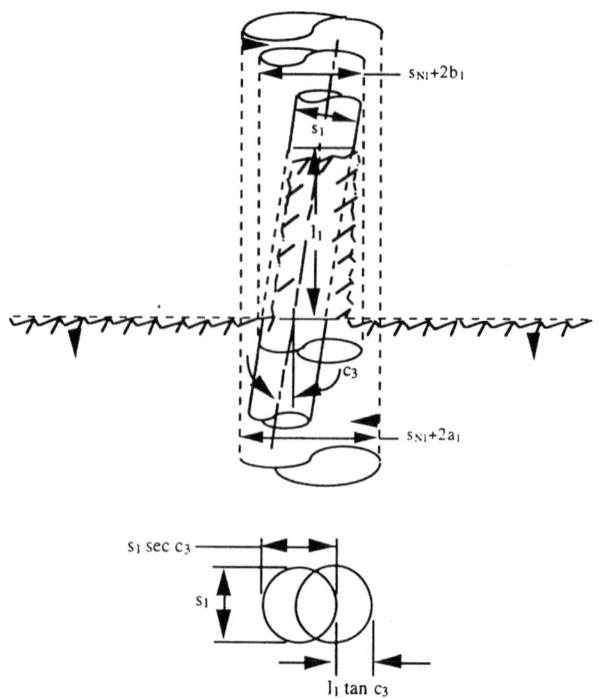

Fig. 8 Orientation tolerancing: assembly requirement for a cylindrical stud.

data and procedures. Data is in the form of a parametric representation, associated datums, and conditional tolerance relastionships that must be maintained for the parameters of the primitive. Procedures include the methods for arriving at values of parameters. Examples of tolerance primitives are shown in Figure 9.

Tolerance primitives can be composed to form more complex tolerance relationships. Intervals and equations are used to describe each tolerance relation. A variety of artificial intelligence techniques, including constraint networks, frame, rule-based reasoning, and dependency tracking are used to support the representation and computation requirements of CASCADE-T.

4. VARIATIONAL MODELING

A variational model is a computer representation of a variational class, and stands for a collection of different instances of the part or assembly. A variational model may incorporate one or more different types of variations. Variational modeling involves applying variations to a computer model of a part or assembly of parts. Any CAD system that permits us to apply variations to a model can be termed a variational modeling system. According to the representation used, we can classify variational modeling schemes as either procedural or declarative.

Procedural modeling systems store a step-by-step plan for constructing the geometric elements that comprise the part or assembly.

In constructive solid geometry-based (CSG) modelers, the system builds a sequential procedure for constructing a part or assembly model in the form of a tree. In such modelers, we can induce variations in two ways:

1. We can change model variables associated with the shape parameters and location parameters of the primitives and reevaluate the construction procedure to produce a new model instance.
2. We can apply variations to the boundaries of the primitives.

In feature-based systems, a part model is expressed in terms of base features and certain feature-forming operations. We can vary the location and shape parameters associated with the feature geometry. The choice of features used and the sequence in which they are applied determine the variational coverage of the model.

With declarative solid modelers, the system builds a representation of each geometric element, together with information about interconnections. In pure

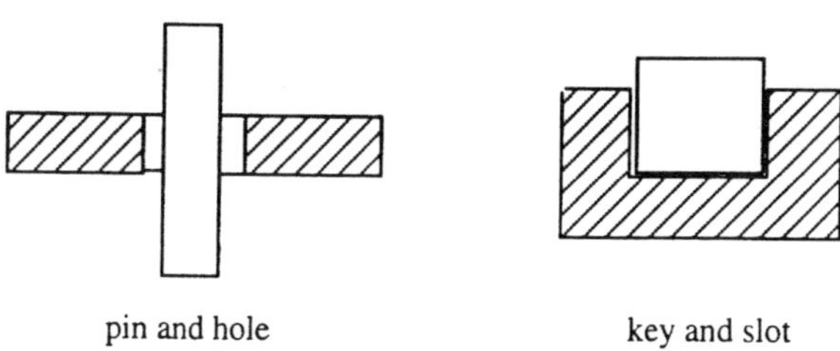

Fig. 9 Tolerance primitives.

boundary representation (B-rep) modelers, we can introduce variations by varying the coordinates of vertices or the positions and orientations of surfaces:

1) If we introduce variations by varying the vertex coordinates, then must define the surfaces relative to the vertex coordinates.
2) If we introduce variations by varying the part surfaces, we then must compute the vertex coordinates from the surface equations.

For procedural or declarative modeling strategies, we want to let the user modify the geometry by defining a set of dimensional constraints on the model. A number of researchers (R. Light and D. Gossard, 1982) (R. C. Hillyard and I. C. Braid, 1978) have explored the use of numerical techniques to simultaneously reevaluate geometric information. Others (B. Aldefeld, 1988) have explored symbolic approaches to recompute geometry. These systems apply artificial intelligence search techniques to reorder that dimensional constraints so that geometric elements can be computed sequentially.

Two important applications benefit from a variational modeling capability:

- *Large variations* We can use a variational model to create nominal models of different sizes and shapes in a variational design system.
- *Small variations* We can use a variational model to represent different manufactured instances for solving tolerancing problems.

4.1. Applications of Variational Modeling to Automated Tolerance Analysis

In the effort of the Automated Tolerancing Project at the Rensselaer Design Research Center (RDRC), the researchers have developed advanced methods for solving tolerancing problems. The work is based on *variational modeling technology* and a *feasibility-space approach.* The goal was to develop software for automated tolerance analysis and synthesis, implemented as part of GEOS, a solid modeler for automated tolerancing developed at Rensselaer, and suitable for integration with corporate CAD systems.

The set of model variables applied to part surfaces defines an n-dimensional Cartesian space called the feasibility space. Each n-tuple of this space corresponds to a particular instance of the variational model. Each tolerance applied to the part as an inequality constraint on the n-tuples of this space is interpreted (see Figure 10). Collectively, the tolerances applied to the part define a feasible region of space. Any n-tuple within this feasible region corresponds to an intolerant part.

4.2. Feasibility-Based Approach

The feasibility-based approach provides a mathematical interpretation of tolerances that we can use as a basis for developing algorithms for tolerance analysis and synthesis. The designer specifies part tolerances and the software determines the total variation of some critical design function. Moreover, in a tolerance synthesis problem, the designer provides acceptable limits of variation of the critical design functions, and the software determines compatible tolerance values.

A design function is expressed as a function of the n-tuples of the feasible region. Numerical methods are used to derive linear approximations of the feasible region's boundaries and to derive a linear expression for the design function in terms of the

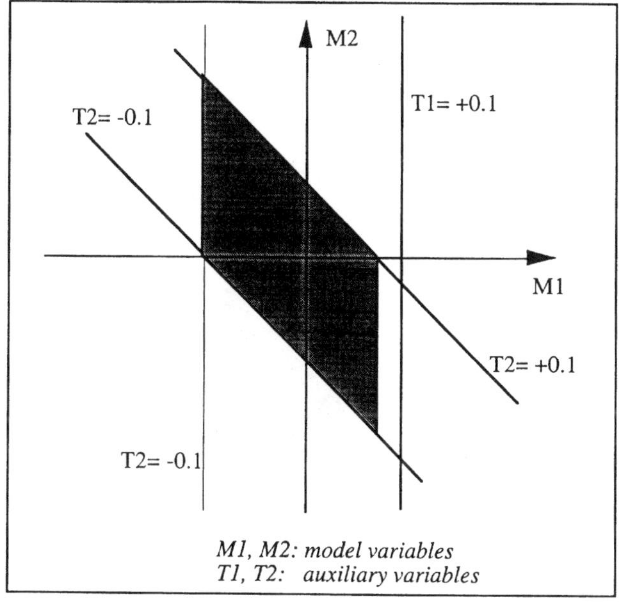

Fig. 10 Feasibility-space representation of a tolerance constraint.

model variables. A linear programming method is then used to perform worst-case or statistical tolerance analysis. This produces upper and lower bounds on the values of the design function (see Figure 11).

The feasibility-space approach to tolerance analysis mainly has two important advantages: (1) The designer is free to use as many or as few tolerances as necessary to control critical design properties. The omission of a function-critical tolerance leads to an unconstrained variation that can be detected and reported back to the designer. Extra tolerances simply add extra inequality constraints; (2) All limit tolerances and geometric tolerances can be hadled and represented.

5. AN ATTEMPT FOR STANDARDIZATION-PDES

Product Design Specification

Product design specification is a key piece of information that flows from the design session to the manufacturing processes. Both design and manufacturing groups use this information for a variety of purposes such as design analysis, process planning and production system design.

The traditional media for communication of mechanical design specification is a set of engineering drawings. A typical drawing contains graphical descriptions of nominal (ideal) parts, plus tolerancing information that defines allowable departures from the nominal objects. In addition, they usually indicate material properties and other design specifications such as internal stress state.

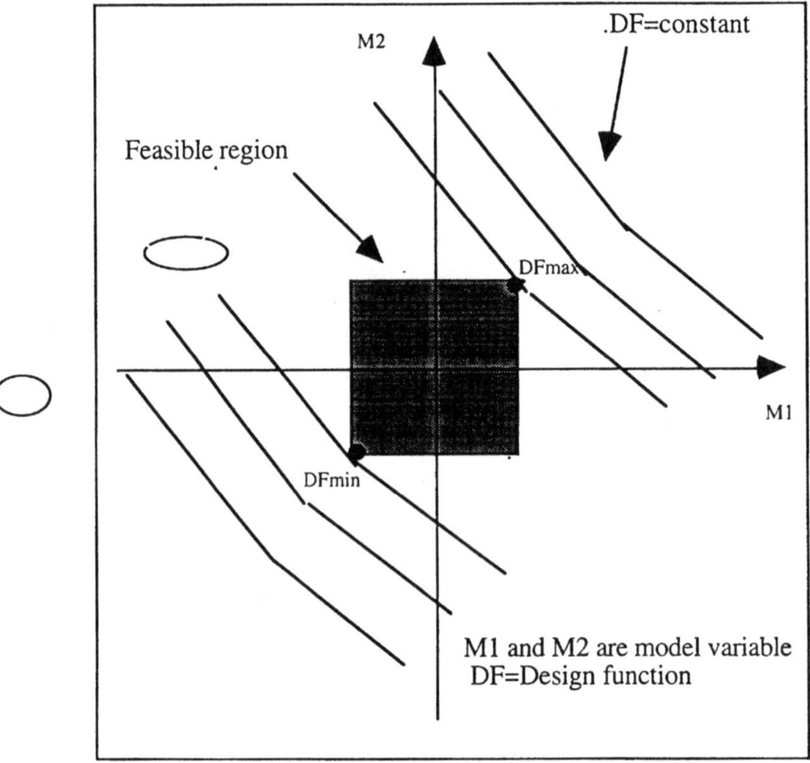

Fig. 11 Worst-case tolerance analysis.

Conventional tolerancing uses "±" to specify dimensional variation from the basic size. Either unilateral tolerance or bilateral tolerance can be used. The basic location where most dimensional lines originate is the reference location, i.e., the datum. For working surfaces that require surface control, control surface symbols are used. In the surface control symbol, several surface characteristics are specified: roughness, height, lay, waviness height, roughness width, etc.

Conventional tolerancing only provides information concerning *size and surface condition, which is not enough* (Chang, Wysk, and Wang 1991).

Geometric Tolerancing

Geometric tolerancing specifies the tolerance of geometric characteristics. According to ANSI's Y14.5M 1982 standard, some basic geometric characteristics are:

Straightness – defines the maximum deviation on the assumed center line over the entire length or a cylindrical component.

Flatness – defines the maximum deviation allowed on a flat surface.

Roundness – defines the irregularity of the diameter at any given cross-sectional location of a cylindrical component.
Cylindricity – is similar to roundness except that it defines the irregularity over the entire length.
Profile of a line/a surface – describe the deviation on the profile.
Parallelism – any feature of a component can be specified as being parallel to any given datum.
Perpendicularity
Augularity – is similar to perpendicularity except that the relationship of a feature to a given datum need not be 90°.
Concentricity – is used to establish a relationship between the axes of two or more cylindrical parts of a component.
Runout – the composite deviation from the desired form of a rotational part during full rotation (360 °) of the part on a datum axis.
True Position – expresses the location of the center line with respect to a feature.

In addition to the geometric characteristics, "modifiers" are used to clarify implied tolerances. Modifiers include:

Maximum Material Condition (MMC)
Least Material Condition (LMC)
Regardless of Feature Size (RFS)

The modifiers are associated with the material bulk requirements.

Tolerancing defined in PDES

Some important assumptions are made in PDES:

- Types of tolerance are described as they pertain to common tolerancing practices which are used to define the form, fit and function of the modeled product.
- To specify certain types of tolerances, independent or dependent geometric constructs may be needed in addition to shape defining elements. Examples of these are hole centerlines and off-part datum.
- All tolerance dimensions are derived from shape or size elements or are explicitly stated as a part of an *Implicit Feature*.

The dimension/tolerance feature (**dt feature**) in PDES includes:

Shape elements (Face, Edge and Vertex in Brep term)
Form feature
Feature of size (a subset of which is form feature)
Geometric derivation

The tolerances described in PDES include Coordinate and Geometric Tolerances. Coordinate tolerances contain "extra" data that were intended to facilitate the determination of the value of the intended dimension. The coordinate tolerances are similar to the conventional tolerancing mentioned in the previous Section.

Geometric tolerances are defined to accommodate typical dimension practices which reference geometry that is not a part of the shape definition of the object, but that is derivable from shape elements. The most common examples are centerlines of holes and the spatial inter-section of two-part surfaces. A geometric tolerance, angularity tole-rance, defined in PDES is given here as an example.

Angularity tolerance: is the condition of a surface or axis at a specified angle (other than 90 degrees) from a datum plane or axis.

```
ENTITY angularity
SUBTYPE OF (geometric_tolerance):
toleranced_ents //a set of elements to which the
                                  tolerance applies
magnitude
material_condition
projection                //a projected tolerance zone which
specifies the addtional height and direction of the projected tolerance zone
outside the feature boundary.
primary_datum
secondary_datum
tertiary_datum
WHERE
secondary_datum < > primary_datum
tertiary_datum  < > primary_datum
tertiary_datum  < > secondary_datum
valid_mlsn (material_condition, toleranced_ents)
END_ENTITY
```

Other geometric tolerances defined in PDES are circular runout, circularity, concentricity, cylindricity, flatness, parallelism, perpendicularity, position, profile line, profile surface, straightness, total runout, datum, conditioned datum, unconditioned datum, feature of size, form feature of size, and size feature. The detailed definitions can be found in (Smith *et al.*, 1988).

6. SOFTWARE FOR TOLERANCE ANALYSIS

Both the parametric tolerancing and the geometric tolerancing modeling approaches are available in one comercial package or another. Commercially available software provides support for analysis of the effect of geometric and parametric tolerances on the functionally significant properties of parts and assemblies (Joshua Turner, 1993). Software for tolerance analysis can play an important role in evaluating design concepts.

Available technology still requires considerable input from the user in building the analysis models, and a thorough understanding of the assumptions of the package. Support for the ANSI/ISO geometric tolerancing standards is limited. However, when used with care and knowledge, the available products can be of considerable value to design engineers in reducing product cost and improving product quality. In this section, the basic capabilities for tolerance analysis of the technologies used in currently available software are discussed. We are concerned with computer software that performs a worst-case or a statistical tolerance analysis of some design function here.

There are three principal techniques in use: techniques based on procedural models, techniques based on dimension-driven models and special-case techniques.

6.1. Procedural Models

The Procedural modeling approach appears to be best suited for the analysis of detailed designs, where assembly sequence plans, fixturing requirements, and process distributions have already been determined. The techniques of procedural models analyze a parametric model in which each model parameter (or a small set of parameters) determines the location of the tolerance features. In some cases the procedural model is explicitly constructed by the user; in other cases the software is capable of constructing the procedural model automatically.

One of the most interesting commercially available procedural modeling techniques is based on a set of 3-D points. The VSA package employs this technique. The procedural model specifies how each of the points is to be constructed. If the procedural model is constructed by the user, then the user must identify the key points of each of the parts, and determine how to construct the points. Parameters are used in the construction procedure specified for each of the points. The software analyzes the influence of each of these parameters on the design function. The user must be careful in selecting the points to be modeled, and the construction procedures to be used in modeling the points. Figure 12 illustrates a user-constructed model which is probably a defective one.

In Figure 12, P1 is constructed at (0,0); P2, P3, and P4 are constructed at P1 + (a,0), P1 + (0,b), and P3 + (a,0) respectively, where a, b and c are the parameters. Because of the way the model has been constructed, the 90 degree angle at the lower left corner of the part cannot vary, and the top and bottom edges of the part will remain parallel. It is likely that significant sources of variation have been overlooked.

In most cases, the procedural model is constructed by an experienced user or a consultant from the toleranced drawings, and a knowledge of the anticipated behavior of the manufacturing processes to be used. If geometric tolerances are to be

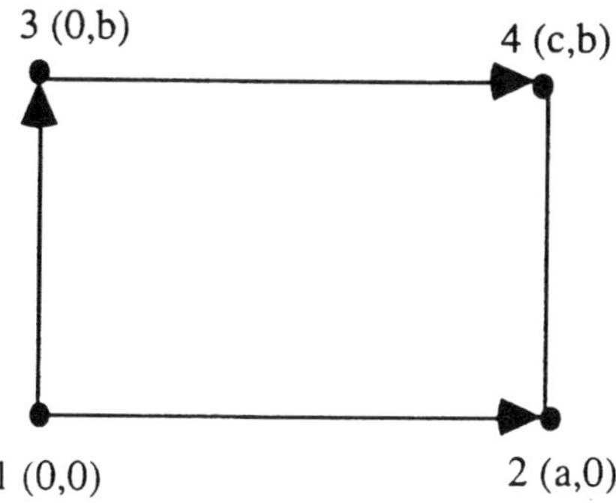

Fig. 12 User-Constructed procedural model.

analyzed, then the user must construct a procedural model that accurately reflects the definition of each of the geometric tolerances given in the standard.

A new version of VSA employs AI techniques to automatically construct the procedural model. The package incorporates a tolerancing expert which attempts to construct a correct procedural model from the geometric tolerance specifications and the manufacturing process specifications.

6.2. Dimension-Driven Models

The dimension-driven approach appears to be best suited for the analysis of design concepts at early stages in the design process. The dimension-driven techniques model analyzes a 2-D dimensioned sketch. If the sketch's dimensions are varied, the software package is able to recompute the sketch geometry so that the geometry is consistent with the new dimensions. A tolerance analysis is performed by analyzing the effects of the variations in the sketch dimensions. Cognition's Mechanical Advantage package (Cognition, 1989) and Analytix (Saltire, 1989) are the pilot commercial tolerancing packages based on dimension-driven technique. SDRC and Intergraph also developed dimension-driven capabilities for use within their CAD system.

A dimension-driven software package normally has the following basic operations:

(1) The user enters a 2-D sketch, and the software creates a geometric model consisting of vertices and lines or curves.
(2) The user adds dimensions, which are used to define algebraic equations on the (x, y) coordinates of the vertices.
(3) The software solves these equations to determine the values of the vertex coordinates.
(4) A tolerance analysis is performed by varying each of the dimensions and determining the effect of these variations on the design function.

In Figure 13, as an example, a set of equations are generated based on the dimensions given in the sketch:

$$X_1 = 0, \quad Y_2 = 0, \quad Y_2 = 0$$

$$D_1 = \tan^{-1}\left(\frac{Y_3 - Y_1}{X_3 - X_1}\right) - \tan^{-1}\left(\frac{Y_2 - Y_1}{X_2 - X_1}\right)$$

$$D_2 = \tan^{-1}\left(\frac{Y_1 - Y_2}{X_1 - X_2}\right) - \tan^{-1}\left(\frac{Y_4 - Y_2}{X_4 - X_2}\right)$$

$$D_3 = \sqrt{(X_2 - X_1)^2 + (Y_2 - Y_1)^2}$$

$$D_4 = \sqrt{(X_3 - X_1)^2 + (Y_3 - Y_1)^2}$$

$$D_5 = \sqrt{(X_4 - X_2)^2 + (Y_4 - Y_2)^2}$$

$$D_6 = \sqrt{(X_3 - X_5)^2 + (Y_3 - Y_5)^2}$$

$$D_6 = \sqrt{(X_4 - X_5)^2 + (Y_4 - Y_5)^2}$$

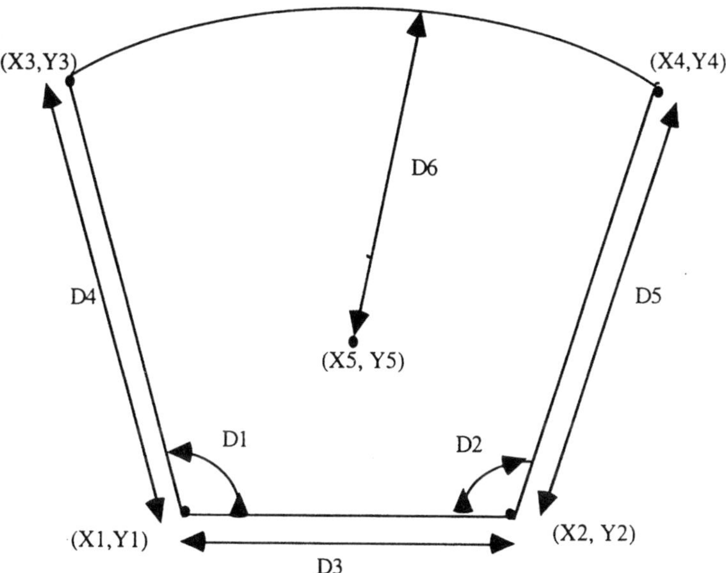

Fig. 13 Dimension-driven model.

The D's are given by the user. The X's and Y's are unknowns. In total, 10 equations in 10 unknowns are generated by the software.

At an early stage in product design, it is useful to sketch up the principle parts and analyze the main effects of the dimensional variations. When an analysis is performed at this stage, it may serve for identifying critical problems with the product concept. However, the dimension-driven approach becomes unwieldy when applied to 3-D problems and the analysis of complex geometric tolerances.

6.3. Special-Case Techniques

The special-purpose approaches may be most efficient in cases where the analysis problem fits the software package capabilities. There are several packages that facilitate the analysis of common special-case situations.

For examples:

(1) ***One-dimensional tolerance analysis*** One-dimensional tolerance analysis refers to a class of problems where all the relevant dimensions stack up in a straight line. When parametric tolerances are used, both worst-case and statistical tolerance analysis are straightforward. Texas Instruments has introduced a 1-D analysis capability (TI/TOL) for use with PRO/Engineer.

(2) ***Part mating situations*** Part mating situations involving the assembly of two parts with mating hole patterns and floating fasteners, or with mating pins and holds, are very common. Both size and location tolerances are usually involved in this situation. The VALISYS package provides capabilities for tolerance assignment and analysis of such situations. Geometrical locating tolerances are supported in the package. The Texas Instruments package apparently contains similar support.

References

B. Aldefeld, Variation of geometries based on a geometric-reasoning method, *Computer-Aided Design*, **20(3)**, 117–126 (1988).

ANSI, Dimensioning and Tolerancing, in **ANSI** Standard Y14.5, American Society of Mechanical Engineers, New York, NY (1982).

T.L. Chang, R.A. Wysk and H.P. Wang, *Computer-aided manufacturing*, Prentice-Hall Inc., Englewood Cliffs, NJ (1991).

Cognition, Inc. Billerica, MA, *Mechanical Advantage Reference Manual* (1989).

Cognition, Inc. Billerica, MA, *Mechanical Advantage Training Manual* (1989).

F. Etesami, On the theory of geometric tolerancing, *Proceeding of the 1987*, ASME Computers in Engineering Conference, New York, 327–334 (1987).

F. Etesami, Tolerance verification through manufactured part modeling, *Journal of manufacturing systems*, **7**(3), 223–232 (1988).

A. Fleming, Geometric relationships between toleranced features, *Artificial Intelligence*, **7**(3), 223–232 (1988).

A. Fleming, Geometric relationships between toleranced features, *Artificial Intelligence*, **37**, 403–412 (1988).

W.H. Greenwood and K.W. Chase, A new tolerance analysis method for designers and manufacturers, Translations of ASME, *Journal of Engineering for Industry*, **109**(2), 112–16 (1987).

R.C. Hillyard and I.C. Braid, Analysis of dimensions and tolerances in computer-aided mechanical design, *Computer Aided Design*, **10**, No. 3, 161–166 (1978).

P. Hoffman, Analysis of tolerance and process inaccuracies in discrete part manufacturing, *Computer Aided Design* **14**(2), 83–88 (1982).

R. Jayaraman, and V. Srinivasan, Geometric tolerancing: I. virtual boundary requirements, **IBM** Journal of Research and Development, **33**(2), 90–104 794 (1989).

U. Joshua Turner Current Tolerancing Packages, CRTD-Vol. 27, *International Forum on Dimensional Tolerancing and Metrology* ASME , 241–248 (1993).

R. Light, and D. Gossard, Modification of geometric models through variational geometry, *Computer Aided Design*, **14**(4), 209–214 (1982).

V.C. Lin, D.C. Gossard, and R.A. Light, Variational geometry in computer aided design, ACM Computer Graphics, **15**(3), 171–177 (1981).

S. C-Y Lu, and R.G. Wilhelm, Automating tolerance synthesis: a framework and tools, *Journal of Manufacturing System*, **10**(4), 279–296 (1991).

B. Smith, and G. Rinaudot, Product Data Exchange Specification : first working draft, NISTIR 88-4004, *National Technical Information Service (NTIS)*, Order PB 89–144 794 (1988).

A. A. G. Requicha, Toward a theory of geometric tolerancing, *The international journal of Robotics research*, vol. **2**, No. 4, 45–60 (1983).

A.A.G. Requicha, Representation of tolerances in solid modeling: issues and alternative approaches, *Solid modeling by computers*, Mary S. Pickett and John W. Boys, Eds. (1984).

A.A.G. Requicha and S.C. Chan, Representation of Geometric features, tolerances, and attributes in solid modelers based on constructive geometry, *IEEE journal of robotics and automation*, Vol. RA-2, **3**, 156–166 (1986).

Saltire Software, Beaverton *OR, Analytix Tutorial and Reference Manual* (1989).

B. Smith, and G. Rinaudot, *Product Data Exchange Specification: first working draft*, NISTIR 88-4004, National Technical Information Service (NTIS), Order PB 89–144 (1988).

V. Srinivasan, and R. Jayaraman, Geometric tolerancing: I. Virtual boundary requirements, *IBM Journal Research Development*, Vol. **33**, No. 2, March, 90–104 (1989).

V. Srinivasan, and R. Jayaraman, Geometric tolerancing: II. Conditional tolerance, *IBM Journal Research Development*, Vol. **33**, No. 2, March, 105–124 (1989).

Suvajit Gupta and U. Joshua Turner Variational Solid Modeling for Tolerance Analysis, *IEEE Computer Graphics & Applications*, Vol. **13**, No. **3**, May, 64–74 (1993).

CHAPTER 5

Integrated Concurrent Design Using a Quantitative Intelligent System

D. XUE and Z. DONG*

Department of Mechanical Engineering, University of Victoria, British Columbia, Canada

1. INTRODUCTION

1.1. Background

The development of a mechanical product undergoes a sequence of processes including conceptual design, detailed design, design analysis, prototype making and testing, production process planning, machining, inspection, and assembly. The conventional practice is to carry out each of these tasks manually and sequentially as illustrated in Figure 1 (a). With the advances in computer technologies, many of

*To whom all correspondence should be addressed.

these design and manufacturing activities have been automated. Computer-Aided Design (CAD), finite element analysis, automated tolerance analysis, Computer-Aided Process Planning (CAPP), Computer Numerical Control (CNC) machining, automated CNC programming, automated inspection using a Coordinate Measuring machine (CMM), and automated assembly by industrial robots have been developed to improve design and manufacturing efficiency and quality of products (Chang *et al.*, 1991; Zeid, 1991).

Although the productivity of each design and manufacturing module has been improved considerably, the new computer-automated product development processes, as shown in Figure 1 (b), present many similarities to the traditional design and manufacturing processes. The design and manufacturing tasks are executed sequentially through a set of isolated "island of automation." In product development, a design often needs to be tested using several prototypes, and it goes through several re-designs before the design can be released for mass production. The isolation between design and manufacturing modules can reduce design efficiency, thereby leading to long product development lead times. It is an obstacle to success in today's highly competitive industrial environment.

To reduce the barrier among computer automated design and manufacturing modules, researches on integration of Computer-Aided Design and Manufacturing (CAD/CAM) system are carried out. A detailed review of relevant work can be found in (O'Grady and Menon, 1986). The main objective for CAD/CAM integration is to extract useful product information from the existing CAD database, or to generate useful product information for downstream manufacturing applications such as production process planning and CNC machining. These include the work on *feature recognition* and *feature-based design*. The former extracts feature geometry, such as a hole, or a slot, to be produced by a certain manufacturing operation from a CAD database (Henderson, 1984; Choi *et al.*, 1984; Voelcker *et al.*, 1986; Dong and Wozny, 1988). The latter constructs new CAD systems using mechanical features as basic building blocks for modeling a design (Pratt 1984; Luby *et al.*, 1986; Chung *et al.*, 1988; Cutkosky *et al.*, 1988; Shah and Rogers, 1988; Tunner and Anderson, 1988). The feature modeling method, presented in this chapter, falls into the second category. Other efforts in this area include the introduction of the product data exchange standard, PDES/STEP (Standard for the

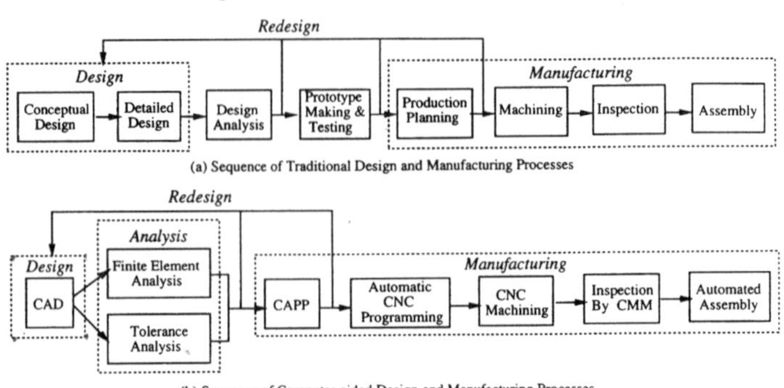

Fig. 1 Traditional and computer-aided design and manufacturing processes.

Exchange of Product Model Data) (Bloor and Owen, 1991), and considerable amount of research, carried out to make PDES/STEP a backbone information system for CAD/CAM systems.

With the computer integration of various degrees, various computer analysis tasks, such as design function and strength analysis using finite element programs, automated tolerance analysis, automated production cost estimation, kinematics simulation and graphical animation, may be carried out in the early design stage, to perfom "soft prototype." Soft prototyping, although not as accurate as real prototype testing, can provide considerable useful information to design, and can significantly reduce the lead times caused by prototype making and testing.

Despite the progress, the product description data in these integrated design and manufacturing systems are mostly transformed in one direction only: from design to manufacturing. The manufacturing information available to guide design and re-design is very modest. To further improve the product development efficiency, a mechanism that can support information flow in the other direction (from manufacturing to design) is desperately required. This mechanism will allow the activities of many product life-cycle aspects, including design, manufacturing, assembly, maintenance, marketing, etc., to be carried out simultaneously, thus leading to a well-balanced design.

Recent researches on concurrent design modelling provide a new approach to integrate various product development modules and to bring in considerations from many downstream manufacturing moduls into the early design stage (Kusiak, 1993). In addition to the capability of soft prototyping, the approach simultaneously carries out the design, analysis, planning, production, and testing tasks that are conducted sequentially in the traditional practice. The method can significantly reduce product development lead times and allow the functional performance of a product to be considered jointly with production costs, customer satisfaction and maintenance costs, thus generating design with better overall life-cycle performance. This chapter will review and summarize the research on developing an integrated and intelligent environment for concurrent design at the University of Victoria.

1.2. Concurrent Engineering and Integrated Concurrent Design

The intent of concurrent engineering is to break the barrier between design and other product development modules, especially manufacturing. Concurrent design (or concurrent engineering design) evaluates a design from various product life-cycle aspects simultaneously, thereby producing a design with balanced design functional performance, production costs, customer satisfaction, maintenance costs, and so on.

It has been estimated that 65–80 percent of the life-cycle cost of a product is driven by decisions made in the first 20 percent of the development effort (O'Grady and Young, 1991; Rosenblatt and Watson, 1991; Dowlatshahi, 1992a; Dowlatshahi, 1992b). In addition, design accounts for only about 5 percent of the total product costs. Therefore, an effective design could be a vital component in improving industrial profitability.

The practice of concurrent engineering was initiated in the automotive industry by forming product development teams in the 1960s (Ward *et al.*, 1994). Members of such teams would usually include representatives from design and manufacturing, as well as possibly those from sales/marketing, accounting, maintenance and the other life-cycle functions. The team may carry out a post-design review, a review at different stages of the design, or be continually involved throughout the design process (O'Grady and Young, 1991). The key issues include selections of the right persons for the team, a harmonic working relationship among team members, and the ability to reach a general design conclusion that reflects the best compromise, considering all contribution factors.

The introduction and adoption of concurrent engineering have led to many archetype industrial successes (O'Grady and Young, 1991; Rosenblatt and Watson, 1991), due to its capability of producing a balanced design and reducing the number of costly redesigns. However, incorporation of relevant information from all product life-cycle functions imposes an even heavier load on designers. A computer software system for assisting designers in systematically considering various life-cycle aspects of a product, and identifying the design with balanced aspect performance, is desperately needed.

Presently, most concurrent design-relevant integrated systems focus on only one non-design life-cycle aspect. Many *design for X* systems, including design for manufacturing (Bralla, 1986; Dong, 1993), design for assembly (Andersen *et al.*, 1983; Boothroyd and Dewhurst, 1983), design for serviceability (Makino *et al.*, 1989; Gershenson and Ishii, 1993), design for reliability (Birolini, 1993), and design for economics (Noble and Tanchoco, 1993), have been developed and implemented. In *a design for X* system, the domain knowledge of a specific product life-cycle aspect, other than design functions, is modeled as guidelines or constraints, leading to design configurations and parameters favourable to the considered product life-cycle aspect. Many advanced computation techniques, including rule-based reasoning (Dixon, 1986), constraint-network (Young *et al.*, 1992), and optimization (Dowlatshahi, 1992a), have been employed for developing these systems.

To further improve the approach, and to introduce the *Design for All* integrated concurrent design methodology, a number of key issues need be addressed. These include: (1) a general concurrent design model for incorporating many product life-cycle aspect considerations simultaneously should be introduced; (2) an integrated database and knowledge base scheme for generating and maintaining the relations among these product aspects should be developed; and, (3) an optimal design identification mechanism for achieving a design with balanced design functional performance, production costs, customer satisfaction, maintenance costs, etc., based upon quantitative evaluations to the different product life-cycle aspects, should be obtained.

In the authors' previous work, a feature-based intelligent design system was developed (Xue and Dong, 1993). Mechanical features are modeled from three perspectives: design, manufacturing, and geometry, representing the different product life-cycle aspects. The database and knowledge base schemes regarding these features follow a design knowledge base and database representation language — Integrated Data Description Language (IDDL), which was originally developed at the Centre for Mathematics and Computer Science (CWI) in Amsterdam and the University of Tokyo (Veth, 1987; Xue *et al.*, 1992). A quantitative intelligent system mechanism was introduced for organizing the qualitative and quantitative relations among the different aspect models (Xue and Dong, 1994a). A design and manufacturing feature coding system has been developed for organizing the feature database and for automated generation of feasible design and planning of production process (Xue and Dong, 1994b). Because the modeling of production costs is a key issue in concurrent design, a number of new production cost models for different production processes were introduced (Dong *et al.*, 1994). Several other models for identifying the design with minimum production costs based upon the analysis of tolerance and production process have also been proposed (Dong, 1990; Dong, 1991; Dong, 1994b).

In this section, a general integrated concurrent design model is introduced. Many sub-systems developed in the authors' previous work are used as functional modules for modeling, generating, and maintaining the data and their relations of the different product life-cycle aspects. In addition, a mechanism for identifying the optimal design with balanced functional performance and production costs, based upon joint evaluation of the two key aspects, is also presented.

2. PRODUCT LIFE-CYCLE ASPECT MODELS AND THEIR REPRESENTATION

2.1. Life-Cycle Aspect Models of a Mechanical Product

The *life-cycle aspect models* are introduced to allow various life-cycle aspects of a mechanical product, including design, manufacturing, assembly, maintenance, etc., to be considered separately, and their performance from these aspects to be modeled and evaluated. These aspect models are associated through an intermediate product description, called *metamodel*. Each aspect model may have its own representation scheme, but its relation to the metamodel is always defined. Due to these links to the metamodel, any changes made in one aspect model will automatically propagate to all others. Figure 2 illustrates the relations among different aspect models and the metamodel.

The concept of metamodel was first introduced by Tomiyama and Yoshikawa (Tomiyama and Yoshikawa, 1987) for representing the evolutionary design database of a General Design Theory (GDT). In GDT, a design is considered as a mapping process from a design specification to a design solution (Yoshikawa, 1981; Tomiyama and Yoshikawa, 1987). The metamodel was improved as a central

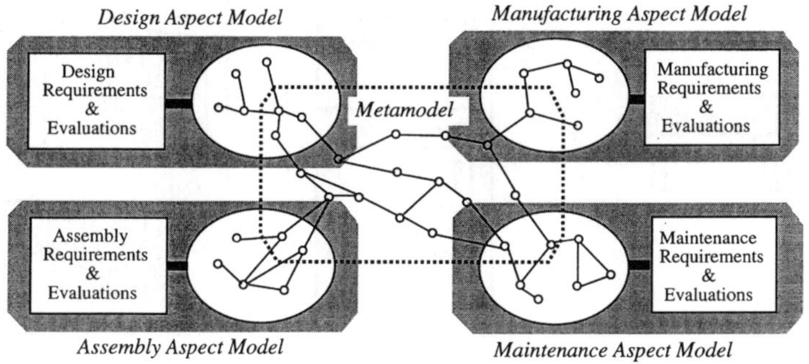

Fig. 2 Aspect models of a product.

database for integrating different design object models (Tomiyama *et al.*, 1989; Kiriyama *et al.*, 1991). In this work, the concept of the metamodel is employed for representing the evolutionary design database, and for maintaining the relations among different product life-cycle aspect models (or concurrent engineering inputs) at a specific design stage.

The dictating requirements for forming each aspect model are a set of constraints and evaluation indices. A *constraint* is a description which the product must satisfy. It can be either a qualitative relation among entities, or a quantitative relation among attributes. An *evaluation index* is a measure of the product quality from a particular evaluation aspect. It is normally based upon a quantitative measure represented by a function of attributes.

2.2. Product Database Representation Scheme

To describe the requirements and database of the product life-cycle aspect models and metamodel, the general product database representation scheme, Integrated Data Description Language (IDDL), is used. IDDL was originally used for an intelligent CAD system called Intelligent Integrated Interactive CAD (IIICAD) (Tomiyama and ten Hagen, 1987; Xue *et al.*, 1992). This language combines logic programming and object-oriented programming techniques for modeling design knowledge base and database. A pair of gears represented using this scheme is illustrated in Figure 3.

In this product database representation scheme, design entities and their relations are represented by *entities* and *facts*, respectively. Facts are described by predicates for symbolic reasoning. Entities are described as the terms of these predicates. For example, *gearPair(g1,g2)*, as shown in Figure 3, is a fact representing a pair of gears. The two terms, *g1* and *g2*, of this predicate are entities representing the two gears. An entity may have several *attributes*. The quantitative relations among attributes are described by *functions*. The entity *g1*, shown in Figure 3, may have attributes including the number of teeth, *z*, module of the gear, *m*, and the nominal diameter, *d*. The built-in predicate $=(d[g1], m[g1]^{*}z[g1])$, is a function for

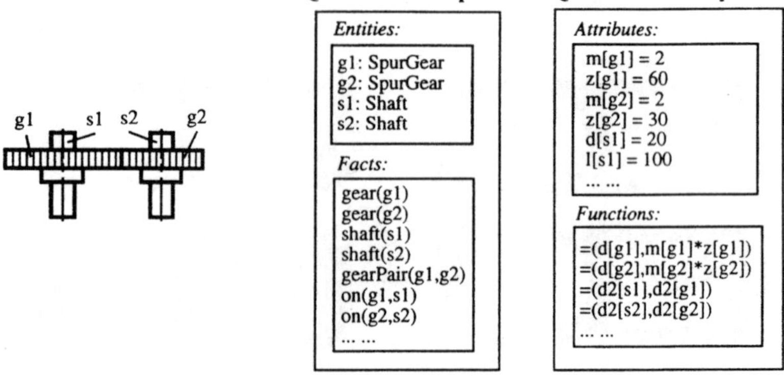

Fig. 3 Product database for representing a pair of gears.

calculating the nominal diameter attribute, $d[gl]$, using related attributes, $m[gl]$ and $z[gl]$.

2.3. Feature-Based Aspect Model Representation

The performance evaluation in different product life-cycle aspects can only be carried out in the physical world, based upon different mechanical features. A feature modeling system is developed to serve that need in this work (Xue and Dong, 1993). Three types of features — design features, manufacturing features, and geometry features — are used for modeling the three most important aspects of a mechanical product. A design feature is a functional primitive in the form of a mechanical component or a mechanism for modeling a mechanical system. A manufacturing feature is formed by the partial component geometry produced by a specific manufacturing operation. A geometry feature is a geometric primitive, used in a 3D CAD system, for representing design geometry. In these feature models, each feature is represented by its type and a number of feature parameters. The relationships among these related, but distinct features for representing a product are illustrated in Figure 4.

The different product life-cycle aspect models are described by the relevant aspect features. The design aspect model allows us to create, evaluate and improve the design, based upon its functional performance. The manufacturing aspect model allows us to examine the manufacturability and production costs of the design. The geometry aspect model integrates the design and manufacturing models into the framework of metamodel for automated model mapping.

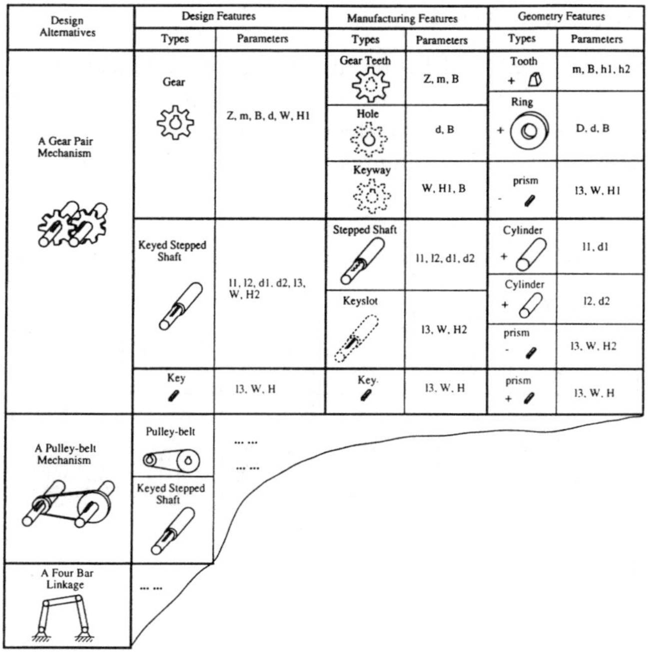

Design Alternatives	Design Features		Manufacturing Features		Geometry Features	
	Types	Parameters	Types	Parameters	Types	Parameters
A Gear Pair Mechanism	Gear	Z, m, B, d, W, H1	Gear Teeth	Z, m, B	Tooth +	m, B, h1, h2
			Hole	d, B	Ring +	D, d, B
			Keyway	W, H1, B	prism -	l3, W, H1
	Keyed Stepped Shaft	l1, l2, d1, d2, l3, W, H2	Stepped Shaft	l1, l2, d1, d2	Cylinder +	l1, d1
					Cylinder +	l2, d2
			Keyslot	l3, W, H2	prism -	l3, W, H2
	Key	l3, W, H	Key	l3, W, H	prism +	l3, W, H
A Pulley-belt Mechanism	Pulley-belt	... ...				
	Keyed Stepped Shaft	... ...				
A Four Bar Linkage	... ...					

Fig. 4 Relations among three types of features for representing a product.

(a) Design Aspect Model

The design aspect model is built up by combining design features. A design feature can be either a mechanical component or a mechanism to deliver a required design function. A number of design features are given in Figure 5.

Design features are described at two levels: class level and instance level, corresponding to general templates and special applications using object oriented modeling (Goldberg and Robson, 1983; Budd, 1991), as shown in Figure 6. Instance design features are generated from class design features. Each class feature is described by a group of data definitions including entities, facts, attributes, and functions. Entities are represented as variables starting with uppercase letters. The *class* itself is an entity and is represented as *Self*. Class design features are organized in a hierarchical data structure. A class feature is defined as a child of a superclass. All descriptions of a super-class are inherited by its sub-classes automatically. When an instance feature is generated using a class feature as its template, all the descriptions of that class feature will be added to the design database automatically. In an instance feature, all variables in its class feature are instantiated by special entities (starting with lowercase letters).

Instance design features are combined during design database modeling. For instance, a linear motion dial indicator, as shown in Figure 7 (a), is modeled by the design features given in Figure 7 (b). The indicator is used for measuring linear motion and displaying the amount of the motion. The indicator is primarily composed of two design features: *rackPinionPair1* and *gearPair1*. These two mechanism design features are connected by a component design feature – *shaft1*.

(b) Manufacturing Aspect Model

The manufacturing aspect model is represented using manufacturing features. Each manufacturing feature is produced by a manufacturing operation. In the authors' work, the manufacturing features produced by various manufacturing processes are studied. A number of manufacturing features, produced by machining, are given in Figure 8.

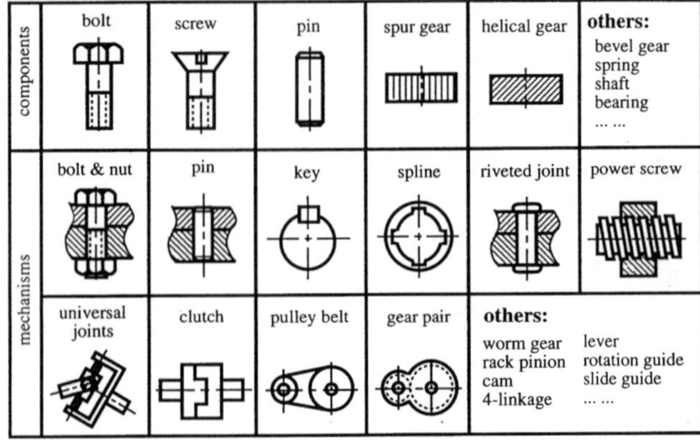

Fig. 5 Design features.

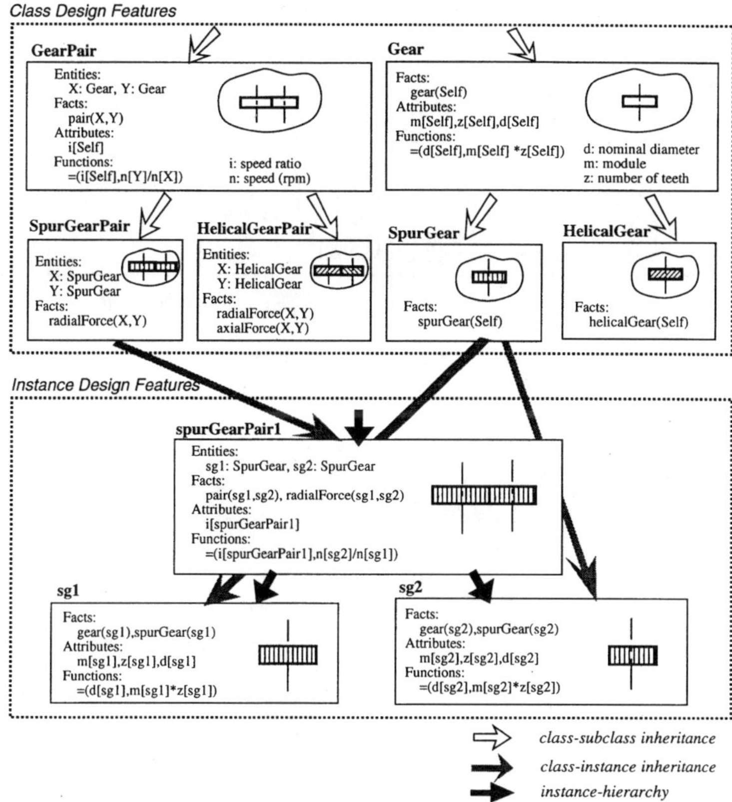

Fig. 6 Representation of class and instance design features.

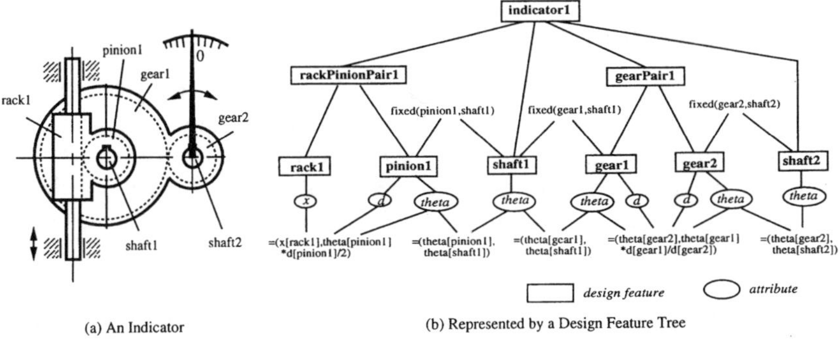

Fig. 7 A design aspect model.

A mechanical component can be represented by a manufacturing feature tree, as shown in Figure 9. Each node of this tree is a manufacturing feature represented by the type of the feature, its location and orientation, and all feature parameters. In manufacturing, these features are produced from higher to lower level. Features at the same level can be machined simultaneously.

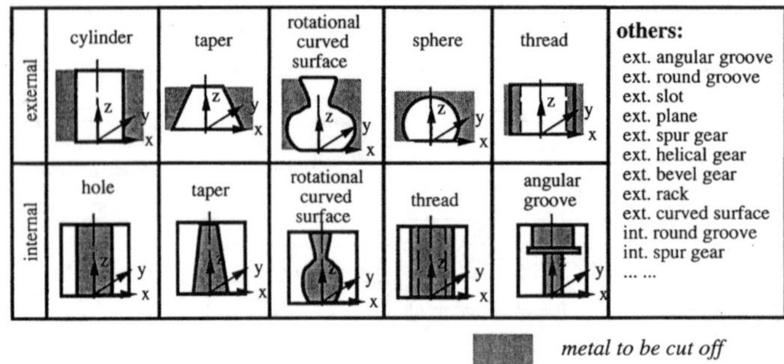

Fig. 8 Manufacturing features.

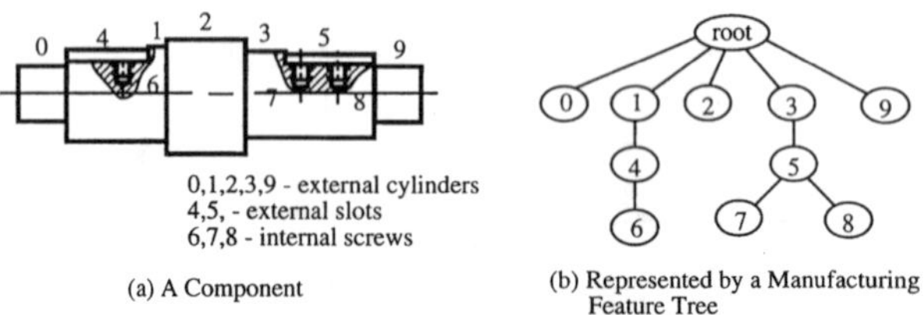

Fig. 9 A manufacturing aspect model.

A manufacturing feature can be either extracted from the 3D geometric model of a CAD system, or defined manually. The generated manufacturing features are used for planning the production process, analyzing manufacturability and costs, and producing the part.

(c) Geometry Aspect Model

The geometry aspect model has minor significance for the life-cycle performance evaluation. However, it serves as the backbone of the metamodel that allows other aspect models to be integrated in the CAD system. A geometry aspect model is represented using geometry features. A geometry feature can be either an elementary geometric primitive such as a prism or a cylinder, or a complex one generated by sweeping a section or extruding a surface. A number of geometry features are given in Figure 10.

The geometry aspect model is constructed using geometry features. The specification of geometry feature is illustrated by a simple mechanical component and its composing geometry features as in Figure 11. The geometry features are organized in a tree data structure. In each geometry feature, the feature type, its location and orientation, and feature parameters including dimensions and tolerances are described (Dong, 1993). Each dimension and tolerance pair is modeled by an arc starting from the reference surface and ending at the dimensioned surface of related geometric features. The dashed lines represent overlapped graph nodes or contacting surfaces.

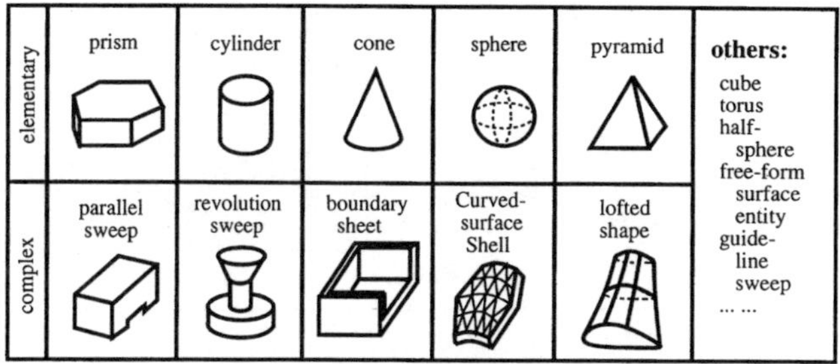

Fig. 10 Geometry features.

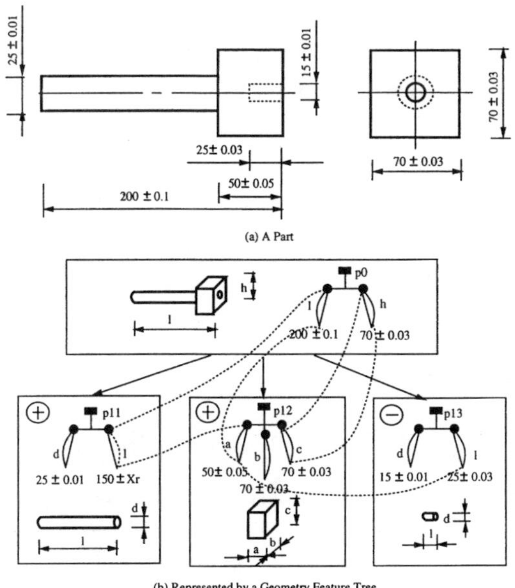

Fig. 11 A Geometry aspect model.

To implement the feature-based geometry representation in a solid modeling system, a hybrid geometric modeler was developed. In the modeler, the geometry features are organized by a symbolic model. Geometric elements, including objects, faces, edges, vertices, and their topological relations, are organized by a boundary representation (B-rep) solid model (Zeid, 1991). The two models are associated through geometric elements, which are entities in the symbolic model and basic data structure in the B-rep solid model. Any change in one model can propagate to another automatically. A pyramid geometry feature represented using the hybrid geometric model is illustrated in Figure 12.

2.4. Integration of Aspect Models

The qualitative and quantitative relations among different aspect models are described by an *entity relation network* and an *attribute relation network* (Xue and

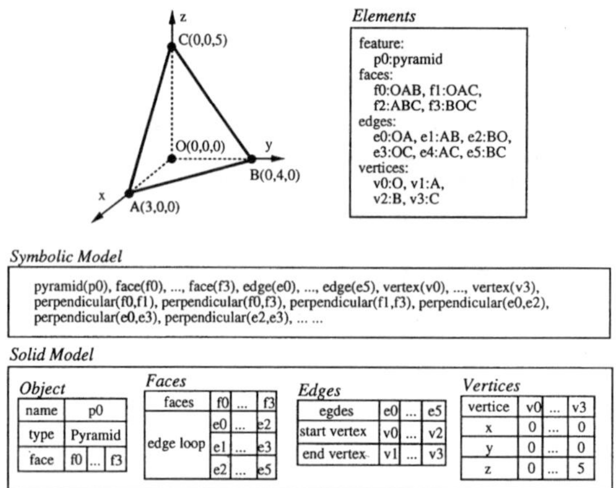

Fig. 12 A hybrid geometric model.

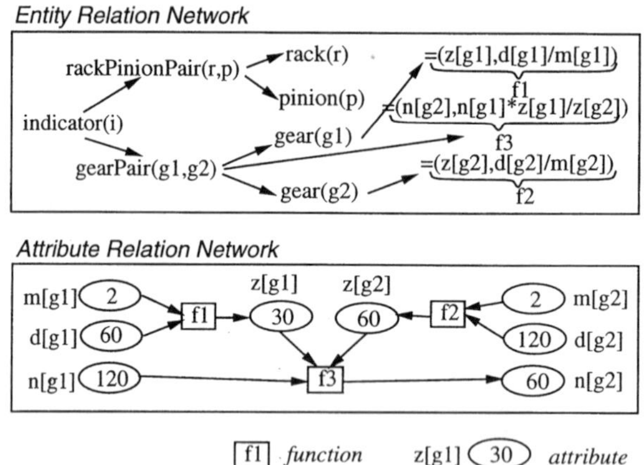

Fig. 13 Qualitative and quantitative relation networks.

Dong, 1994a), as shown in Figure 13. The entity relation network, represented by entities, facts, attributes, and functions, is created by symbolic reasoning. The generated qualitative relations are maintained using an Assumption-based Truth Maintenance System (ATMS) (de Kleer, 1986). In ATMS, dependent relations among data are described. When part of the database is altered, all other related data are updated. The generated quantitative relations described by functions are then transformed into an attribute relation network. This network can be used for calculating the attribute values, subject to the constraints defined in the attribute relation network.

3. AUTOMATED GENERATION OF ASPECT MODELS

The feature modeling mechanism provides a platform for representing and maintaining the data and their relations in the different product life-cycle aspect models. Although this mechanism has improved the efficiency of concurrent design considerably, many decision-making tasks, such as generation of feasible design candidates, selection of appropriate geometry, and planning of the production process, still need to be automated. To carry out these tasks in feature-based, integrated concurrent design, more advanced computer automation techniques are required. The recent advances in artificial intelligence techniques have proved most effective for developing intelligent design and manufacturing systems (Kusiak, 1992; Dong, 1994a). In this work, relevant artifical intelligence techniques, including knowledge representation, rule-based reasoning, fuzzy data clustering and pattern identification, and contradiction resolution, are applied.

3.1. Knowledge Representation Scheme

The knowledge for generating life-cycle aspect models follows the scheme of the knowledge base and database representation language — IDDL (Veth, 1987; Xue *et al.*, 1992). In IDDL, a piece of knowledge is decribed by a scenario. A scenario is composed by a collection of rules. Each rule has the IF – THEN structure for matching and accessing the introduced product database. Both the condition part and the result part are described by a set of predicates connected with the logical-and (&). Rules can be used in *forward reasoning* and *backward reasoning*. An example scenario describing the knowledge of gears is given in Figure 14.

The scenario-based inference system, as shown in Figure 15, was developed based upon a computable design process model (Takeda *et al.*, 1990). In this model, design knowledge is represented at two different levels: meta level and object level, corresponding to general knowledge on design process activities and the specific knowledge for solving specific problems. In the scenario-based inference system, meta level scenarios are used for representing meta level knowledge K_m, which describes the relations between database conditions (C) and possible operations (O) to the database. The possible operations include selection of object level knowledge, execution of object level inference modules, and display of the present product database status. Object level scenarios are used for representing object level knowledge K_o, which describes the relations between design objects D_s and their

gears

```
BEGIN
IF gear(X) THEN =(d[X],m[X]*z[X]).
IF gear(X) & =(angle[X],0) THEN spurGear(X).
IF gear(X) & >(angle[X],0) THEN helicalGear(X).
IF gear(X) & gear(Y) & pair(X,Y) THEN gearPair(X,Y).
IF gearPair(X,Y) THEN =(pair[X],Y) & =(pair[Y],X)
    & =(dist[X],(d[X]+d[Y])/2) & =(dist[Y],dist[X]).
IF pair(X,Y) & ~gear(X) THEN ~gearPair(X,Y).
... ...
END
```

Fig. 14 A scenario.

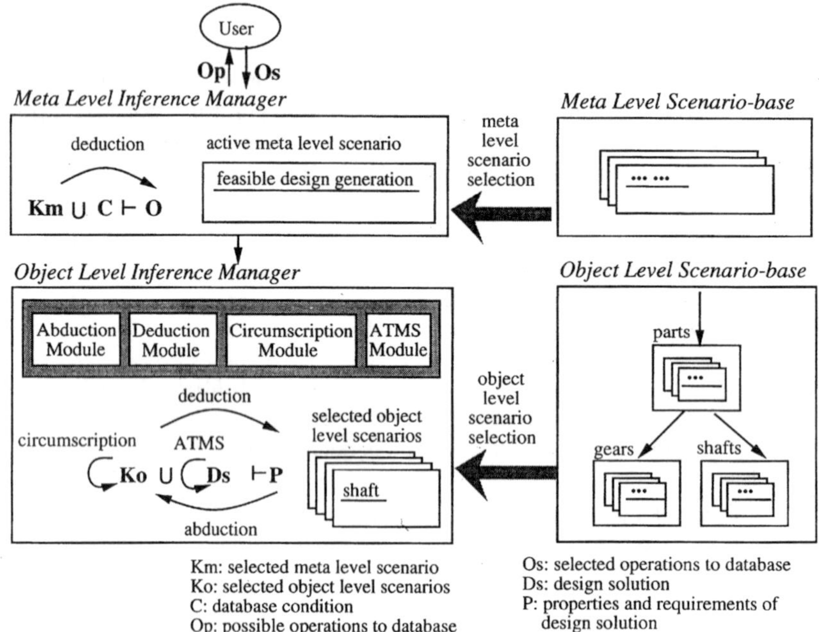

Fig. 15 Scenario-based inference system.

functions and properties P. During a design, meta level scenarios are first selected and used for observing the condition of the database and for giving the user the available possible operations O_p. The user then selects one of the O_s to execute an object level inference module. The object level inference modules include an abduction module (Fann, 1970) for generating possible problem solutions through backward reasoning, a deduction module for deriving more information related to the solution through forward reasoning, a cirumscription module (McCarthy, 1980; Nakagawa and Mori, 1987) for solving a contradiction by modifying the incomplete design knowledge, and an ATMS (de Kleer, 1986) for solving a contradiction by changing part of the product database. In this work, the fundamental modules of the scenario-based inference system are used for generating the life-cycle aspect models.

3.2. Generation of Feasible Design

As discussed previously, a design aspect model is built up by selecting and combining appropriate design features. The task is carried out by scenario-based reasoning.

Design is usually considered as a process to obtain a design solution for a given set of functional requirements (Yoshikawa, 1981). During a design, a required function is decomposed into a number of sub-functions. Each is to be accomplished by a feasible design feature. The selected features are combined to form a candidate design. The design is evaluated using required functions as the criteria. The scenarios serve as the knowledge for function decomposition, candidate identification, design synthesis and evaluation.

The process of feasible design generation is illustrated in Figure 16. The task is to design the linear motion measurement indicator shown in Figure 7 (a). At first, two entities, a linear motion rod *r1* and a rotation motion indicator *i*, are given in a database called *world1*. The design objective is to find a mechanism to achieve the function of transmitting small linear motion into large rotation motion, *LRLarge-Trans(r1, i)*. Since this function cannot be achieved by any single design feature directly, the function is decomposed into two sub-functions, a linear-to-rotation (LR) transmission function and a rotation-to-rotation (RR) magnification transmission function. A component design feature, shaft *s*, is used to connect these two functions. The design database is then evolved as *world2*. These two sub-functions are replaced by two design features, a rack-pinion pair and a gear pair. By linking these generated design features, the original design objective is accomplished.

The selection of appropriate design features to satisfy a given design function can also be carried out using a design feature coding system (Xue and Dong, 1994b). This coding system was originally developed for organizing the numerous design features. In this system, design functions are classified into 13 major categories as shown in Table 1. Each category consists of several function items. A function code is represented by a set of bits corresponding to all the function items of the 13 categories. Each bit of the code is a number between -1 and $+1$, representing one aspect of design functions. The numbers -1 and $+1$ represent the absolute negative and positive cases, respectively. Any other number presents a fuzzy statement. The number 0 represent an "unknown" status.

Design features are classified into many clusters according to their function codes using the fuzzy c-means (FCM) pattern clustering method (Bezdek, 1981). The

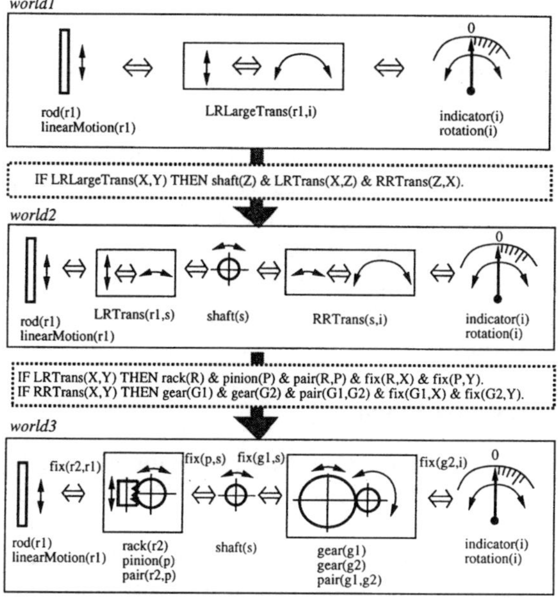

Fig. 16 Feasible design generation through scenario-based inference.

TABLE 1
Design feature coding system.

(1) Main Shape [1.1 – 1.3]
(2) Accuracy Level [2.1 – 2.3]
(3) Major Dimension [3.1 – 3.4]
(4) Form of Power Transmission [4.1 – 4.11]
(5) Form of Kinematic Movement [5.1 – 5.4]
(6) Form of Kinematic Transmission [6.1 – 6.8]
(7) Shape of Connected Components [7.1 – 7.3]
(8) Connection - Removability [8.1 – 8.4]
(9) Transmission Magnitude [9.1 – 9.2]
(10) Supporting Component [10.1 – 10.4]
(11) Hydraulic Component[11.1 – 11.4]
(12) Relation between Input\Output Shafts [12.1 – 12.3]
(13) Form of Motion Guide [13.1 – 13.3]

clustering is conducted at multiple levels to decrease the number of design features in each group. The tree data structure of design feature clusters is illustrated in Figure 17. Each cluster has a centroid function code vector C_i. Elements of the bottom layer clusters are represented by design features.

The search of a design feature for a given design function starts with coding this design function using the introduced coding system. Because the number of design features is huge, matching between the required function and design features is conducted through the hierarchical data structure of the design feature clusters. In Figure 17, the required design function X is first compared with three clusters *C2*, *C3*, and *C4*. Since the cluster *C4* is more similar to X than the two others, this cluster is selected for further comparing. Using this method, the design feature *F8* is finally selected.

3.3. Specification of Design Geometry

The specification of design geometry is accomplished by associating the component design features with the geometry features using relevant knowledge. Locations and orientations of these geometry features can either be obtained from design descriptions, or assigned manually by the user. The geometry aspect model represented by geometry features is called the symbolic geometry model. The symbolic model is linked with a B-reps solid model, using the hybrid geometry modeler introduced previously. The fundamental idea regarding design geometry creation is given in Figure 18.

3.4. Production Process of a Design

An accurate estimate to the production costs and manufacturability of a design can only be accomplished when the production process of the design is determined. In addition, only the optimal production process with minimum production costs can truly reflect the production costs of a design. Automated generation of a minimum cost production process for a given design therefore becomes essential. It is worthwhile to mention that computer automated process planning is a very broad and active research area. Readers may find in-depth discussions of the technology in

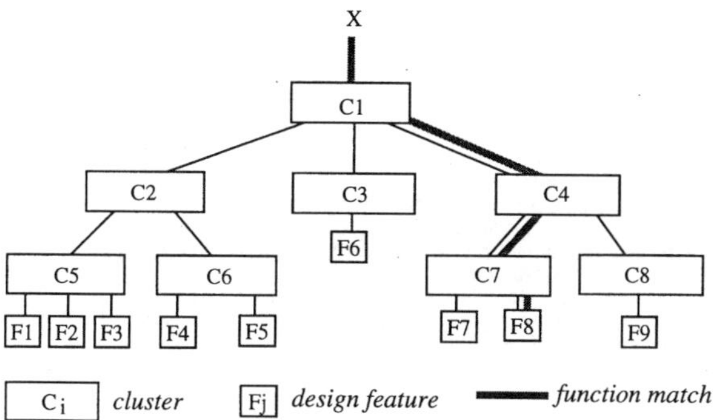

Fig. 17 Design feature clusters and design feature search.

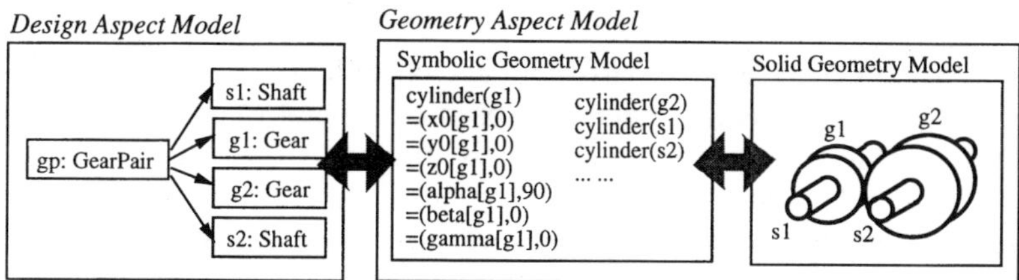

Fig. 18 Creation of design geometry.

(Chang, 1990; Chang *et al.*, 1991; Zhang and Alting, 1993). The focus of the authors' work is not to develop a general process planning system, but to find a means for accurately estimating the production costs of a design.

The production process of a product relies heavily on the manufacturing features of the design. The first step of production process generation (or planning) is to represent a mechanical design using manufacturing features. As discussed previously, manufacturing features can either be extracted from the 3D CAD databases or defined by a user manually. During the last decade, many research projects on feature extraction have been carried out (Henderson, 1984; Choi *et al.*, 1984; Voelcker *et al.*, 1986; Dong and Wozny, 1988). In this work, a new feature extraction approach, based upon scenario-based reasoning using the hybrid geometry model, is developed. A scenario for extracting a "hole" manufacturing feature is given in Figure 19. The feature extraction is conducted through three steps: (1) pre-processing of the CAD databases to represent geometric primitives using IDDL entities; (2) creation of manufacturing features based upon the created IDDL entities and; (3) post-processing of manufacturing features to add feature attributes. Extraction of the hole manufacturing feature from a given design is carried out by finding the internal edge loop on the face, selecting an edge in the loop, and calculating the angle between the two adjacent faces of the edge. This procedure is used because a circle is represented by a polygon which is also considered as an edge loop in most solid modelers.

The extracted or defined manufacturing features of a component are represented in a tree data structure, as illustrated in Figure 9. To organize these manufacturing

```
[rule 1]: IF face(F) & hasInternalEdgeLoop (F)
            THEN generateInternalEdgeLoopEntity(F,EL) & internalEdgeLoop(EL) & has(F,EL),
[rule 2]: IF internalEdgeLoop(EL)
            THEN generateFirstEdgeEntity(EL,E) & edge(E) & has(EL,E),
[rule 3]: IF face(F) & internalEdgeLoop(EL) & edge(E) & has(F,EL) & has(EL,E) & <(angle[E],180)
            THEN generateHoleEntity(H) & hole(H) & has(F,H) & topEdgeLoop(H,EL) & =(topEdgeLoop[H],EL),
[rule 4]: IF hole(H) & =(bottomEdgeLoop[H],nil)
            THEN generateBottomEdgeLoop(H,EL) & bottomEdgeLoop(H,EL) & =(bottomEdgeLoop[H],EL),
[rule 5]: IF hole(H) & topEdgeLoop(H,EL1) & bottomEdgeLoop(H,EL2) & sameShape(EL1,EL2) & parallel(EL1,EL2)
            THEN =(crosssectionProperty[H],'constant') & =(height[H],distance[EL1,EL2]),
[rule 6]: IF hole(H) & topEdgeLoop(H,EL) & circularShape(EL) & =(crosssectionProperty[H],'constant')
            THEN =(crosssectionShape[H],'circular') & =(diameter[H],diameter[EL]),
[rule 7]: IF hole(H) & topEdgeLoop(H,EL) & polygonShape(EL) & =(crosssectionProperty[H],'constant')
            THEN =(crosssectionShape[H],'polygon'),
[rule 8]: IF hole(H) & topEdgeLoop(H,EL) & =(crosssectionProperty[H],'constant') & edge(E) & has(EL,E)
          >(angle[E],89.5) & <(angle[E],90.5)
            THEN =(direction[H],'straight'),
... ...
```

Fig. 19 A scenario for extracting a hole manufacturing feature.

features in an efficient manner, as well as schedule the production process, a manufacturing feature coding system has been developed (Xue and Dong, 1994b). The manufacturing feature coding system is similar to the design feature coding system. In this coding system, manufacturing functions are classified into nine major categories as shown in Table 2. Each category consists of several items. A manufacturing feature is described by a manufacturing feature code that forms a bit vector corresponding to the nine categories.

The manufacturing properties of a component are described by a manufacturing code which is obtained from the composing manufacturing features. The components are classified into many groups according to their manufacturing codes. This approach is similar to the group technology (GT) method (Mitrofanov, 1959; Chang *et al.*, 1991), in which the parts are classified according to their similarities using a coding system, such as Opitz (Opitz, 1970) and KK-3 (Japan Society for the Promotion of Machinery Industry, 1980) systems. In this work, composing manufacturing features are coded directly, instead of the component at a higher level. This method has two advantages. First, it is comparatively straightforward and simple to code manufacturing features than to code the component. Second, the composing manufacturing features are changed and updated constantly during design. The component manufacturing codes should also be updated dynamically. Production process planning of the components using their manufacturing codes serves as *multiple component production process planning*.

Since a manufacturing feature is produced by a set of production operations, a manufacturing feature tree can then be transformed into a production operation tree. The production operation tree, shown in Figure 20 (a), is obtained from the manufacturing tree shown in Figure 9 (b). This tree is then transformed into a production operation sequence. A software tool has been developed for optimizing the sequence of the production operations (Xue and Dong, 1994b). The procedure and result of production process planning are shown in Figure 20 (b). The generation of the production process, or optimization of the production sequence is aimed at the prediction of production costs and manufacturability of design. The task involves a single component only, and forms a *single component production process planning* process.

TABLE 2
Manufacturing feature coding system.

(1)	Main Shape [1.1 – 1.3]
(2)	Cross Section Shape [2.1 – 2.2]
(3)	External Surface [3.1 – 3.14]
(4)	Internal Surface [4.1 – 4.9]
(5)	Raw Material Form [5.1 – 5.6]
(6)	Type of Material [6.1 – 6.8]
(7)	Major Dimension (L) [7.1 – 7.4]
(8)	Length-to-diameter Ratio (L/D) [8.1 – 8.6]
(9)	Accuracy Level [9.1 – 9.3]

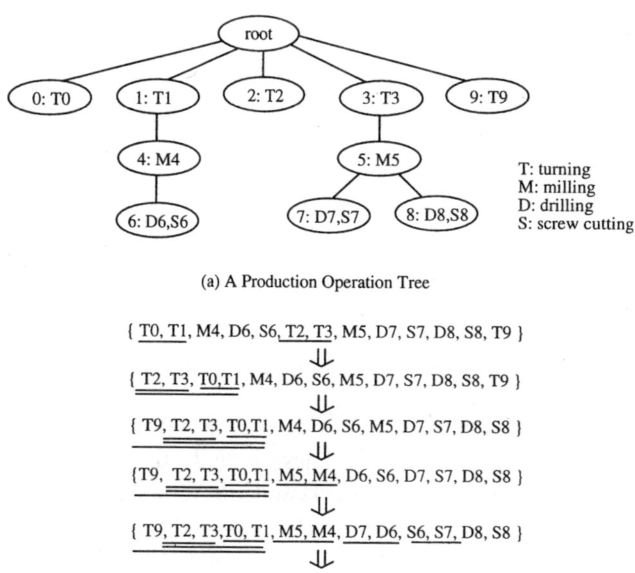

Fig. 20 Production process planning for a single component.

4. INTEGRATED CONCURRENT DESIGN BASED UPON LIFE-CYCLE PERFORMANCE OPTIMIZATION

The methods of feature-based representation and automated generation of product life-cycle aspect models provide a basis for modeling various concurrent design activities. The key issue of a concurrent design — identification of optimal design with the best overall life-cycle performance — is carried out using the developed modeling tools.

4.1. Life-Cycle Performance of a Product

The ultimate goal of concurrent design is to form a design with the best overall life-cycle performance. This goal can only be reached by evaluating all the related product life-cycle aspect performance, and conducting global optimization on the product life-cycle performance. The evaluation and optimization, in turn, are based upon appropriate modeling of all product life-cycle aspect performance, using either theoretical models or empirical data.

The essence of concurrent design, if represented as the global optimization of product life-cycle performance, can be expressed as

$$\max_d \mathbf{I}(\mathbf{d})_i = \lambda_F \mathbf{I}_i^{(F)} - \lambda_C \mathbf{I}_i^{(C)} + \lambda_S \mathbf{I}_i^{(S)} - \lambda_M \mathbf{I}_i^{(M)} + \ldots - \ldots \tag{1}$$

$$\lambda_F + \lambda_C + \lambda_S + \lambda_M + \ldots - \ldots = 1$$

where, $\mathbf{I}$ represents the product life-cycle performance; $\mathbf{d}$ includes all design variables for the design configuration i; $\mathbf{I}_i^{(F)}$ and $\mathbf{I}_i^{(S)}$ are measures of functional performance and customer satisfaction, respectively; $\mathbf{I}_i^{(C)}$ and $\mathbf{I}_i^{(M)}$ are measures of manufacturing and service (maintenance) costs, included as negative performance or performance loss; and λ_i are application dependent weighting factors.

In this work, we narrowed down the scope of our consideration to the two most representative terms of the formulation — design functional performance and production costs.

4.2. Formulation of Functional Performance and Production Costs

A design at a specific product development stage is represented by a n-dimensional parameter vector $\mathbf{d} = (d_1, d_2, \cdots, d_n)^T$. These include discrete parameters, such as design configuration (type of feature) and material, as well as continuous parameters, such as feature dimensions and tolerances. Parameters regarding the three aspect models — design, manufacturing, and geometry — are represented using three vectors, $\mathbf{d}^{(D)}$, $\mathbf{d}^{(M)}$, and $\mathbf{d}^{(G)}$, respectively. A parameter, d_i, used for describing the product in one aspect model can also be used in other aspect models. Only the design parameters that have influence on both design and manufacturing aspects are considered.

Functional performance is a quality measure of the product from the design aspect. The functional performance, $F(\mathbf{d}^{(D)})$, is represented by a group of functional performance indices, $F_i(\mathbf{d}^{(D)})$. For instance, motion transmission accuracy is a functional performance index for a gear-pair. A functional performance index is associated with design aspect parameters through design features. In general, a design with a higher functional performance measure has better design quality. The overall functional performance of a design can thus be represented as

$$F(\mathbf{d}^{(D)}) = f(F_1(\mathbf{d}^{(D)}), F_2(\mathbf{d}^{(D)}), \cdots, F_p(\mathbf{d}^{(D)})) \tag{2}$$

Similarly, production cost is a quality meausure of the design from the manufacturing (or manufacturability) aspect. Production costs are associated with manufacturing aspect parameters through manufacturing features. Suppose the cost for producing a manufacturing feature (j) of a design is represented as $C_j(\mathbf{d}^{(M)})$, the overall production costs of the design can then be calculated by adding the costs

$$C(\mathbf{d}^{(M)}) = \sum_{j=1}^{q} C_j(\mathbf{d}^{(M)}) \tag{3}$$

Given the basic functional performance requirements and maximum acceptable production costs (or manufacturability), a region of feasible design can be defined. The functional performance requirements, maximum acceptable production costs and manufacturability limit serve as constraints from design and manufacturing aspects. In this work, design evaluation and optimization are conducted only for feasible designs.

The selection of design parameters, and formulation of functional performance and production costs are illustrated using a design example. In this example, we are to design a multiple spindle drill head. For ease of illustration, we will focus on only two design variables of the drill head — the size tolerance, δ_D, and location tolerance, δ_{xy}, of the hole on the drill head case for the spindle, as shown in Figure 21. The design objective is to identify the optimal values of these two tolerances, which lead to the best life-cycle performance of the drill head.

The life-cycle performance of the spindle hole is examined from design and manufacturing aspects. The hole can be modeled as a design feature "spindle/shaft hole" for supporting spindle rotation and maintaining accurate spindle location. The spindle hole is also manufacturing feature "hole" to be produced by drilling and reaming. In the geometric aspect model, this hole is created by applying the "difference" Boolean operation to the geometric model of the case using a cylindrical primitive. In practice, the spindle holes are machined after the installation of the journal bearings.

The size tolerance of the spindle hole, δ_D, influences the clearance between the spindle and the journal bearings. Two functional performance measure, power loss

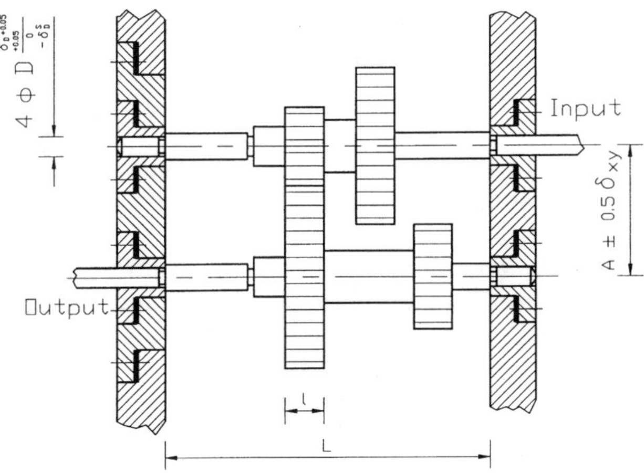

Fig. 21 Spindle hole of a multiple-spindle drill head.

variation ΔPL and spindle case working temperature variation ΔT, are related to the clearance change. The location tolerance of the spindle hole, δ_{xy}, and the size tolerance of the spindle hole, δ_D, both influence the alignment of spindle and other shafts. Misalignments between two shafts with mating gears will change the distribution of the contact stress on the tooth of the gear. Stress concentration will reduce the designed life-time of the gears.

The manufacturability and production costs are sensitive to both the hole location tolerance, δ_{xy}, and the hole dimension tolerance, δ_D. The production cost for machining the hole to an accurate size, C_D, and the production cost for producing the hole at an accurate location, C_{xy}, can be modeled using empirical production cost-tolerance data.

Following the identification of the design parameters, the next step is to formulate the functional performance and production cost indices using these design parameters.

The original functional performance data with regard to: (a) clearance vs. power loss; (b) clearance vs. temperature; and (c) gear life vs. maximum stress, can be obtained from mechanical design textbooks and handbooks (Faires 1965; Shigley and Mischke, 1989). The clearance between the spindle and the bearing depends upon the tolerances of the hole δ_D and the tolerance of the shaft δ_D^S, as illustrated in Figure 21. The minimum clearance is selected as 0.05 *mm*. In practice, because it is much easier to achieve a higher accuracy in shaft machining than in hole machining, the tolerance of the shaft is selected as half that of the tolerance of the hole.

A larger tolerance δ_D can introduce a larger variation of clearnace Δ, thereby increasing the variations of power loss and temperature, and leading to poor product quality in mass production. The functional performance measures are defined as

Functional Performance I:

$$\Delta PL(\delta_D) = -|PL(\Delta_{max}) - PL(\Delta_{min})| \tag{4}$$

Functional Performance II:

$$\Delta T(\delta_D) = -|T(\Delta_{max}) - T(\Delta_{min})| \tag{5}$$

The direct relations between the hole diameter tolerance, δ_D, and the power loss variation, ΔPL, as well as the temperature variation, ΔT, are plotted in Figure 22 (a) and (b).

Both the size and location tolerances, δ_D and δ_{xy}, contribute to the alignment of the shafts. The misalignment can introduce stress concentration on the surface of the gear, thereby reducing the life-time of the gear. The life-time of the gear is defined as the third functional performance to be evaluated:

Functional Performance III:

$$N(\delta_D, \delta_{xy}) = N_0(\sigma_{max}) \tag{6}$$

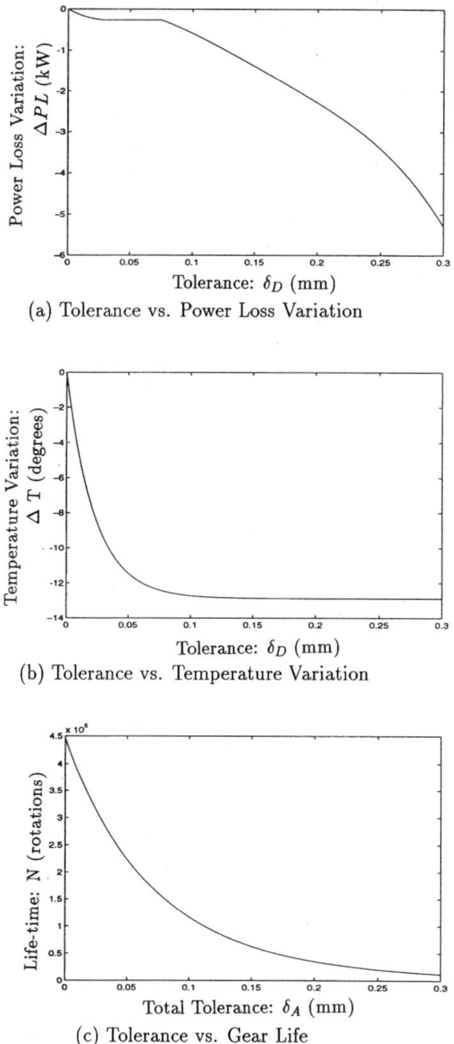

(a) Tolerance vs. Power Loss Variation

(b) Tolerance vs. Temperature Variation

(c) Tolerance vs. Gear Life

Fig. 22 Hole tolerances and functional performance measures.

When other design parameters of the multiple spindle drill head are chosen, as listed in Table 3, the relation between gear life-time and size/location tolerances can be obtained as shown in Figure 22 (c). The δ_A in this figure is calculated using

$$\delta_A = 0.5\delta_{xy} + 0.5(1.5\delta_D + 0.05) \tag{7}$$

The production costs for machining the spindle hole will vary according to the size and location tolerances specified in the design. High accuracy and small tolerance need more manufacturing effort and require higher costs. The production cost-

TABLE 3
List of selected design parameter values.

Parameters	Symbols	Values
Tangential load to the gear surface	W_t	$1.2 \times 10^3 N$
Pressure angle of the gear	ϕ	$20°$
Width of the gear	l	$25.4 \times 10^{-3} m$
Pitch diameter of the first gear	d_{p1}	$101.6 \times 10^{-3} m$
Pitch diameter of the second gear	d_{p2}	$371.5 \times 10^{-3} m$
Elasticity module of the first gear	E_1	$207.0 \times 10^9 Pa$
Elasticity module of the second gear	E_2	$100.0 \times 10^9 Pa$
Poission's ratio of the first gear	V_1	0.292
Poission's ratio of the second gear	V_2	0.211
Module of the gear	m	$4 \times 10^{-3} m$
Length of the shaft	L	$762 \times 10^{-3} m$

tolerance relations were obtained from machine shop and experiments (Trucks, 1976), and modeled in the authors' earlier work (Dong *et al.*, 1994). These relations are illustrated in Figure 23. The production cost-tolerance models have the following forms:

Production Cost I:

$$C(\delta_D) = a_o + a_1\delta_D + a_2\delta_D^2 + a_3\delta_D^3 + a_4\delta_D^4 + a_5\delta_D^5 \tag{8}$$

where, the parameters $a_0, a_1, a_2, a_3, a_4, a_5$ are $112.3, -1.061 \times 10^3, 5.833 \times 10^3, -1.534 \times 10^4, 1.845 \times 10^4, -8.269 \times 10^3$, respectively (Dong *et al.*, 1994).

Production Cost II:

$$C(\delta_{xy}) = b_0 + b_1\delta_{xy}{}^{-b2} + b_3 e^{-b_4\delta_{xy}} \tag{9}$$

where, the parameters b_0, b_1, b_2, b_3, b_4 are $1.000 \times 10^{-5}, 5.326, 0.4475, 6.652, 11.72$, respectively (Dong *et al.*, 1994).

4.3. Feasible Design Space

The feasible design space of this simplified problem is defined by the minimum functional performance and manufacturability limitations imposed on the design. These constraints are summarized in Table 4.

The feasible design space, if plotted with respect to the two selected design variables, δ_D and δ_{xy}, has a closed area as illustrated in Figure 24.

4.4. Identification of Optimal Design

Three types of life-cycle design optimization are considered in this work. They are: (a) production cost priority design; (b) functional performance priority design; and, (c) balanced functional performance and cost design.

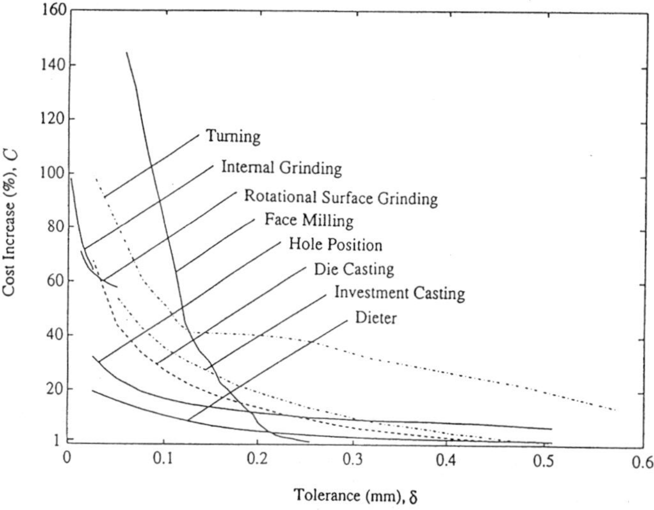

Fig. 23 Relations between production operations and production costs.

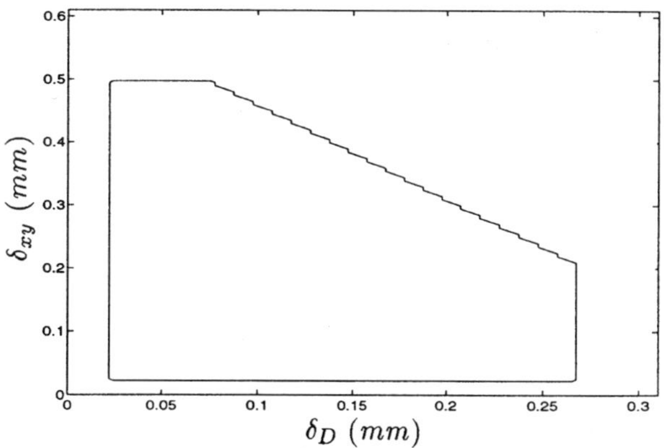

Fig. 24 Feasible design space.

TABLE 4
Design and manufacturing constraints for the spindle hole tolerance design.

Design Constraint I	$\Delta PL \leq 4\ kW$
Design Constraint II	$\Delta T \leq 13°$
Design Constraint III	$N \geq 5^4$ *rotations*
Manufactring Constraint I	$\delta_D \in$ [0.02 mm, 0.5 mm]
Manufacturing Constraint II	$\delta_{xy} \in$ [0.02 mm, 0.5 mm]

(a) Production Costs Priority Design

Production cost priority design is carried out by minimizing the total production costs, subjecting to all given functional performance constraints:

$$\min_d C(\mathbf{d}) \tag{10}$$

subject to

$$F(\mathbf{d}) \geq F_0$$

Production cost priority design is widely used in industry for producing commercial products. In developing a complex mechanical product with many components, the overall design functions are often decomposed into many sub-functions. These sub-functions are the minimum design requirements to be satisfied by different departments and suppliers. Since production cost is a critical measure of product competitiveness, the reduction of production costs, while satisfying the required functional performance, becomes the major goal of all participating departments or suppliers.

In the multiple-spindle drill head design, the total production costs for producing the holes are calculated using

$$C(\delta_D, \delta_{xy}) = 4(C(\delta_D) + C(\delta_{xy})) \tag{11}$$

The related total production costs vary for different values of δ_D and δ_{xy}. Cost variation can be illustrated by plotting the equal cost contours over the feasible region, as illustrated in Figure 25. The optimal design based upon the production cost priority design principle leads to the optimal design parameters of $(\delta_D, \delta_{xy})^T = (0.225\text{ mm},\ 0.270\text{ mm})^T$.

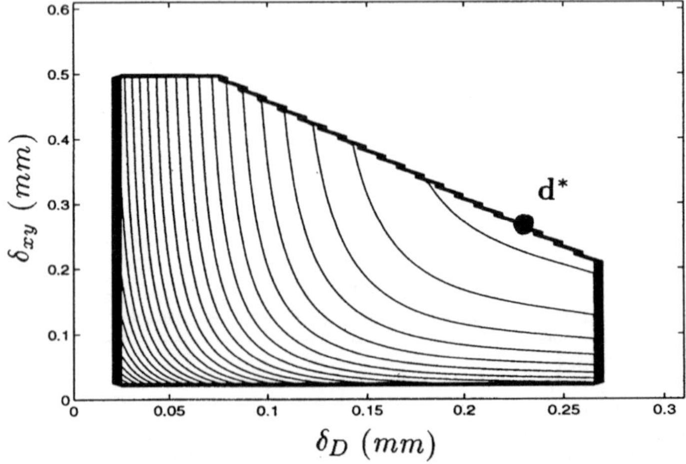

Fig. 25 Contour map of related production costs in the feasible design space.

(b) Functional Performance Priority Design

Functional performance priority design is carried out by maximizing the overall functional performance, subject to given manufacturing cost constraints. The optimization problem, if modeled in the standard minimization format, has the form:

$$\min_d \; -F(\mathbf{d}) \tag{12}$$

subject to

$$C(\mathbf{d}) \leq C_0$$

Functional performance priority design is employed in the case where the functional performance is considered more important. For instance, in the design of mechanical components for aerospace applications, volume and power consumption are overriding considerations rather than costs, and in the design of artificial human organs, reliability is more critical than costs.

Because the functional performance measures have different units, these measures need to be transformed into a comparable measure to form the overall functional performance and to associate with production costs. In this work, the functional performance increase compared with a reference design is used as the common measure, or is called functional performance index.

In the multiple-spindle drill head design, the design with the minimum production costs is selected as the reference design $(\delta_{D0}, \delta_{xy_0})$. The three selected functional performance measures, power loss variation, temperature variation, and gear lifetime, with respect to the reference design are represented as ΔPL_0, ΔT_0, and N_0, respectively. The three functional performance indices are obtained using the following equations:

$$I^{(F)}_{\Delta PL}(\delta_D) = \frac{\Delta PL(\delta_D) - \Delta PL_0}{|\Delta PL_0|} \tag{13}$$

$$I^{(F)}_{\Delta T}(\delta_D) = \frac{\Delta T(\delta_D) - \Delta T_0}{|\Delta T_0|} \tag{14}$$

$$I^{(F)}_{N}(\delta_D, \delta_{xy}) = \frac{N(\delta_D, \delta_{xy}) - N_0}{|N_0|} \tag{15}$$

Here the absolute values, $\Delta PL_0, \Delta T_0$, and N_0, are used to transform a better functional performance into an index with higher reading. The overall functional performance index is calculated using:

$$I^{(F)}(\delta_D, \delta_{xy}) = \frac{I^{(F)}_{\Delta PL}(\delta_D) + I^{(F)}_{\Delta T}(\delta_D) + I^{(F)}_{N}(\delta_D, \delta_{xy})}{3} \tag{16}$$

Identical weighting factors are used in this calculation. The overall functional performance is to be maximized. The contour map of the overall functional

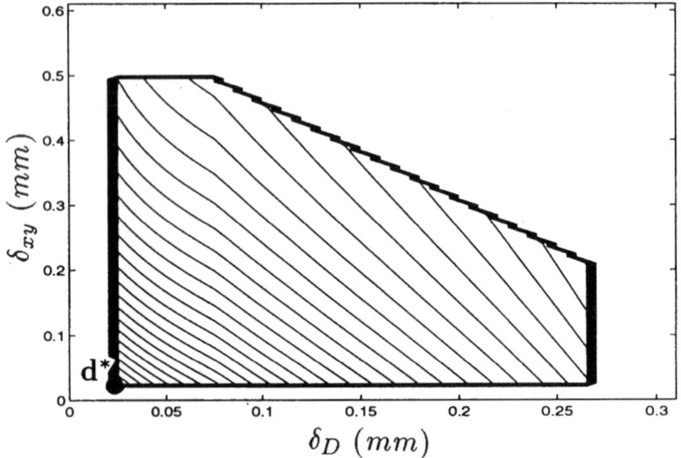

Fig. 26 Contour map of overall functional performance in the feasible design space.

performance index, if plotted within the feasible design space, is shown in Figure 26. The design with the tolerances of $(\delta_D, \delta_{xy})^T = (0.02 \text{ mm}, 0.02 \text{ mm})^T$ is the optimal design solution with best functional performance.

(c) Balanced Performance and Cost Design

Balanced performance and cost design is aimed at identifying the best trade-off between these two aspects. To associate the two "performance" measures of distinct nature, the functional performance and production costs are transformed into a comparable form — functional performance index, $I^{(F)}(\mathbf{d})$, and production cost index, $I^{(C)}(\mathbf{d})$, respectively.

A design with balanced functional performance and production costs can be achieved by:

$$\min_{d} -I(\mathbf{d}) = -\lambda_F \cdot I^{(F)}(\mathbf{d}) + \lambda_C \cdot I^{(C)}(\mathbf{d}) \tag{17}$$

$$\lambda_F + \lambda_C = 1$$

Where $I(\mathbf{d})$ is the overall product life-cycle performance index; and λ_F and λ_C are application dependent weighting factors.

This equation is a short form of Eq. (1), considering only functional performance and manufacturing costs. The formulation is a good representation of the concurrent engineering principle. An ideal design can be accomplished by improving the design with more functional performance increase and less production cost increase, rather than turning to one of the two extremes.

The functional performance index is calculated using the method introduced in 4.4 (b). In the multiple-spindle drill head design, the design with the minimum production costs, represented by $(\delta_{D0}, \delta_{xy_0})^T = (0.225\ mm, 0.270\ mm)^T$, is used as the reference design. The obtained functional performance index is shown in Eq. (16).

The production cost index is formed as

$$I^{(C)}(\delta_D, \delta_{xy}) = \frac{C(\delta_D, \delta_{xy}) - C_0}{C_0} \tag{18}$$

The product life-cycle performance index, represented in Eq. (17), is then transformed as

$$I(\delta_D, \delta_{xy}) = I^{(F)}(\delta_D, \delta_{xy}) - I^{(C)}(\delta_D, \delta_{xy}) \tag{19}$$

The optimal design with the best product life-cycle performance is identified by maximizing the objective function of Eq. (19). The identified optimal design is at $(\delta_D, \delta_{xy})^T = (0.105\ mm, 0.100\ mm)^T$. The contour map of the product life-cycle performance is illustrated in Figure 27.

4.5. Comparison with Present Design Practice

Present design practice follows a quite different approach from the discussed life-cycle performance optimization approach. In general, a designer would first specify a rough design target in terms of functional performance. Based upon the determined design objective, the values of design parameters are determined according to the recommendation of design handbooks and/or experience.

Three major problems lie within this design practice. First, the values of design parameters are determined following the "one size fit many" design handbook. These values will seldom be optimal. Secondly, the actual functional performance and production costs of a design can hardly be known before the design is manufactured and tested, and the appropriate balance of these two product life-cycle aspects is very difficult to accomplish. Thirdly, the design and redesign process is carried out in darkness. It requires many trials and a great deal of experience to improve a design to a satisfactory form.

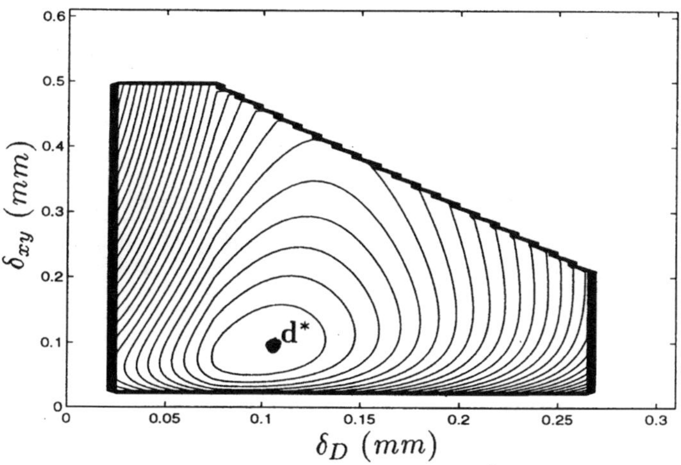

Fig. 27 Contour map of product life-cycle performance in the feasible design space.

For the multiple spindle drill head design, the values of the two considered design variables, the spindle size tolerance and spindle hole tolerance, are chosen as $(\delta_D, \delta_{xy})^T = (0.0840\ mm,\ 0.115\ mm)^T$ according to the recommendation of the American National standards Institute (ANSI) tolerance grades IT10 and IT9, respectively.

The functional performance, production cost, and life-cycle performance readings of the previously discussed four different design approaches are listed in Table 5. The functional performance and production cost oriented designs are strongly biased with poor life-cycle performance readings, although optimization is used. The manual design approach, which is generated through very careful study, presents a good reading of the life-cycle performance index. The balanced performance and cost design leads to the highest life-cycle performance rating.

5. A PROTOTYPE INTEGRATED CONCURRENT DESIGN SYSTEM

A prototype integrated concurrent design system has been developed based upon the discussed methods. Many software modules are implemented for generating life-cycle aspect models, maintaining their relations, evaluating life-cycle performance, and identifying optimal design. The architecture of the presently developed system is shown in Figure 28.

At the conceptual design stage, the design aspect model is generated by selecting and combining appropriate design features, according to the given design requirements, using the scenario-based inference module and the functional code-based design feature search module. The evolutionary design database is represented by a sequence of metamodels. Only one of these metamodels is *active* at a time for design modification and evaluations. From design descriptions, the symbolic geometric aspect model is created using geometry features, and is associated with a B-rep type research solid modeler for representing geometry details. Manufacturing features are extracted or defined from design geometry. Each component of the product is described by a manufacturing feature tree. Because each manufacturing feature is produced by a known sequence of production operations, the production process needed for manufacturing the product can be directly obtained. This further

TABLE 5
Performance increase of different design models.

Performance Items	Production Cost Priority Design	Functional Performance Priority Design	Manual Design	Balanced Performance and Cost Design
$I^{(F)}(\delta D, \delta_{xy})$	0	2.39	1.14	1.28
(rank)	(4)	(1)	(3)	(2)
$I^{(C)}(\delta_D, \delta_{xy})$	0	2.41	0.62	0.67
(rank)	(1)	(4)	(2)	(3)
$I(\delta_D, \delta_{xy})$	0	−0.02	0.52	0.61
(rank)	(3)	(4)	(2)	(1)

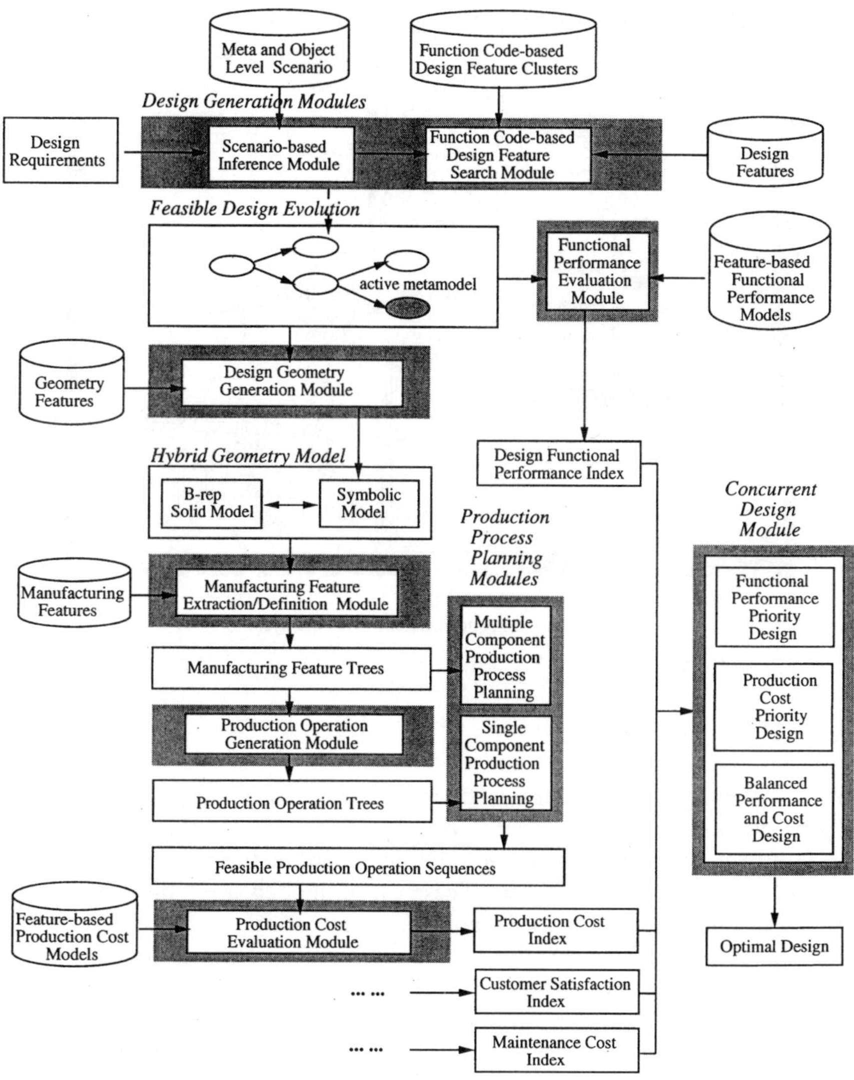

Fig. 28 Architecture of the prototype integrated concurrent design system.

allows the optimal production sequence to be identified, and the production costs estimated. Based upon the evaluation of functional performance and production costs using relevant features and knowledge, various optimal designs, including the functional performance priority design, production cost priority design, and balanced performance and cost design, can be identified.

The system is implemented using Smalltalk-80, an object-oriented programming language (Goldberg and Robson, 1983). Smalltalk-80 is used because of its good programming environment for rapid software prototyping. A snapshot of the developed system is given in Figure 29.

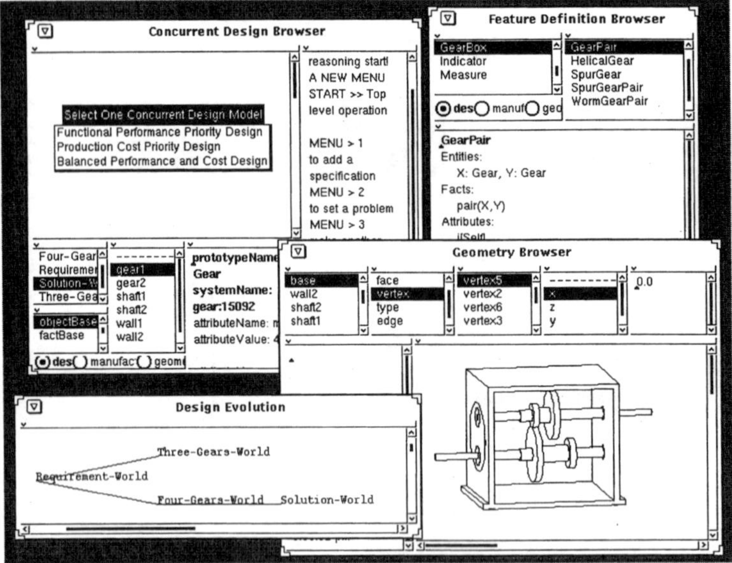

Fig. 29 A snapshot of the prototype system.

6. SUMMARY

A prototype integrated concurrent design system, based upon the quantitative intelligent system approach, was developed. In this system, the three most important life-cycle aspect of a product, namely design, manufacturing, and geometry, are described by relevant aspect models. These aspect models are built up using aspect features. An intelligent system is used for generating the aspect models and maintaining their relations. Design and manufacturing features are also used for associating the product evaluation measures, including functional performance and production costs, with design parameters. Based upon the evaluation, the design with the best life-cycle performance can be identified. The introduced methods are illustrated using a number of design examples and compared with the present manual design practice. The study serves as a basis for developing the next generation CAD systems with concurrent design functions.

Acknowledgements

Financial support from the Natural Science and Engineering Research Council (NSERC) of Canada is gratefully acknowledged.

References

M.M. Andersen, S. Khler and T. Lund, *Design for Assembly*. IFS Publications Ltd., Bedford, UK (1983).

J.C. Bezdek, *Pattern Recognition with Fuzzy Objective Function Algorithms*, Plenum, New York (1981).

A. Birolini, Design for reliability, Chapter 12 in *Concurrent Engineering: Automation, Tools, and Techniques*, (ed A. Kusiak), John Wiley and Sons, Inc., pp. 307–347 (1993).

M.S. Bloor and J. Owen, CAD/CAM product-data exchange: the next step. *Computer-aided Design*, **23**(4), 237–243 (1991).

G. Boothroyd and P. Dewhurst, *Design for Assembly: A Designer's Handbook*, Boothroyd Dewhurst Inc., Wakerfield, RI (1983).

J.G. Bralla (ed) *Handbook of Product Design for Manufacturing*, McGraw-Hill Inc (1986).

T. Budd, *An Introduction to Object Oriented Programming*, Addison-Wesley (1991).

T.C. Chang, *Expert Process Planning for Manufacturing*, Addison-Wesley Pub.Comp. (1990).

T.C. Chang, R.A. Wysk and H.P. Wang, *Computer-Aided Manufacturing*, Prentice Hall (1991).

K. Choi, M. Barash and D. Anderson, Automatic recognition of machined surfaces from a 3-D solid model. *Computer-Aided Design*, **16**(2), pp. 81–86 (1984).

J.C. Chung, R.L. Cook, D. Patel and M.K. Simmons, Feature-based geometry construction for geometric reasoning, in *ASME Computers in Engineering Conference*, San Francisco, CA (1988).

M. Cutkosky, J. Tenenbaum and D. Miller, Features in process based design, in *Proceedings of ASME Computers in Engineering Conference*, San Francisco, CA (1988).

J. de Kleer, An assumption-based TMS. *Artificial Intelligence*, **28**, pp. 127–162 (1986).

J.R. Dixon, Artificial intelligence and design: a mechanical engineering view, in *Proceedings of Fifth National Conference on Artificial Intelligence*, AAAI-86/2 (1986).

X. Dong and M. Wozny, Feature extraction for computer-aided process planning, in *Proceedings of the Third International Conference on Computer-Aided Production Engineering*, University of Michigan, Ann Arbor, Michigan (1988).

Z. Dong, Design for automated manufacturing Chapter 9 in *Concurrent Engineering:Automation, Tools, and Techniques*, (ed A. Kusiak), John Wiley & Sons Inc., pp. 207–234 (1993).

Z. Dong (ed) *Artificial Intelligence in Optimal Design and Manufacturing*, PTR Prentice Hall (1994a).

Z. Dong Automated generation of minimum cost production sequence, in *Artificial Intelligence in Optimal Design and Manufacturing*, (ed Z. Dong), PTR Prentice Hall, pp. 153–172 (1994b).

Z. Dong and W. Hu, Optimal process sequence identification and optimal process tolerance assignment in computer-aided process planning. *Computer in Industry*, **17**, pp. 19–32 (1991).

Z. Dong, W. Hu and D. Xue, New production cost-tolerance models for tolerance synthesis. *Journal of Engineering for Industry, Transaction of ASME*, **116**, pp. 199–206 (1994).

Z. Dong and A. Soom, Automatic optimal tolerance design for related dimesion chains. *Manufacturing Review*, **3**(4), pp. 262–271 (1990).

S. Dowlatshahi, Product design in a concurrent engineering environment: an optimization approach. *International Journal of Production Reaserch*, **30**(8), pp. 1803 – 1818 (1992a).

S. Dowlatshahi, A comparison of approaches to concurrent engineering. *Internaltional Journal of Advanaced Manufacturing Technology* (1992b).

V. Faires, *Design of Machine Elements*, MacMillan and Company, NewYork (1965).

K.T. Fann, *Peirce's Theory of Abduction*, Martinus Nijhoff, The Hague, The Netherlands. (1970).

J. Gershenson and K. Ishii, Life-cycle serviceability design, Chapter 14 in *Concurrent Engineering: Automation, Tools, and Techniques*, (ed A. Kusiak), John Wiley & sons, Inc., pp. 363–384 (1993).

A. Goldberg and D. Robson, *Smalltalk-80: The Language and Its Implementation*, Addison-Wesley (1983).

M.R. Henderson, Extraction of feature information from three dimensional CAD data, Ph.D. Dissertation, Purdue University (1984).

Japan Society for the Promotion of Machinery Industry. *Group Technology*, University of Tokyo Press, Tokyo (1980).

T. Kiriyama, T. Tomiyama and H. Yoshikawa, The use of qualitative physics for integrated design object modeling, in *Proceedings of ASME Design Theory and Methodology*, — DTM' 91, pp. 53–60 (1991).

A. Kusiak, (ed) (1992) *Intelligent Design and Manufactring*, John Wiley & Sons, Inc.

A. Kusiak, (ed) *Concurrent Engineering: Automation, Tools, and Techniques*, John Wiley & sons, Inc (1993).

S.C. Luby, J.R. Dixon and M.K. Simmons, Creating and using a features database. *Computers in Mechanical Engineering*, **5**(3), (1986).

A. Makino, P. Barkan and R. Pfaff, Design for serviceability, in *Proceedings of the 1989 ASME Winter Annual Meeting*, San Francisco, CA (1989).

J. McCarthy, Circumscription — a form of non-monotonic reasoning, *Artificial Intelligence*, **13**, pp. 27–39 (1980).

S.P Mitrofanov, *Scientific Fundamentals of Group Technology*, USSR (1959).

H. Nakagawa and T. Mori, Computable circumscription in logic programming, *Transactions of Information Processing Society of Japan*, **28**(4), pp. 330–338, in Japanese (1987).

J.S. Noble and J.M.A. Tanchoco, Design for economics, Chapter 16 in *Concurrent Engineering: Automation, Tools, and Techniques*, (ed A. Kusiak), John Wiley & Sons, Inc., pp. 401–435 (1993).

P. O'Grady and U. Menon, A concise review of flexible manufacturing systems and FMS literature. *Computer in Industry*, **7**, pp. 155–167 (1986).

P. O'Grady and R.E. Young, Issues in concurrent engineering systems. *Journal of Design and Manufacturing*, **1**, pp. 27–34 (1991).

H. Opitz, *A Classification System to Describe Workpieces*, Elmsford, NY: Pergamon (1970).

M.J. Pratt, Solid modeling and the interface between design and manufacturing. IEEE *Computer Graphics and Application Magazine*, pp. 52–59 (1984).

A. Rosenblatt and G.F. Watson, Concurrent engineering, *IEEE Spectrum*, **28**(7), pp. 22–37 (1991).

J.J. Shah, Assessment of features technology. *Computer Aided Design*, **23**(5) (1991).

J.J. Shah and M.T. Rogers, Functional requirements and conceptual design of the feature-based modeling system. *Computer-Aided Engineering Journal,* **5**(1), pp. 9–15 (1988).

J.E. Shigley and C.R. Mischke, *Mechanical Engineering Design,* McGraw-Hill Book Company (1989).

H. Takeda, P.J. Veerkamp, T. Tomiyama and H. Yoshikawa, Modeling design processes. *AI Magazine*, **11** (4), pp. 37–48 (1990).

G.P. Tunner and D.C. Anderson, An object oriented approach to interactive, feature based design for quick turn around manufacturing, in *Proceedings of ASME Computers in Engineering Conference,* San Francisco, CA (1988).

T. Tomiyama, T. Kiriyama, H. Takeda, D. Xue and H. Yoshikawa, Metamodel: a key to intelligent CAD systems. *Research in Engineering Design*, **1**(1), pp. 19–34 (1989).

T. Tomiyama and P.J.W. ten Hagen, The concept of intelligent integrated interactive CAD systems, CWI Report No. CS-R8717, Centre for Mathematics and Computer Science, Amsterdam, The Netherlands (1987).

T. Tomiyama and H. Yoshikawa, Extended general design theory, in *Design Theory for CAD, Proceedings of the IFIP Working Group 5.2 Working Conference 1985 (Tokyo)*, (eds. H. Yoshikawa and E.A. Warman), North-Holland, Amsterdam, pp. 95–130 (1987).

H.E. Trucks, *Designing for Economical Production*, Society of Manufacturing Engineers, Dearborn, Michigan (1976).

B. Veth, An integrated data description language for coding design knowledge, in *Intelligent CAD Systems I: Theoretical and Methodological Aspects* (eds. P.J.W. Ten Hagen and T. Tomiyama), Springer-Verlag, Berlin, pp. 295–313 (1987).

H.B. Voelcker, R.J. Marisa, E.E. Hartquist, A.A. Requicha and N.L. Lawrence, CNC machining: simulation, verification, programming, planning, communication and control, Technical Report, University of Rochester (1986).

A.C. Ward, J.K. Liker, D.K. Sobek and J.J. Christiano, Set-based concurrent engineering and Toyota, in *Proceedings of ASME Design Theory and Methodology Conference,* DE-Vol. **68**, pp. 79–90 (1994).

D. Xue and Z. Dong, Feature modeling incorporating tolerance and production process for concurrent design. *Concurrent Engineering: Research and Applications*, **1**, pp. 107–116 (1993).

D. Xue and Z. Dong, Developing a quantitative intelligent system for implementing concurrent engineering design. *Journal of Intelligent Manufacturing*, **5**, pp. 251–267 (1994a).

D. Xue and Z. Dong, Coding and clustering of design and manufacturing features for concurrent design, in *Proceedings of 1994 ASME Design Automation Conference: Advances in Design Automation 1994*, DE-Vol. 69-1, Minneapolis, Minnesota, pp. 533–545 (1994b).

D. Xue, H. Takeda, T. Kiriyama, T. Tomiyama and H. Yoshikawa, An intelligent integrated interactive CAD, in *Intelligent Computer Aided Design*, (eds. M.B. Waldron, D. Brown, and H. Yoshikawa), North-Holland, Amsterdam, pp. 163–192 (1992).

R.E. Young, A. Greef and P. O'Grady, An artificial intelligent-based constraint network system for concurrent engineering, *International Journal of Production Research*, **30**(7), pp. 1715–1735 (1992).

H. Yoshikawa, General design theory and CAD systems, in *Man-machine Communication in CAD/CAM,* (eds.T. Sata and E. Warman), North-Holland, Amsterdam, pp. 35–58 (1981).

I. Zeid, *CAD/CAM Theory and Practice,* McGraw-Hill (1991).

H-C. Zhang and L. Alting, *Computerized Manufacturing Process Planning Systems*, Chapman & Hall (1993).

CHAPTER 6

Rapid Prototyping of Hardware and Software

UNNY MENON

Department of Industrial and Manufacturing Engineering
Cal Poly State University, San Luis Obispo, California 93407, USA

1. INTRODUCTION

In this chapter we present rapid prototyping from a concurrent engineering viewpoint. First, we examine the definition of concurrent engineering and then we explore the role of rapid prototyping for both hardware and software, to enable *Concurrent Design of Products, Manufacturing Processes and Systems*. The range of currently available rapid prototyping technologies are presented with guidelines for choosing the right process, for a given design-validation-need. Strategies for successful deployment of rapid proto-typing are presented throughout this chapter and contemporary research initiatives in-progress are outlined.

In this chapter the terms *Concurrent Design of Products, Manufacturing Processes and systems* and Concurrent Engineering (CE) will be regarded as equivalent with CE as the general abbreviation. The CE approach has become the normal methodology for product development in industry. This "team-approach" encourages design engineers to interact with all other organizational functions and take into account all factors that affect the product from a total-life-cycle viewpoint (see O'Grady and Young (1991), Pugh (1991) and Rzevski (1992) for more details).

This Concurrent Engineering product-development-process, encourages the team to take into account the following product-life-cycle considerations: assemblability, manufacturability, maintainability, reliability, testability, recyclability, disposability, portability, etc often referred to as the "-ilities." The design process should be iterative and use prototyping technologies to reduce the number of design changes occurring in later phases of the product cycle. "Design Maturity" is a qualitative term used to describe the degree of certainty which the design team can provide to others that the current stage of the design has reached a certain level of "stability" and we have gone past the "uncertain stage" which is very common on new product developments with no prior track history. "Design Maturity" is not a binary metric in terms of absolutely firm or completely fluid, but a continuum between those two extremes, and it is highly desirable to have at least some subjective index of design maturity which enables members of the design team and others interacting with the product development team to know when we can begin to make major commitments and investments based on the knowledge that the "design has stabilized" and unlikely to change in a major way.

In some cases "rapid prototyping techniques" can help to accelerate the "design maturity process" by providing opportunities to test the speculative design options and identify the most viable product design and configuration. Thus we have now begun to focus on design maturity as a metric to be monitored by the product development team, to influence product design philosophies and to control design actions. In particular the management of "design creativity" and "acceptability/un-acceptability of design changes" will now become subject to considerations of design maturity in the time domain of our development sequence, whereby changes are preferred early in the cycle and will require strong justification if requested late in the development cycle. The notion of design maturity will also allow the Project Leader to promote creativity and speculative high-risk pursuits by the design team when the design is regarded as "fluid" and at a later stage, the Project Leader can stipulate that major revisions will not be permitted, when the design maturity is close to the "absolutely firm" stage in the development cycle. In this engineering management strategy, creativity is encouraged and could be rewarded during the early stages while it is strongly discouraged when the design has matured and is close to final release for volume production.

2. CONCURRENT ENGINEERING (CE)

Concurrent Engineering **(CE)** has become the new norm for organizing and managing all aspects of the product design and development process. This approach is also referred to by a number of other synonyms: Integrated Product Development, Simultaneous Engineering, Life Cycle Engineering, Parallel Engineering, and Team Engineering. In this paper we will use Concurrent Engineering and the abbreviation **CE** in referring to this approach. The generally accepted definition of CE as formulated by the Institute for Defense Analysis and documented by Winner *et al.*, (1988) is as follows:

> "*Concurrent Engineering is a systematic approach to the integrated, concurrent design of products and their related processes, including manufacture and support. This approach is intended to cause the developers, from the outset to consider all elements of the product life cycle, from conception through disposal, including quality, cost, schedule and user requirements.*"

This approach requires a high level of teamwork and simultaneous involvement of all company functional disciplines, very early in the product-concept-design process. This could ensure that all necessary modifications are made when it is easy to do so and development teams are empowered with more autonomy to enhance the overall product life cycle. Effective implementation of CE can benefit companies with greater customer satisfaction, lower cost, higher quality and impressive reductions in time-to-market cycles (see Pawar, Menon and Riedel 1994) from concept through to full-scale volume production. Many companies now regard CE as being essential to remain competitive and for the U.S. Defense industry, the Department of Defense (DoD) now expects all contractors to use this integrated product development approach. Comprehensive discussions of CE including some industrial case histories can be found in the July special edition of IEEE Spectrum edited by Rosenblatt & Watson (1991), with more detailed descriptions in Allen (1990), Hartley (1992), Kusiak(1993), Shina (1991), Syan & Menon (1994) and the influential Institute for Defense Analysis report R-338 documented by Winner *et al.* (1988).

3. TRADITIONAL DESIGN PARADIGMS AND NEW TRANSITIONS

If CE is the new and preferred approach, what then is the old approach which we are seeking to replace and what was wrong with it? In comparative discussions, the "traditional approach to Engineering Design" has been assigned the following self-descriptive labels: Serial Engineering, Over-the-wall Engineering, Sequential Engineering, etc. It is assumed that in the old approach a designer translated his/her perception of customer requirements into a concept design and final detail design, which was tossed "over-the-wall" to Manufacturing Engineering and other functional disciplines who were required to overcome any obstacles in translating the design to a satisfactory product, which conforms to all customer specifications and expectations.

A good metric for the consequences of this style of design in product development is the "Engineering Change Order – ECO," which is a documentation of "imperfections in the design process." Of course there are many reasons for originating an ECO, but a very substantial majority of ECO's are attributable to poor design decisions which in most cases could have been avoided, if there had been more discussions during the formative periods of the preliminary design, between the designer and other "down-stream functional disciplines." In many cases where highly innovative concepts and processes are part of the product design, even such dialogue among the product development team would be insufficient and iterative cycles of "prototyping at many levels of the product hierarchy" may be necessary, to enable a "right-first-time product design" or "to compress time-to-market" (see Pawar, Menon and Riedel (1994) for illustrative case examples).

This arguably simplistic premise of the "old approach to design" assumes that many designers do not consult all "requisite downstream functional disciplines" and/or do not "prototype-to-trouble-shoot" the design concept. Hence, we find the consequence that the number of ECO's that became necessary, are at a much higher level than is justifiable. The comparative metric for this premise is that if we compare our industry to Japan, we find the following contrasts:

- Japan has much shorter concept-to-market development cycles.
- Japanese products have fewer ECO's issued after product launch.

- The frequency distribution of ECO's over time is left-skewed for Japanese products and right-skewed for U.S. products which indicates that their higher level of teamwork and early prototyping contrasts with our "over-the-wall" discover-problems-late in the product cycle and hence we suffer higher costs per ECO, since they occur late in the development cycle, when changes are more expensive.

U.S. and European industry have recognized that we have to change our approach to product development and we must find ways to compress the time-to-market, if we are to remain competitive in global markets. Thus, we are beginning to see significant changes in industry and the emergence of new paradigms for organizing the product development process which reflects a Concurrent Engineering approach with teamwork and greater emphasis on prototyping to identify design modifications.

In the subsequent sections of this chapter we will explore how we can deploy concurrent engineering oriented prototyping to enable the *Concurrent Design of Products, Manufacturing Processes and Systems* approach which forms the basis of this textbook.

4. RAPID PROTOTYPING TO FACILITATE CONCURRENT ENGINEERING

Consider the premise that many late-design-changes could be avoided if the development team makes effective use of preliminary prototypes to identify problems of "form-fit-function" which could not be caught, without the "physical part" versus just drawings or CAD (Computer Aided Design) views of the part. This premise is gaining widespread acceptance and some influential customers of engineering products are beginning to expect it for some products; the Department of Defense Directive (DoDD) 5000 states the following:

> "DoDD 5000.1 Program plans... must provide for a systems engineering approach to the simultaneous design of the product and its associated manufacturing, test and support processes, aggressive prototyping (including manufacturing processes) are to be used to reduce risk."

> "DoDD 5000.2 Technology Transition and Prototyping of critical manufacturing processes and software systems and sub-systems shall be conducted to reduce risk,... The quality emphasis will be on preventing product deficiencies rather than on detecting and correcting defects."

One of the setbacks to prototyping in the past has been the time needed to come up with a prototype of the current product design. This time-limitation is alleviated with the emergence of "Rapid Prototyping **(RP)** Processes" whereby it is now possible to produce a "three-dimensional-hardcopy" of the physical part from a CAD solid model of the product design. Similarly, fourth generation software now enables rapid construction of software-prototypes for computer dependent sub-systems in new products (see Britton & Doake 1993).

All of the Rapid Prototyping (RP) processes for physical parts "grow" a three-dimensional part by stacking thin layer cross-sections of the part-geometry using coordinate data obtained from the CAD solid model (see Figure 1). The newer term "layered manufacturing" conveys this basic principle of rapid prototyping which is common to all the RP processes.

- Some **RP** processes create the physical part by using the characteristic of certain photo-polymer resins to cure in those regions which are exposed to ultra-violet light (usually from a laser source for a precise beam). This concept using for example, "Stereolithography Apparatus **(SLA)**" as the specific RP process is illustrated in Figure 2 (see Jacobs (1992) for comprehensive details of the Stereolithography Process). SLA was the first commercial rapid prototyping process for layered manufacturing which became available in 1987 from 3-D Systems inValencia, California, based on the patented invention of Charles W. Hull.
- The same principle of building a rapid prototype in an "additive manner" (versus the subtractive processes of machining), is used in a number of alternative processes to SLA which have emerged in recent years, in this emergent new technology for rapid prototyping. In Selective Laser Sintering **(SLS)** a layer of powder is "sintered" (i.e. particles bonded with around 90% physical density) by a laser beam (see Figure 3 and for the original ideas of SLS process see Deckard and Beaman (1988)).
- In Fused Deposition Modeling **(FDM)** a filament of wax or nylon is melted and resolidified in layers (see Figure 4 and for more details on FDM as well as other layered manufacturing processes see the text-book by Burns (1993)).
- In Laminated Object Manufacturing **(LOM)** the cross-section-edge of each layer is "burned-out by a laser beam" from a continuous sheet of adhesive-backed-material (see Figure 5 with more details in Burns (1993)).

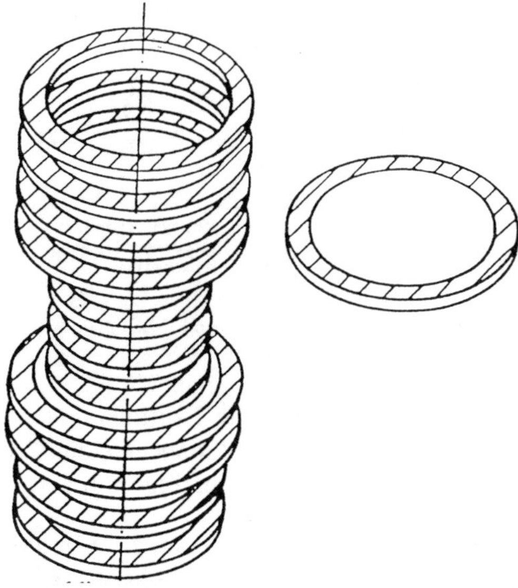

Fig. 1 Layer-by-layer additive process from 3-D CAD.

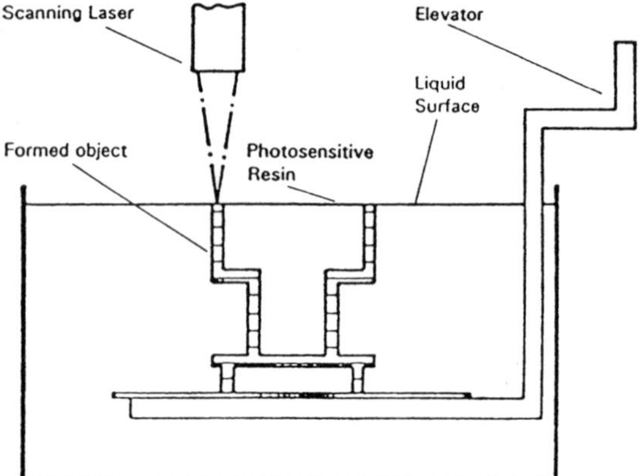

Fig. 2 Stereolithography apparatus and process.

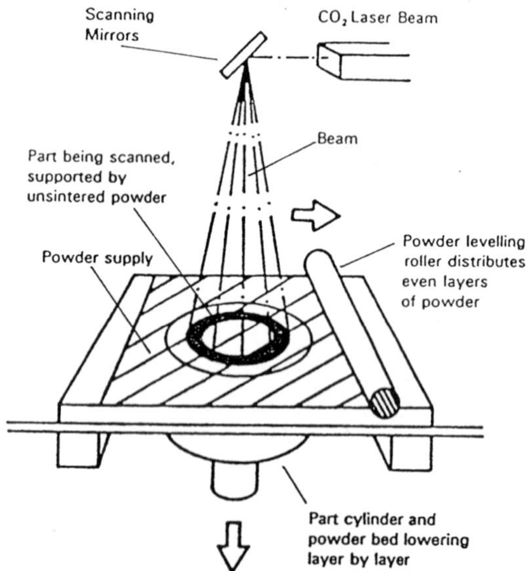

Fig. 3 Selective laser sintering (SLS) process.

- Research completed at the University of Nottingham led by Phil Dickens, was directed towards the development of "Shape Welding" where a robot builds metallic layers by welding each layer (see Figure 6 and Dickens *et al.*, (1992) for more details).
- In "3-D Printing" developed at MIT by Sachs *et al.*, (1991), concepts from ink-jet printer technology have been adapted to "selectively bind" layers of powder with adhesive injected on to powder to produce the physical part (see Figure

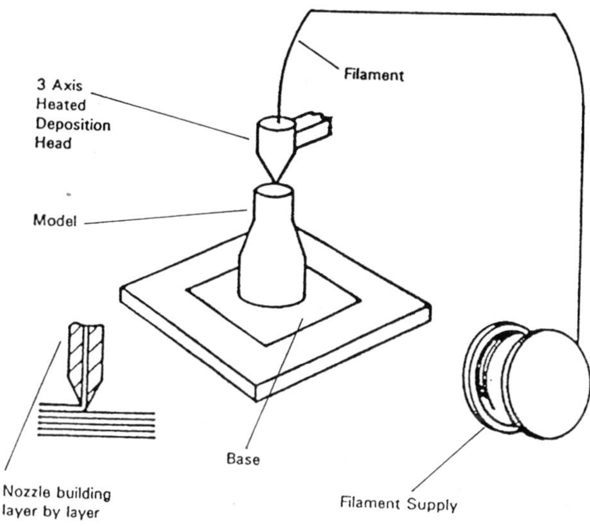

Fig. 4 Fused deposition modeling (FDM) process.

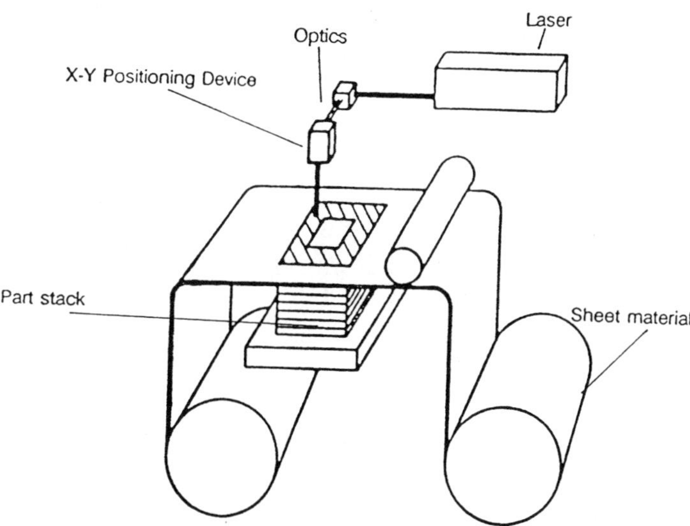

Fig. 5 Laminated object manufacturing (LOM) process.

7). This "3-D printing concept" developed by Sachs *et al.*, (1991) at MIT has now been licensed to Soligen Inc. in California, to produce commercial versions which have found unique applicability in investment casting foundries where the ceramic shell of the "part cavity" enables liquid metal to be solidified giving precise investment-cast metal prototypes.

For more details of all these RP processes the reader is referred to the following comprehensive reviews in Ashley (1991), Burns (1991, 1993), Deckard and Beaman

Fig. 6 Shape welding of solids.

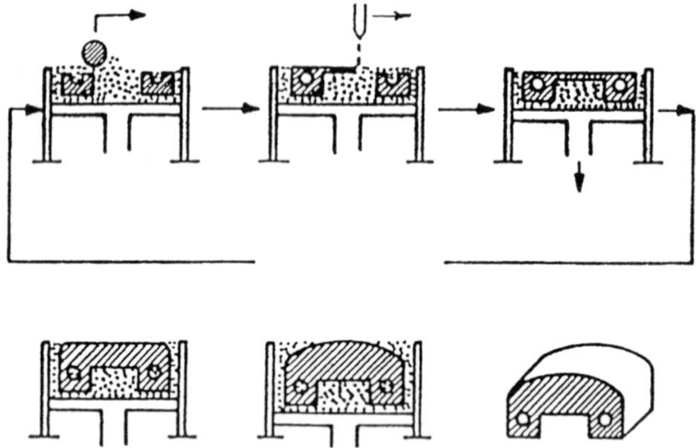

Fig. 7 3-D Printing of solids.

(1988), Jacobs (1992), Menon and Dickens (1992), Sachs (1991), Sprow (1992), Syan and Menon (1994, Ch.8.).

From a user viewpoint there are several important considerations in selecting the "right" RP process for producing the "physical 3-D prototype." These considerations include a) whether the RP process introduces "in-process" temporary additions to the basic part for "producibility," b) type of material that is feasible for a given RP process, c) accuracy d) surface smoothness properties and e) other physical properties. Space limitations do not permit comprehensive discussions here in this chapter and the reader is referred to Burns (1993, pp. 157–183).

Thus, we now have the option to generate a physical prototype of CAD representations, to enable design-reviews and validation of "form-fit-function" and to make any needed corrections in the iterative process of evolving a mature product design. We can also use the prototype to validate uncertainties related to product and process implications of specific designs. Although we have focussed on prototyping physical parts, it is the concept of prototyping which we are advocating which is applicable to the entire spectrum of engineering systems developments, including software, for complex systems and overall product or project configuration.

5. CONCURRENT ENGINEERING ORIENTED PROTOTYPING

To provide insight into some of the key issues in this review of "Concurrent Engineering Oriented Prototyping" of physical hardware, using the layered manufacturing processes, we can list the following major elements, derived from the broader "engineering management perspective" and the *Concurrent Design of Products, Manufacturing Processes and Systems* theme of the textbook:

- The primary purpose of physical prototyping is to facilitate iterative design in a rapid manner, whereby the design team is able to work with a "three-dimensional hardcopy" and hence able to use it for validation of "form-fit-functionality" and to try and detect any potential downstream-problems or objections from other functional disciplines (e.g. customer response tests by Marketing).
- Rapid Prototyping processes for layered manufacturing, are an evolving and young maturing technology (circa 1987). While prices are coming down, deployment of RP is still expensive compared to the traditional manufacturing processes (although they are generally much slower compared to RP). The analogy of RP being the equivalent of Express Mail Delivery and traditional manufacturing processes being similar to First Class mail is useful in trying to compare the two approaches with respect to cost and time.
- Industrial users of rapid prototyping have confirmed that deployment of rapid prototyping does enable early detection of potential problems and the "physical 3-D model" appears to have significant advantages over other forms of conceptual representations of the design, presented as drawings or Computer-Aided-Design 3-D visualizations, even when sophisticated computer graphics are deployed for realistic rendering on the computer monitor; people seem to want to touch and feel the product and seem to be better at detecting problems, when they have a physical model which they can touch and feel, versus paper or computer-screen representations.
- Given the fundamentally different mode of manufacture in rapid prototyping, where "layered manufacturing" is the means deployed to produce the part, versus conventional manufacturing (e.g. the subtractive approach in machining or mold-filling in net-shape processes), it is possible to readily produce complex shapes which may be difficult or in some cases impossible to produce as single-piece using most machining processes, where the cutter may not be able to reach certain locations. This type of manufacturing complexity, arises when we are trying to rapidly produce the physical model of "complex intertwined surfaces" defined by high-order mathematical equations (e.g. as deployed in some bio-medical and aerospace applications).
- Thus, the role of layered manufacturing processes for rapid prototyping is to provide a means of validating design ideas, from a variety of functional perspectives, at the product and process development phase of the overall

development cycle. Given this as the primary purpose for "rapid prototyping technologies," layered manufacturing does not seek to compete with respect to cost, against conventional manufacturing processes, but will almost always be able to compete on "rapid cycle time," to produce the physical part, compared to most conventional manufacturing processes.

6. SOFTWARE PROTOTYPING

The software component of modern engineering products of the current era is substantial and should be assigned "equal consideration and importance" as given to mechanical and electrical sub-systems of the total product. For example even a cursory review of automobile sub-systems or modern appliances, would reveal a large proportion of micro-processor/software controlled devices which have replaced electro-mechanical sub-systems of earlier models. This overlapping role of mechanical/electronic/software sub-systems of most modern products has resulted in the emergence of the new field of "mechatronics" (see Auslander *et al.*, (1993) and Bradley *et al.*, (1991) for more details on mechatronics) which addresses all three disciplinary interests, in an integrated manner in accord with the concurrent engineering philosophy. This means that we must have prototyping capability not only for mechanical and electrical systems but also for the often overlooked area of system-software to enable validation of error-free final products which rely on the synergy of all three sub-systems for effective performance.

Software prototyping can take one of the following general forms: a) Storyboard Prototyping or Static Screen Layouts, b) Disposable Prototyping, c) Incremental Prototyping and d) Evolutionary Prototyping. Each of these forms of prototyping serve a different purpose and is particularly applicable for specific types of software development. In this chapter the reader is provided an overview of the major issues and is directed to the following publications for detailed information on all forms of software prototyping (see Andriole 1989, Britton and Doake 1993, Connell and Schafer 1989, Hekmatpour and Ince 1988, Syan and Menon, Ch.9 1994).

6.1. Storyboard Prototyping or Static Screen Layouts

This approach is particularly suitable for detecting any "divergence from user needs" and to facilitate the "congruence of perceptions" by the user and software developers by presenting a storyboard screen-layout of the principal input-output functionality of a system under development. It is produced from the results of the user requirements definition and by presenting the user with a proposed implementation (albeit static screen template) which provides an opportunity to correct any mis-communicated perceptions of what the user "really wanted" and if there are discrepancies this static prototype could initiate the "software design iteration" well before major resources are used in developing the final version. Storyboarding is a fast low-cost option for software prototyping (for examples see Andriole 1989).

6.2. Disposable Prototyping

This type of software prototype is a "means-towards-the-end" in that a model of the proposed system is developed very quickly to simulate the functionality of the desired system enabling the client to see what it will do, but it is not supported by structured design or robustness-details of the production version. Speed is of the essence in this type of prototyping and is made possible by fourth-generation computer languages which can build such rapid prototypes for concept validation (or specification-prototyping or specification-by-example). This type of software prototype is useful not only for refining system specifications, but is equally applicable for exploring alternative system designs and testing the feasibility of new design concepts.

6.3. Incremental Prototyping

Incremental prototyping starts with significant initial effort to validate and consolidate the total system design at a "conceptual level" and freeze the "overall design of the total system," followed by incremental software development of each sub-system one section at a time with incremental prototypes devised and tested one at a time. Incremental prototyping has the advantage of sustaining discipline and stability from the "consistent structural stability" of the overall design frozen at the beginning vis-a-vis the scope for adaptation and incorporation of evolving concepts which is possible in the evolutionary prototyping approach; thus incremental and evolutionary are not synonymous as often implied by naive observers of these alternative design philosophies; who perhaps fail to grasp the subtle difference.

6.4. Evolutionary Prototyping

This approach is particularly suited to applications subject to rapidly changing requirements where there is a need to change both "during and after" the development cycle. The evolutionary prototypes are best suited to environments where the installed system attains maturity and then "invalidates the original requirements," necessitating a new generation with different functional capabilities which must evolve to fulfill that need. Additional modification and adaptation, are an integral part of this mode of software development where the prototype evolves to a production version and design maturity is subject to an iterative loop. This form of prototyping would also be invaluable in software developments where the end-user is unable to articulate requirements clearly at the start or is unable to freeze the initial specifications for some justifiable organizational reasons, whatever they may be to the client.

The conceptual infrastructure for software prototyping include several useful formalisms including state transition diagrams to model system behavior, formal grammars for specifying human-computer interactions, user interface prototyping, simulators, screen generation and language supported facilities (for details see references cited in Hekmatpour and Ince 1988).

7. SOFTWARE PROTOTYPING TOOLS

The reader looking for existing systems which can provide the prototyping functionality discussed in this section will find that there are many providers of commercial systems who can supply turnkey systems and CASE (Computer Aided Software Engineering) tools to meet specific user needs. In this chapter we will mention just two CASE systems: STATEMATE (see Syan and Menon, Ch. 9, 1994) from I-Logix Corporation and EPROS outlined by Hekmatpour & Ince (1988).

STATEMATE © is a commercial software prototyping and development system produced and sold by I-Logix Inc. (22 Third Ave, Burlington MA 01803, USA). It consists of the following sub-systems: Kernel, Analyser, Prototyper, Documentor and Dataport. The Kernel is used to create the Statemate Specification which is an executable model of the system. The user formulates all "key structural and behavioral elements" of the system architecture in terms of graphical charts on an advanced workstation. The system designer depicts the user needs and system elements using STATEMATE © running on an advanced engineering computer workstation (e.g. SUN or SGI), in each of three key domains on the appropriate chart: Entity-State Transition Chart (see Figure 8), Activity Chart (see Figure 9) and Module Chart (see Figure 10). In this conceptual and schematic representation of the TOTAL SYTEM, we have captured system behavior in the entity-state chart, data flow and data control in the activity chart which captures system processing capability using hierarchic decomposition and the hardware/software components are depicted in the module chart.These three co-domains of the system exist in "multiple-related windows" of the computer screen, allowing the user to work on them concurrently, based on evolving and progressive information available to the system designer. STATEMATE © continually monitors both syntax and logic, which keeps the system designers focussed on the crucially important logic of the system; detecting errors and invoking the iterative design process to develop an

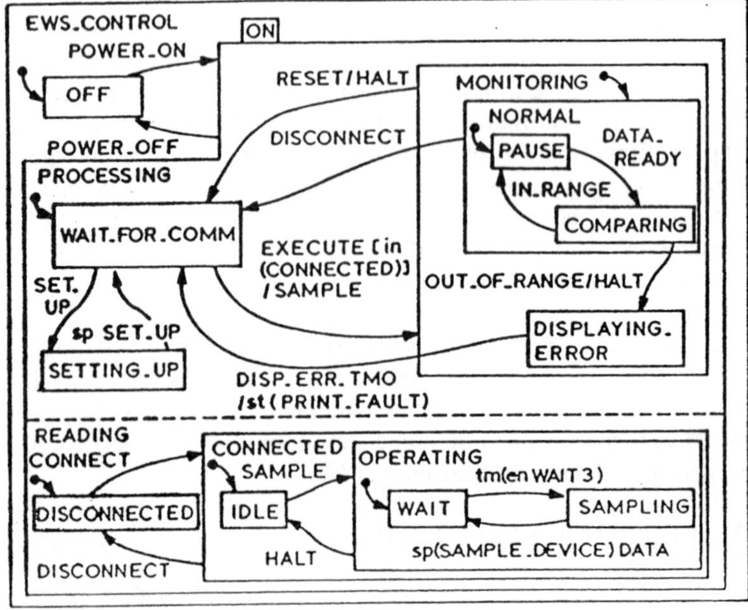

Fig. 8 STATEMATE © entity-state transition chart.

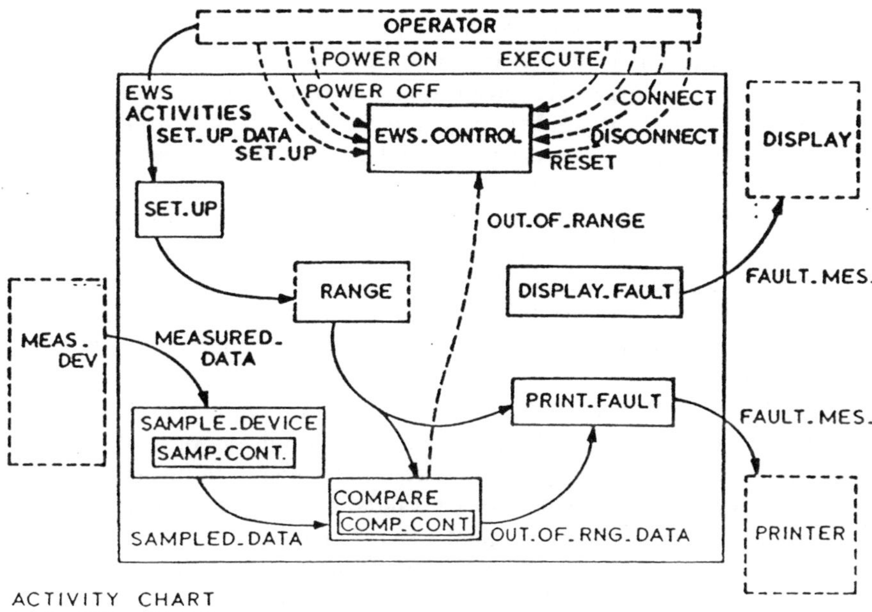

Fig. 9 STATEMATE © activity chart.

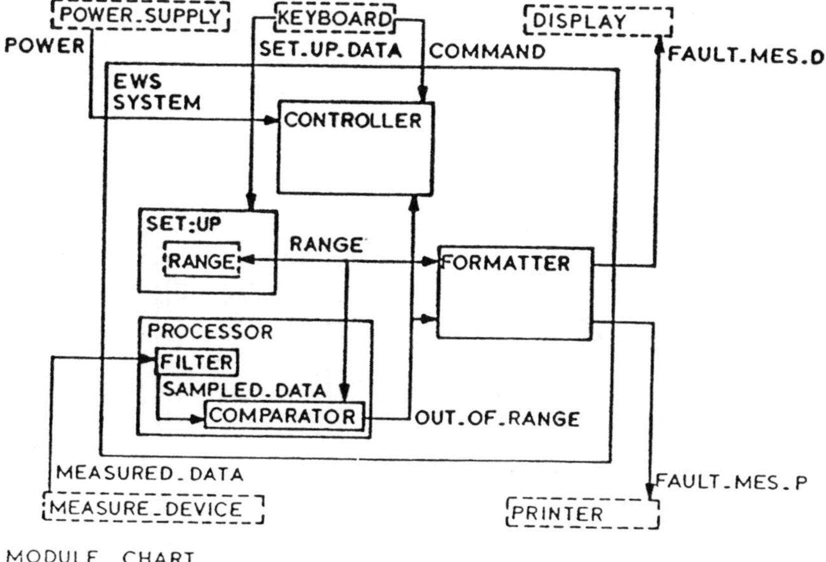

Fig. 10 STATEMATE © module chart.

error-free system. At that stage, STATEMATE © can automatically generate "prototype software" in high-level computer languages including ADA, "C" or VHDL (VLSI (very largescale integration) Hardware Definition Language) computer source code, to simulate system behavior and create fully functional input-output on computer screens which the end user can thus "see and confirm" if it is exactly "What they wanted". For example in avionics, the pilots could use the simulated cockpit screens (see Figure 11) and verify if it is acceptable. These STATEMATE © generated signals can also be connected to full-scale advanced flight simulators normally deployed for pilot training, for further validation of system response and real-time feedback on the performance of the prototype software with experienced pilots responding to the design configuration, well before any substantial capital investment has been made on the proposed system. In addition STATEMATE © also provides accurate and complete software documentation. This is an example of a commercially available software prototyping and system development environment, where prototyping is an integral part of the system design process. This enables formal validation of system viability and error-free operations, "before building the final system" and thus avoiding the potentially high-cost modifications, if they are discovered only after the final version of the system has been completed. STATEMATE © would be applicable to either incremental or evolutionary prototyping as outlined in this chapter.

EPROS © is a software development system with prototyping capabilities developed and published by Hekmatpour & Ince (1988). In EPROS a system is developed in a top-down manner, from the abstract to the detailed. The system evolves in an iterative manner, with every cycle providing a self-contained formal description of the system. This description, no matter how abstract or detailed is always executable and always converted into a working prototype. EPROS © deploys four approaches to prototyping: executable specifications, state-transition diagrams, language supported facilities and reusable software. EPROS provides an executable textual notation for system description and language supported features, which simplify the task of constructing user interfaces. EPROS can be used for all three forms of prototyping outlined earlier (throw-away, incremental or evolutionary).

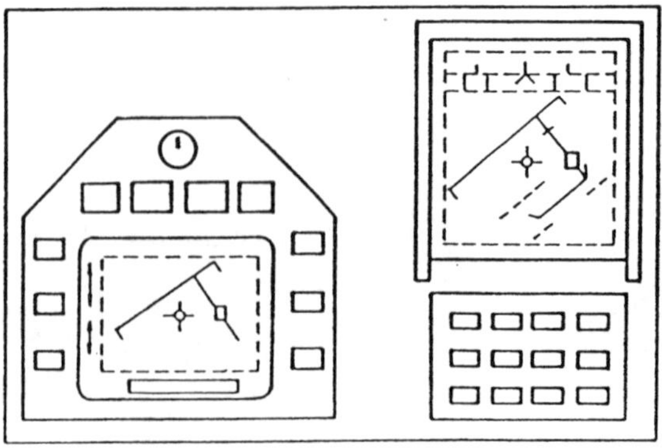

Fig. 11 STATEMATE © prototype of simulated cockpit screens.

8. EXAMPLES OF RAPID PROTOTYPING RESEARCH INITIATIVES

The following are a few illustrative examples of research challenges in the field of rapid prototyping, being addressed at the time that this chapter was written (circa 1996) and are included in this chapter to give the reader some insight into evolving approaches to technological constraints and seeking solutions to meet evolving needs of applications based on layered manufacturing concepts:

- Ballistic particle manufacturing (BPM): In this process we deposit a stream of tiny adhesive droplets on successive layers of material to build up the part. William Masters obtained the first BPM patent and the process is being commercialized by BPM Technology (Greenville, South Carolina). While BPM deploys a single-droplet stream to build the part, a variation of the concept being developed by the research team at Carnegie-Mellon University in Pennsylvania, deploys a "recursive mask and deposit" approach (which they call MD-Star). This process is analogous to spraying paint through a stencil. Other researchers are exploring variations of the "metal-spraying process" for layered manufacturing. For further details see Burns (1993) pp. 95–100 and Fussell and Weiss (1990).
- Embedded sensors: The concept of inserting some "foreign object" into the "product-build" of a layered manufacturing process was discussed by Pomerantz (1993) at a special workshop of leading researchers held in Israel on the topic of "Heterogeneous Models" during March 30–April 1, 1993. The Cubital SOLIDER process, is very amenable to the introduction of objects like sensors or LED's during the part-build, which results in an embedded heterogeneous component. Similarly the Fused Deposition Process (FDM) outlined earlier is also amenable to the introduction of some "foreign object" into the part so that it is embedded within and could provide additional functionality.
- Deployment of ceramics and other materials: Collaborative research being conducted by a large research team at Rutgers University in New Jersey, sponsored by the U.S. Department of Defense (ARPA/ONR) and Allied-Signal Research and Technology Labs of New Jersey, is addressing the challenges to produce certain types of rapid prototype parts made from ceramic materials (see Aggarwala *et al.* (1995) for more details. The 3-D Printing Process developed at MIT (as depicted in Figure 7), which is currently commercialized by Soligen Inc. of California, also results in ceramic parts for use as shell-molds for casting. The full potential for rapid-prototyped ceramic components especially hybrids for micro-electronics using layered manufacturing has yet to be developed, and we are likely to see innovative solutions emerging during the next decade.
- Hybrid-Prototyping Systems: Professor Fritz Prinz and his research team at Stanford University, California are conducting research to develop hybrid-systems which combine layered manufacturing processes with traditional manufacturing systems (multi-axis Computer Numerical Control machining centers, Electrical Discharge Machining (EDM), Surface Treatment via shot-peening, etc.) all within a robot controlled cell to provide much broader

functionality for rapid prototype parts than available from the other processes on their own. For more details see Prinz *et al.*, (1995).

- Virtual Prototyping: While there is merit in advocating the merits of a "physical prototype" for a variety of validation needs, before one proceeds to deploy physical prototyping, it is worth asking the question: "Is it possible to carry out the testing/validation we need by some non-physical means ?" If the answer is yes, then we have a case for considering "virtual prototyping" as a possible lower-cost alternative which is some cases may be equally effective. Virtual prototyping relies on the capability of making very high-fidelity replications of the real system within a computer-aided environment for highly-accurate representation of the product or system to be prototyped and validated. Unpublished research on virtual prototyping, is known to be in progress within various branches of U.S. Defense services and other commercial organizations in Europe and United States. For insights into the contemporary state-of-the-art of virtual prototyping, the reader may wish to visit the following world-wide-web sites on the Internet:
- Virtual Prototyping Research Group of the Fraunhofer Institute, Stuttgart, Germany with web-site is at: http://www.iao.fhg.de/Library/rp/Virtual. html
- Virtual Prototyping Journal published by MCB Press, England, web-site at http://www.mcb.co.uk/services/portfoli/liblink/vpj/jourinfo.htm
- Virtual Reality Interface for Prototyping: While we are not aware of any specific research projects deploying virtual reality for rapid prototyping, in progress at this time, it is nevertheless a futuristic notion that has been discussed at many research forums by leading research leaders looking at the future of rapid prototyping at the level of "blue-sky research possibilities." The concept is that an end-user of any product or the product designer is able to fully visualize all aspects of the product using a "virtual-reality holographic visualization" including the ability to touch and feel the product with complete sensory accuracy to the end-user. Furthermore, the user should have the capability to "manipulate the product" and change it to suit evolving needs without access to the CAD system. The changes so made are captured by the virtual-reality interface and conveyed to the layered manufacturing process for the production of the revised physical prototype for further testing or validation. Some preliminary thoughts are outlined in brief in Burns (1993) pp. 315.

9. CONCLUSIONS

In this chapter we have presented an overview of rapid prototyping processes for both hardware and software, in terms of the essential characteristics and how it could be applied in a concurrent engineering based product and process development environment. The goal of concurrent engineering is to produce robust product designs which can be translated to robust products of high quality in very short development cycles and with very few engineering changes after product release. This development strategy requires sound mechanisms to detect errors early in the product development cycle, where prototyping is one proven way of facilitating that goal. It also promotes creativity and exploration of innovative concepts through rapid prototyping for iterative design improvement and optimization. We thus promote creativity, because risky options can be explored, and we regard it as healthy to accept many engineering changes at the preliminary concept phase and subsequently to gradually increase the penalty for engineering changes as the product design gains maturity. It should be stressed that the most important message we wish to convey, is not about the rapid prototyping processes *per se*, but rather our strong advocacy of "prototyping" itself as an important self-discipline to keep validating ideas and to deploy iterative design as the norm for integrated product, process and enterprise design. A prototype can be as simple as a cardboard cutout or a highly accuratere presentation; what matters the most is that you are willing to restrain your ego and self-confidence; by validating the design continually, deploying as many prototypes as needed. While physical-rapid-prototyping is somewhat expensive to use at this time, it is cost-justified in most instances. With the likelihood of low-cost rapid prototyping machines by about the year 2005, such low-cost-future-RP systems will join the color laser printer, in most engineering offices, adjacent to the computer workstation, enabling engineers to easily deploy the concurrent engineering oriented rapid prototyping philosophy advocated by the author of this chapter. We have much to gain by adopting this philosophy with nothing to lose, making it a win-win development strategy for product, process and enterprise design.

Acknowledgements

The work reported in this chapter was funded by the National Science Foundation-ILI Leadership in Laboratory Development Grant Number USE-9150291 during 1991–93 and an ARPA-DURIP grant DAAH04-95-1-0064 funded by the U.S. Army Research Office 1995. The NSF grant enabled the author to examine concurrent engineering oriented rapid prototyping and the ARPA grant has facilitated the acquisition of RP research equipment to validate our assumptions on the value of RP in iterative design.

References

M.K. Aggarwala, *et al.* (14 co-authors), *Structural Ceramics by Fused Deposition of Ceramics*, Solid Freeform Fabrication Symposium Proceedings, University of Texas-Austin, pp. 1–8 (1995).

C.W. Allen (Editor), *Simultaneous Engineering: Integrated Manufacturing and Design*, Society Manufacturing Engineering, Dearborn, MI (1990).

S.J. Andriole, *Storyboard Prototyping a New Approach to User Requirements Analysis*, QED Information Sciences Inc.,Wellesley, Mass (1989).

S. Ashley, Rapid Prototyping Systems – Special Report, *Mechanical Engineering*, April pp. 34–43 (1991).

D.M. Auslander *et al.*, "Tools for Teaching Mechatronics," ASEE National Meeting Proceedings, Champagne, IL, (published by ASEE, 11 DuPont Circle Suite 200, Washington D.C., 20036-1207), July 20–24, pp. 1280–1285 (1993).

D.A. Bradley, D. Dawson, N.C. Burd and A.J. Loader, *Mechatronics – Electronics in Products and Processes*, Chapman & Hall, London (1991).

J. Braham, Where are the Leaders?, *Machine Design*, pp. 58–62, October 10 (1991).

C. Britton and J. Doake, *Software System Development a Gentle Introduction*, McGraw-Hill, London and N.Y. (1993).

M. Burns, *Rapid Prototyping: System Selection and Implementation Guide*, Management Roundtable Inc, Boston (1991).

M. Burns, *Automated Fabrication: Improving Productivity in Manufacturing*, Prentice-Hall, N.J. (1993).

D.E. Carter and B. Stillwell Baker, *Concurrent Engineering: the Product Development Environment for the 1990's*, Addison Wesley, Reading AM. (1992).

J.L. Connell and L. Schafer, *Structured Rapid Prototyping: An Evolutionary Approach to Software Development*, Yourdon Press / Prentice-Hall, N.J. (1989).

C. Deckard and J.J. Beaman, Process and Control Issues in Selective Laser Sintering, *ASME Publication* PED-V-33, pp. 191–197 (1988).

P.M. Dickens, M. Pridham R. Cobb I. Gibson and G. Dixon, *3D Welding*, Proceedings of 1st European Conference on Rapid Prototyping, University of Nottingham, July, pp. 81–93 (1992).

F.B. Prinz, L.E. Weiss C.H. Amon and J.L. Beuth, *Processing Thermal and Mechanical Issues in Shape Deposition Manufacturing*, Solid Freeform Fabrication Symposium Proceedings, University of Texas-Austin, pp. 118–129 (1995).

J.R. Hartley, *Concurrent Engineering*, Productivity Press, Cambridge, AM (1992).

P.S. Fussell and L.E. Weiss, *Steel-Based Sprayed Metal Tooling*, Solid Freeform Fabrication Symposium Proceedings, University of Texas-Austin, pp. 107–113 (1990).

S. Hekmatpour and D. Ince, *Software Prototyping, Formal Methods and VDM*, Addison-Wesley, Reading, AM. (1988).

Institute for Defense Analysis, *The Role of Concurrent Engineering in Weapons Systems Acquisitions*, IDA report R-338, available from NTIS, # ADA-203615. (1988).

P. Jacobs, *Rapid Prototyping & Manufacturing: Fundamentalsof Stereolithography*, SME publication, Dearborn, MI. (1992).

Kusiak, Andrew *Concurrent Engineering : Automation, Tools*, John Wiley, N.Y. (1993).

U. Menon and P.M. Dickens, *Rapid Prototyping: Role inConcurrent Engineering*, working paper MEOM Dept., University of Nottingham, England (1992).

P. O'Grady and R.E. Young, Issues in Concurrent Engineering, *Journal of Design and Manufacturing*, Vol **1**, No 1, pp. 27–34 (1991).

K. Pawar, U. Menon and J.C.K.H. Riedel, Time to market: getting goods to market fast - and first, *Integrated Manufacturing Systems - An International Journal*, Vol. **5**, No 1, pp 14–22 (1994).

I. Pomerantz, *Discussions on Embedding of Devices in Cubital Solid-Ground-Cured Parts*, Workshop on Heterogeneous Models, Cubital Ltd., 13 Hasadna St, Raanana, Israel (1993).

S. Pugh, *Total Design – Integrated Methods for Successful Product Engineering*, Addison Wesley, Reading MA (1991).

A. Rosenblatt and G.F. Watson, Concurrent Engineering -special Report, *IEEE Spectrum*, Vol. **28**, No 7, pp. 22–37 (1991).

G. Rzevski, Strategic Importance of Engineering Design, *Journal of Design and Manufacturing*, Vol. **2**, No 1, pp. 43–47 (1992).

E. Sachs *et al.*, Three-Dimensional Printing: Rapid Tooling and Prototypes Directly From CAD Representation, *NSF Grantees Conference Proceedings*, Austin Texas, Jan 91, SME publication, Dearborn MI (1991).

S.G. Shina, *Concurrent Engineering & Design for Manufacture of Electronic Products*, Van Nostrand Reinhold, N.Y. (1991).

E.E. Sprow, Rapid Prototyping: Beyond the Wet Look,*Manufacturing Engineering*, November, pp. 58–64 (1992).

C.S. Syan and U. Menon, *Concurrent Engineering: Concepts, Implementation and Practice*, Chapman & Hall, London (1994).

R.I. Winner, J.P. Pennell H.E. Bertend and M.M.G. Slusarczuk, The Role of Concurrent Engineering in Weapon System Acquisition, IDA Report R-338, Institute for Defense Systems Analysis, Alexandra VA, available from NTIS, #ADA-203615 (1988).

8. LIST OF ABBREVIATIONS

ADA	A high-level computer language named after Lady Ada Lovelace
ARPA	Advanced Research Projects Agency, U.S. Department of Defense
BPM	Ballistic Particle Manufacturing
C	A high-level computer language based on the C compiler
CAD	Computer Aided Design
3-D CAD	Three-Dimensional Computer Aided Design
CASE	Computer Aided Software Engineering
CE	Concurrent Engineering
DoD	Department of Defense, U.S.A.
DoDD	Department of Defense Directive
DTM	Desk-Top Manufacturing
DURIP	Defense University Research Instrumentation Program
ECO	Engineering Change Order
FDM	Fused Deposition Modeling
IEEE	Institute of Electronic and Electrical Engineers, U.S.A.
IPPD	Integrated Product and Process Development
IPPED	Integrated Product, Process and Enterprise Development
LOM	Laminated Object Manufacturing
NSF	National Science Foundation, U.S.A.
ONR	Office of Naval research, U.S.A.
RP	Rapid Prototyping
SLA	Stereo-Lithography Apparatus
SLS	Selective Laser Sintering
VHDL	VLSI (very large scale integration) Hardware Definition Language

CHAPTER 7

Rapid Prototyping

DIANE SCHAUB[a] and KOU-REY CHU[b]

[a]*Industrial and Systems Engineering, University of Florida 303 Weil Hall, Gainesville, Florida 32611 USA*
[b]*Phoenix Analysis and Design Technologies, 1465 N. Siesta Blud., Ste. 102, Gilbert, AZ 85234*

1. INTRODUCTION

Manufacturability of product designs has never been more important, given current business trends stressing speed to market of new or improved products. In the effort to create new and improved products, however, design engineers face the challenge of communicating their ideas of form, fit and function to other engineers designing adjoining component parts, to manufacturing personnel charged with producing the parts, and to customers who will ultimately decide the product's competitive success. In turn, manufacturing engineers need to reduce the cycle time required to produce these new products. An important tool to assist in addressing these demands is rapid prototyping (RP).

RP is an *enabling technology* for integrated product, process and enterprise design (IPPED), and is a means of creating a three-dimensional (3D) solid model from an analytical three-dimensional computer screen image, without a mold and within a matter of hours. Typically these models are built layer by layer, but technologies such as laser milling can also be effective (see Hsu and Copley (1990) for more information on laser milling).

RP has been around for less than a decade, but has revolutionalized the world of product research and development (R&D). By having a precise solid model of the designed object, all members of an integrated product design team can touch, feel, test, evaluate and make comments and suggestions to various aspects of the product. This results in a very efficient communication and meaningful discussion during various product design activities.

2. NEED FOR A RAPID PROTOTYPING MODEL

Rapid prototyping models can serve many purposes. According to Jacobs (1992) these models can serve functions as diverse as:

- visualization-allows a view of what the actual part will look like from all sides.
- verification-determines if features are as desired, allowing geometric measurements for intuitive, as well as analytical feedback to the design team.
- iteration-RP models can be created, checked, have errors corrected, and successive models created in as many repetitions as it takes to get a successful design.
- optimization-with the relatively low cost of the RP models as compared to the traditional methods of prototyping, superior designs rather than merely acceptable designs can be created for relatively low cost.
- fabrication-depending on the needed properties of a part, RP can be used as a manufacturing process for short-run parts, or for the creation of patterns, tools and molds.

Many different types of industries have found RP to be invaluable to their R&D process. Some of the first industries to take advantage of RP technology are the aerospace industry, the automotive industry, and bio-medical companies. Examples of applications within these three industries follow.

In the aerospace industry, RP models have been used in many stages within the development process. For example, the creation of an aircraft engine airfoil is quite a complex process, with the first step being the analytical modeling of critical performance characteristics. Figure 1 shows sample airfoils created through Stereolithography, a resin-based RP process. In one case design engineers were surprised to find that calculated aerodynamic model data was imported incorrectly into a CAD model for an airfoil, which would have resulted in the production of a part that was too thin and undersize. By examining the RP model, the design engineers were able to quickly see the error and correct if before incurring expensive retooling costs, and encountering weeks of delay (Hyduke and Chu, 1993). RP models of airfoils have

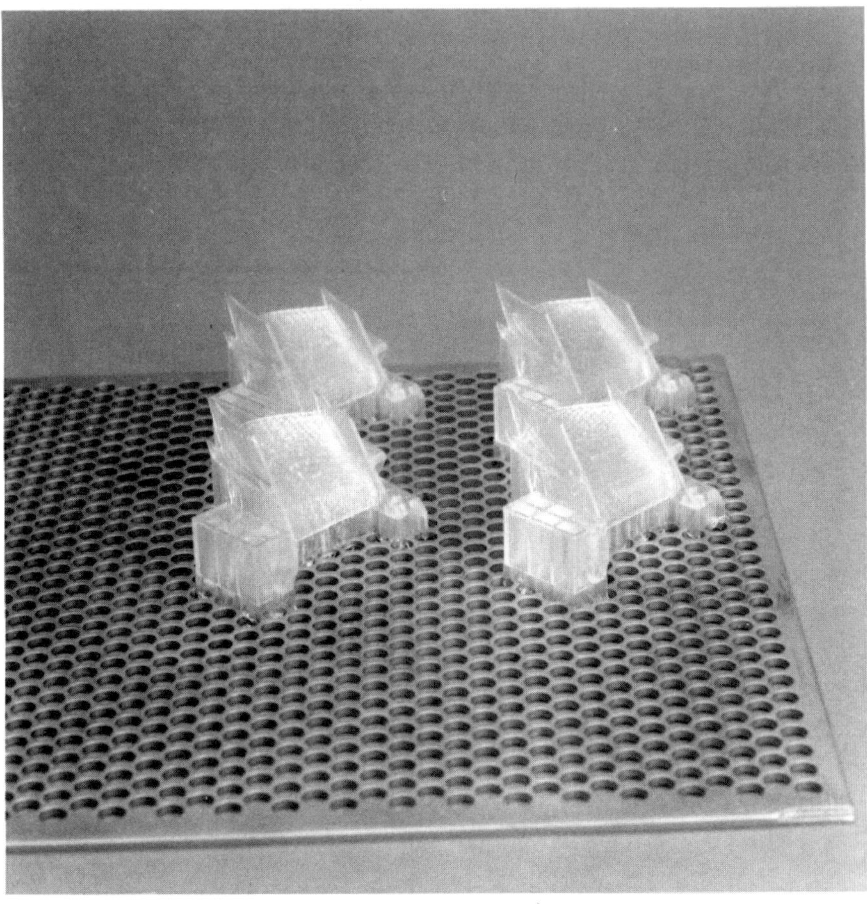

Fig. 1 Sample airfoils created through stereolithography.

also been used for low temperature airflow tests of a laboratory rig, providing simulated test results and verification of the design. Once the design engineers are satisfied with their design, the RP models have then been used to aid in the creation of a mold to produce the actual metal airfoils, as well as to aid discussion with machining and casting vendors by providing a means to intuitively determine the manufacturability of a design.

In the automotive industry, RP models have been used to be evaluate alternate designs quickly. In one example a 3D car gear shift handle model was created from a computer model. When evaluation showed it was too large for small hands to grasp, scaled-down SL models were created that were 8%, 10%, and 12% smaller. The designs helped the design team pick the size handle they wanted, which then went into production (Jacobs, 1992). Figure 2 shows an example of an automotive part created through Laminated Object Manufacturing, a cellulose-based RP process.

With biomedical companies, RP models allow customized prosthetic limbs and joints to be created. Since hip joints need to be custom fit to the recipient, the medical team will take an image of the recipient's current joint and a 3D computer model is created. From the image, an RP model of a mold for a replacement hip joint can be created. Figure 3 shows an investment casting shell of both a hip and knee replacement created through Fused Deposition Modeling, a wax-based RP process.

These are but a few examples of important contributions made possible through RP. New applications are continually being devised, as more and more companies

Fig. 2 Form, fit and function can be verified with RP parts.

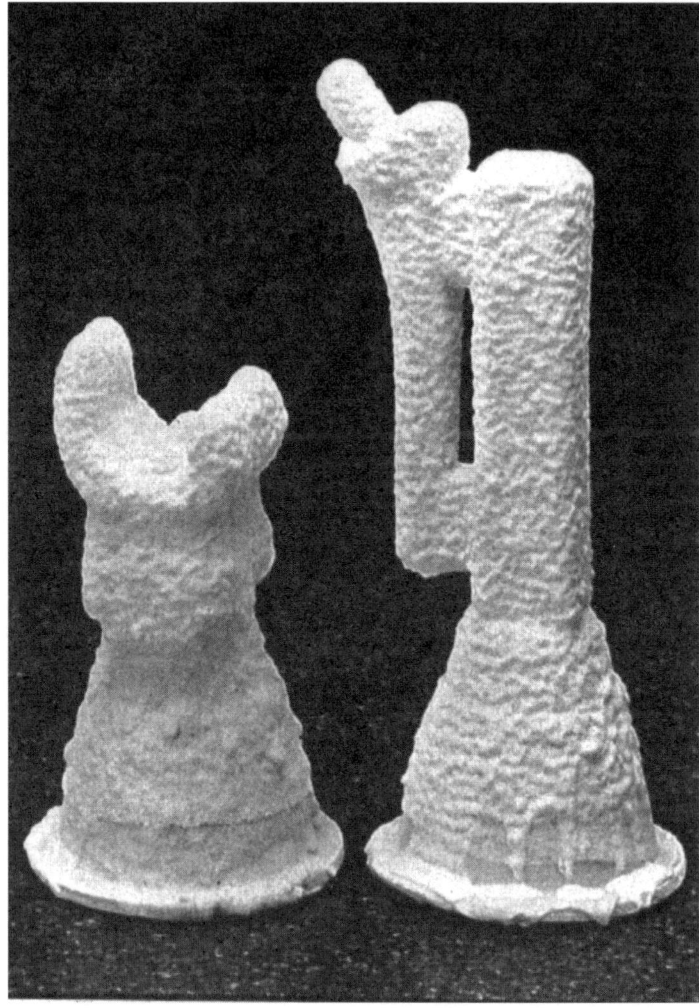

Fig. 3 Investment casting shells created from parts made of stratasys investment casting wax.

come to find the advantages of RP, and as the RP technology matures. Through RP, development times and costs have been reduced, while simultaneously improving design quality.

RP has revolutionalized the R&D process. Concurrent engineering efforts especially benefit from RP. Product design development cycles reduce from the traditional, as seen in Figure 4. Prototypes created early in the design process:

- create a common understanding between team members,
- provide a means of collecting early test evaluation data,
- and help decide the manufacturability of the final product.

With RP, blueprints are virtually unnecessary except for final documentation, after the product is designed, similar to a final report (Ulerich, 1992). In place of blueprints,

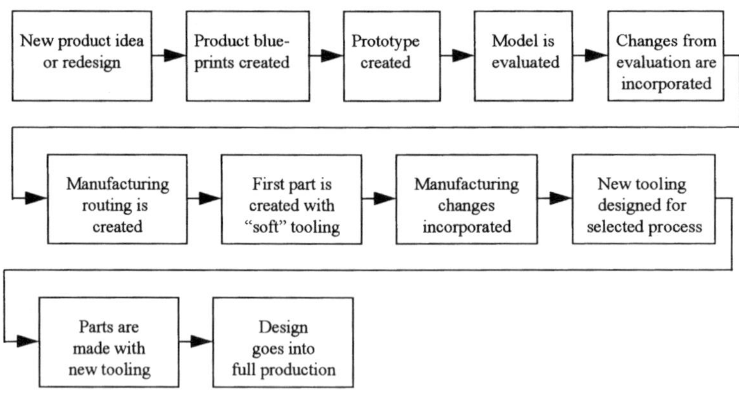

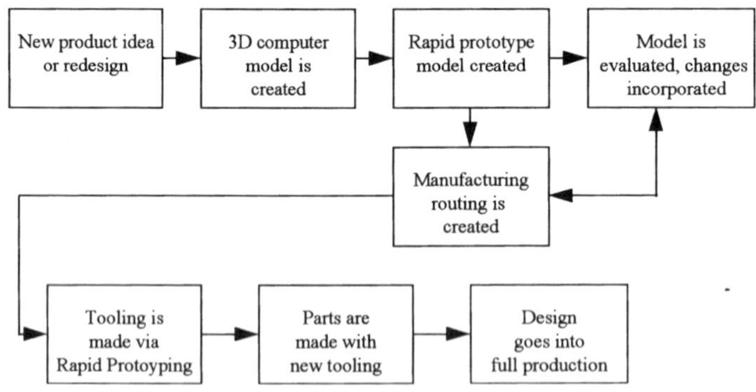

Fig. 4 Rapid prototyping streamlines the product design cycle.

designs based on computer models are used instead, which are utilized to make RP models. The entire functional group can use these models to plan an entire assembly, making revisions as needed earlier in the design cycle.

Savings resulting from use of RP models vs. the traditional methods, such as carving a model or working from drawings, have been impressive. Many of these cost savings are associated with finding design flaws early in the design cycle when they are less expensive to correct, rather than when the product is close to full scale production. It is not unusual for companies to refer to savings in terms of multiples, rather than percentages, as for instance, cycle times are cut from 20 weeks to weeks, by the use-of RP. Many companies start with one RP machine and go on to greatly expand their capacity. RP machines are often set up to run 24 hours/day to keep up with demand.

Experimental design techniques can help optimize the efficiency of the RP equipment, examining key variables with respect to improving, say dimensional accuracy while also improving throughput, see Schaub and Montgomery (1997). The number of variables involved is quite vast, but by using a strategy such as that proposed in Coleman and Montgomery (1993) a list such as that shown in Table 1 can be reduced down to a reasonable number of experiments. This list of variables was developed as a result of brainstorming with RP process experts, part design experts, and a statistician. A large number of these variables affecting the creation of a stereolithography turbine engine airfoil were classified as "held constant" or "nuisance" variables. The resulting experiment for this part required only thirteen tests and yielded results that improved throughput while producing a part with better surface finish.

3. THE THREE-DIMENSIONAL COMPUTER MODEL TO SOLID MODEL

From an input-output point of view, a rapid prototyping system is just a sophisticated output device for computers, the three-dimensional analog of a computer printer. In order to produce output, the computer must be supplied with certain type of data. For rapid prototyping processes, the input data is normally a three-dimensional computer model. This section will present various aspects of 3D computer models intended as input to rapid prototyping systems.

3.1. Three-Dimensional Computer Model

Rapid prototyping systems currently in use require nonambiguous data representation of the part geometry. Simply stated, the computer model must have closed boundaries and contain surface normal information. In other words, if the

TABLE 1
SL variables list.

1a) Resin type	20b) Solvent contact duration	34b) Tray removal
1b) Resin lot miture	21) E_c Factor	35) Input file format
2) Build style	22) Exposure of liquid/solid	36) Laser focus
3) Spot overlap (Hatch spacing)	23a) Part design residual stress	37) Vibration
4) Resin temp.	23b) Prototype residual stress	38) Part size
5) Shrink factor	24) Part geometry	39) Part/vat size ratio
6a) Room temp	25) Part creep	40) Quantity of parts
6b) Humidity	26) D_p Factor	41) Beam shape
7) Age of laser	27) Deep dip	42) Resin volume
8) Beam diameter	28) Wiper velocity	43) Resin height
9) Laser power	29) Wiper gap	44) Optical quality
10) Scan speed	30) STL file resolution	45) Trapped volumes
11) Layer thickness	31) Support structure	46) Part wall thickness
12) Overcure depth	31a) Amount	47a) Beamwidth comp.
13) Power	31b) Location	47b) Shrinkage comp.
14) Part orientation	31c) Overlaps	48) Skin Fill
15) Laser mode	32) Polymer strength	49) MSA
16) Scan pattern	33) Beam positioning accuracy	50) Time bet. prod'n & cure
17) Vat positioning	34) Hand finishing	51a) SL operator
18) Resin viscosity	34a) Support removal	51b) Hand finish operator

computer model is sliced in thin cross-sections, each cross-section would contain one or more closed loops and a surface normal to indicate the orientation. Three type of computer models are most commonly used. They are solid CAD model, 3D surface CAD model and 3D shape digitizers.

3.1.1. Solid CAD Model

A solid CAD model can be defined as a geometric representation of a bounded volume using computer graphics technology. This volume is represented graphically, via curves and surfaces, as well as nongraphically through a topological tree structure which provides a logical relationship that is inherent only with solid models. The topological data defines and maintains the connective relationships between the various faces and surfaces of the geometry.

Most existing high end CAD systems offer solid modeling capability. With parametric and variational technologies, today's solid modeling systems are easy to use and extremely powerful. Complex solid models can be easily created, modified and updated. This is the best choice for creating part geometries to feed into RP systems.

3.1.2. 3D Surface CAD Model

3D surface CAD models are often used to describe aerodynamic and aesthetic shapes. A major functional difference between solid CAD systems and surface CAD systems is the absence of topological data connecting the surfaces. A surface model also lacks the capability to describe the interior of the part.

Typically, the user develops two-dimensional wireframe profiles that are revolved, extruded, swept, or blended together to form the desired surface. The surface entities can be simple, such as planes and tabulated cylinders. Complex entities require definition by mathematical interpolation schemes such as bicubic splines. Other schemes utilize polynomial expressions such as Bezier curves. Most advanced systems utilize NURBS defined entities. NURBS definition bestows optimum control over the surface shape, while allowing editing to be easily accomplished.

The individual surfaces are assembled to form the desired design. Sometimes gaps will exist between surfaces which results in an non-closed volume. This often causes many problems when feeding the data into a RP system. This may require additional work to fix the "gap". However, they have the ability to describe some complex surfaces where many solids-based systems have trouble. It should be noted that users of advanced CAD systems should take advantage of both solid and surface modeling capabilities to create the desired part geometry.

3.1.3. 3D Shape Digitizer

Sometimes the desired rapid prototyping system output is not a new design, but an accurate replication of an existing object. In this case, a number of digitizers can be used to generate a series of point data. These digitized data would normally be fed into a CAD system for reconstruction of the 3D shape in a CAD system. Once a complete CAD model is created, this can then be converted into STL format.

3.2. Rapid Prototyping Input Data

From computer model to rapid prototyping solid model requires an intermediate input data. The de-facto standard input format is known as STL (STereoLithography).

Figure 5 shows an example of the STL file format. It is based on a normalized, tessellated approximation similar to a finite element mesh representation. Solid CAD model generally produce files that correctly follow the STL specification. This is so because they inherently satisfy the vertex to vertex rule, among others contained within the STL structure.

Each CAD system requires its own translator to convert the internal CAD database into the standard STL format. The process of performing the translation varies. The level of user control over the STL resolution also varies from system to system. Every CAD systems has its own STL translation procedure and recommended parameters, and the details of any system under consideration should be investigated.

4. RAPID PROTOTYPING PROCESSES

In this section all the commercially available rapid prototyping processes are described. These processes are grouped into six categories based on their part fabrication principles. The basic mechanics of each fabrication process will first be analyzed and some background and selected specifications of commercially available systems will be provided. However, this is not intended to be a comprehensive study of these processes and systems. For further studies of these rapid prototyping processes, a list of references is provided at the end of this chapter for the interested readers.

4.1. Layer-Additive Resin Polymerization by Laser Scanning

The basic mechanics of this process is based on the principle that photocurable liquid resin solidifies under the influence of light in a specific range of wavelength. In general, this type of systems combines four core technologies: three-dimensional computer aided modeling, lasers, polymer chemistry, and laser scanning. The process

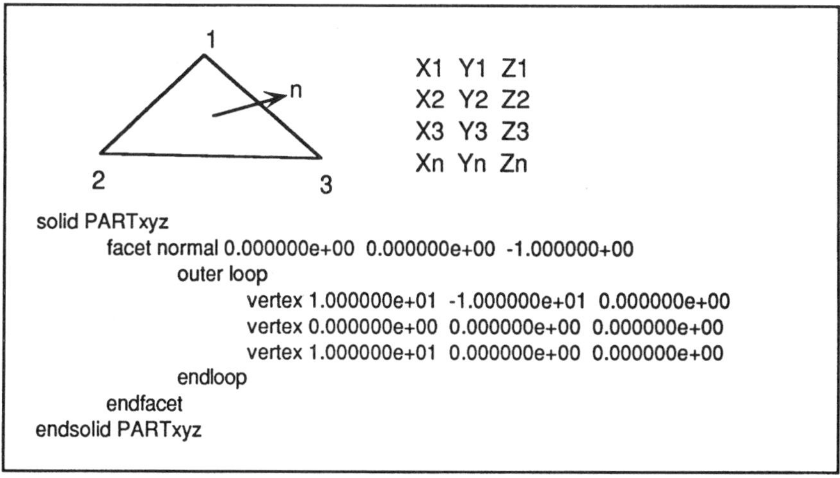

Fig. 5 Example of STL file format.

starts with a 3-D computer model that is horizontally sliced into a stack of 3-D cross sections. Each slice is physically constructed by using a laser beam to solidify the top layer of a photocurable resin bath. The laser beam only draws the pattern that defines the two-dimensional shape of the cross section of the object to be built. Once a single cross section is formed, a fresh layer of liquid resin is coated over the last solidified layer and the process is repeated, with the new layer adhering to the previous one. When the final layer has been built, the object is complete.

4.1.1. ***StereoLithography (SL):*** *3D Systems, Inc. Valencia, California, U.S.A.*

3D Systems was founded in 1986 by inventor Charles W. Hull and entrepreneur Raymond S. Freed. The company is a pioneer and the leading manufacturer of rapid prototyping equipment. 3D Systems coined the term "StereoLithography" to refer to its method of building solid object by laser curing liquid resins.

3D Systems manufactures and sells their StereoLithography Apparatus (SLA) in four configurations, SLA-190, SLA-250, SLA-400 and SLA-500. The model numbers of the apparatus represent the approximate widths of the build envelopes in millimeters. Figure 6 depicts major components of a SLA-250.

The process of building an object with stereolithography normally begins with an STL file, the de facto standard format for all rapid prototyping systems. The STL file consists of triangular facets which approximate the surfaces of the object to be built. In addition to the object STL files, SL process requires some external support structures to attach the object to the build platform. Supports are also needed to provide support to overhang features to hold the object in place while layers are being built. The object and support STL files are then software sliced into a set of

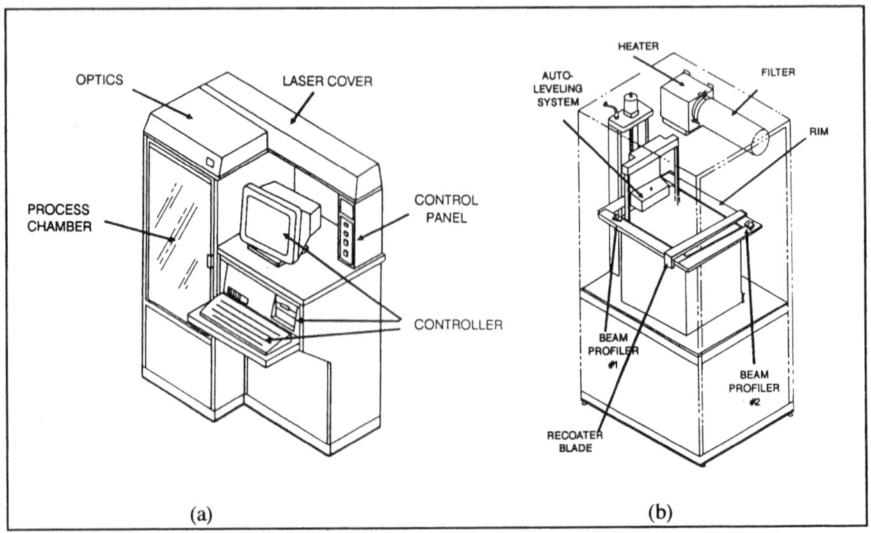

Fig. 6 (a) Major components of a 3D systems' SLA-250 stereolithography apparatus. (b) Major components of SLA-250 process chamber.

two-dimensional cross sections. The slice thickness is defined by the user and can be fixed or varied throughout the entire object. Slice thickness is normally ranging from 0.0625 mm to 0.508 mm (0.0025″to 0.020″). Once all STL files are sliced, the slice data will be merged together into a buildable vector file containing cross-sectional data. Other data files containing control parameters that govern the building process also must be prepared.

The build process takes place within the process chamber. The first step is to check the resin level and replenish any resin used in building the previous part. The chamber must also be heated to the resin-specific operating temperature range. The elevated temperature generally reduces the resin's viscosity and the settling time required to level each layer.

The object is built onto a perforated steel platform which is mounted on a computer controlled elevator. The platform is first immersed in the resin, then is positioned slightly above the resin surface so that it is covered by only a thin film of liquid. At the top of the process chamber is a set of galvanometer scanning mirrors. Under computer control, these mirrors move the focused laser beam to draw the first, bottom, layer of the build file on the resin surface. Figure 7 shows a schematic of a 3D Systems SLA-250 Stereolithography Apparatus.

After the first layer is solidified, the following sequence repeats for each subsequent layer:

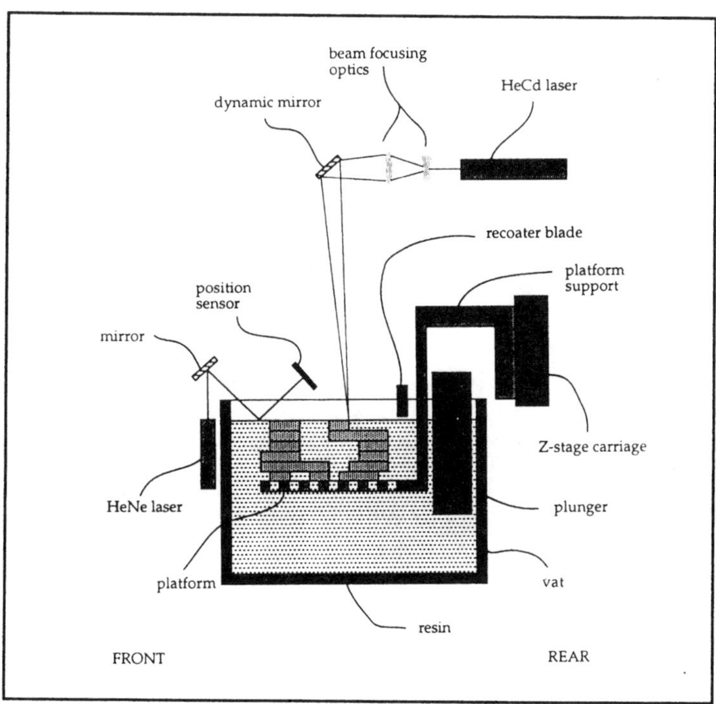

Fig. 7 Schematic a 3D systems' SLA-250 stereolithography apparatus.

1) The platform descends, allowing fresh resin to coat over the top of the previous layer.
2) Then the platform rises to a specified height and a "recoater blade" sweeps the excess resin from the top surface of the previous layer to leave a liquid resin layer that is about the thickness of the next layer to be built.
3) Upon completion of sweeping, the platform moves to the next layer build position. When it comes to rest, the top of the layer of recoated resin is at the free surface level of the resin in the vat. Due to finite surface tension effects, some pertubations happen around the solid to liquid interface. This requires a short period of pause to allow the new layer to flatten.
4) Then the scanning mirrors move the laser across the resin surface, drawing the cross section for the new layer.

After all layers are built, the platform and part are raised to let excess resin drain back into the vat. At this point, the SL part is not fully cured which is called a "green" part. The part with platform is then cleaned and rinsed with resin-stripping solvent. This step will remove all uncured resin. Once this is completed, the part is removed from the platform. Sometimes additional cleaning is necessary after the part is removed from the platform. The next step is to place the "green" part in a post curing apparatus. Here, the part is exposed to Ultra-Violet flood radiation to complete the polymerization process and to improve the final mechanical strength of the part.

Lastly, the SL part is hand-finished to provide a desirable surface finish. Secondary operations such as machining or coating may also be performed at this time. Figure 8 shows the steps of the SL process.

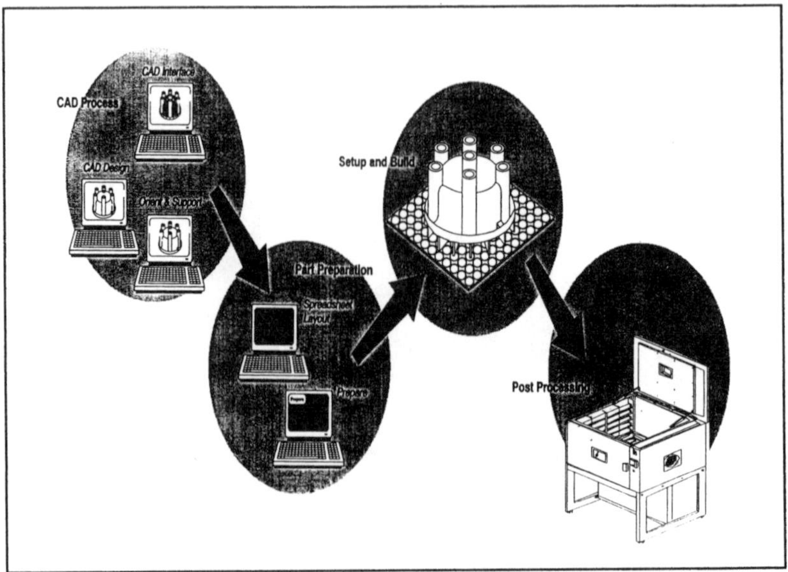

Fig. 8 The stereolithography process steps.

4.1.2. Stereos: *EOS (Electro Optical Systems) GmbH – Munchen, Germany*
The Stereos rapid prototyping systems are manufactured and marketed by EOS, in Germany. The core of the process is the same as 3D Systems' Stereolithography process. Three Stereos models are available, which can build models up to 610 mm × 610 mm × 407 mm (24″ × 24″ × 16″). One specific feature to be noted in the Stereos systems is the resin recoating method. It is called the "active recoating" system which is different from the SLA's "deep dip" technology. The active recoating method reduces recoating time, bubble formation, and greatly improves the quality of parts containing trap volumes. Another feature of the Stereos systems is the interchangeable resin vats. With the ability to exchange vats in different sizes, users can build models using different resins with minimum resin.

4.1.3. SOUP (Solid Object Ultraviolet Plotter): *CMET, Inc. (Computer Modeling and Engineering Technology) – Tokyo, Japan*
The SOUP system is manufactured and marketed by CMET Inc. The technology was developed by Professor Yogi Marutani at the University of Osaka. Currently, CMET's SOUP machines are available in five configurations. Machine work envelopes range from 407 mm × 407 mm × 407 mm (16″ × 16″ × 16″) up to 610 mm × 864 mm × 508 mm (24″ × 34″ ×20″). All models build objects in the same manner as the SLA machines. The significant differences are in the laser beam delivery mechanism. SLA machines use galvanometer mirrors, while SOUP machines use either galvanometer mirrors or gantry plotter mechanisms.

4.1.4. Solid Creation System *(SCS): D-MEC Ltd. (Design-Model Engineering Center) – Tokyo, Japan*
The Solid Creation System rapid prototyping technology was developed by SONY and marketed by D-MEC Ltd. in Japan. The SCS JSC-3000 from D-MEC is the largest capacity rapid prototyping system currently available on the market. The JSC-3000 has a work envelop of 990 mm × 787 mm× 508 mm (39″ × 31″ × 20″). The functional principle and general features are the same as the SL process and 3D Systems SLA machines.

4.1.5. Soliform: *Teijin Seiki Company Ltd. – Kanagawa, Japan*
The Soliform system is based on the SOMO (SOlid MOdeling System) laser curing process developed by DuPont Imaging Systems of New Castle, Delaware, U.S.A. Teijin Seiki currently manufactures and sells Soliform machines in two models which can build prototypes up to 508 mm × 508 mm ×508 mm (20″ × 20″ × 20″). Soliform machines use the same resin curing by laser scanning principle as used in SL process. One distinction in the Soliform process is that the system exposed the resin layer using a raster-scanning mode, rather than a vector scan, with a modulated fast shutter to define the cross-section boundaries.

4.1.6. COLAMM *(Computer Operated Laser Active Modeling Machine): Mitsui Engineering & Shipbuilding Company Ltd. – Tokyo, Japan*
The COLAMM rapid prototyping system is developed and marketed by Mitsui Engineering in Japan. The basic principle of the COLAMM process is similar to the SL process. However the COLAMM machine is unique in that the laser scanner is

installed at the bottom, and the laser beam is directed upwards through a glass plate. In the COLAMM process, objects are built from the top down, rather than bottom up.

4.2. Layer-Additive Resin Polymerization by Masked Lamp Radiation

The general principle of resin polymerization by masked lamp radiation is the same as using laser scanning which is based on that photocurable liquid resin solidifies under the influence of light. The essential difference between the two approaches is that the surface is not illuminated in a point-to-point mode but by a complete lighting and hardening of the entire surface, using specially prepared masks. Generation of the masks uses ionography technology. Each cross section geometry will be developed on a charged glass plate. The layer of liquified resin is exposed through the masks by a strong UV light source. At the illuminated areas the photopolymer hardens, and the entire object can be built layer by layer.

4.2.1. Solid Ground Curing *(SGC). Cubital Ltd. – Raanana, Israel*
Solid Ground Curing (SGC) system is developed and marketed by Cubital, Ltd., located in Raanana, Israel. The company was founded by the inventor Itzchak Pomerantz in 1986. Currently Cubital offers two models of the SGC system. The Solider 5600 has a 508 mm × 356 mm × 508 mm (20 ″ × 14″× 20″) work envelop, and the Solider 4600 has a 356 mm × 356 mm × 356 mm (14″ × 14″ × 14″) work envelop.

Cubital's SGC system builds parts in a multi-step process which utilizes an optical mask system to selectively expose layers of photocurable resin. A thin layer of liquid resin is spread onto the workpiece and then exposed to ultraviolet light through a patterned mask having transparent areas corresponding to the cross section of the part being built. Light passing through the mask solidifies areas of the resin. The remaining unexposed resin, still liquid, is then removed with a suction device, and replaced with wax. The wax acts to support the part as it is being built. The layer, now consisting of both solidified resin and wax, is then milled to a precise thickness, forming a flat substrate on which the next layer is built.

The SGC process consists of mask generation and model building cycles. The sequence of operation of the SGC process is illustrated as follows: (Figure 9 illustrates the SGC process.)

Mask Generation:

1) Transfer sliced CAD data to the mask genrator.
2) Charge the glass mask plate through an ionographic process.
3) Develop the charged image with electrostatic toner.
4) Expose the resin with a strong flash of UV light passing through the mask. All the exposed areas in the layer are fully cured instantly.
5) Erase the exposed mask physically and electrostatically, then the process repeats.

Model building:

1) A thin layer of photopolymer is spread on the support carriage.
2) Support carriage is moved to the exposure station. A strong UV light radiates through the mask.

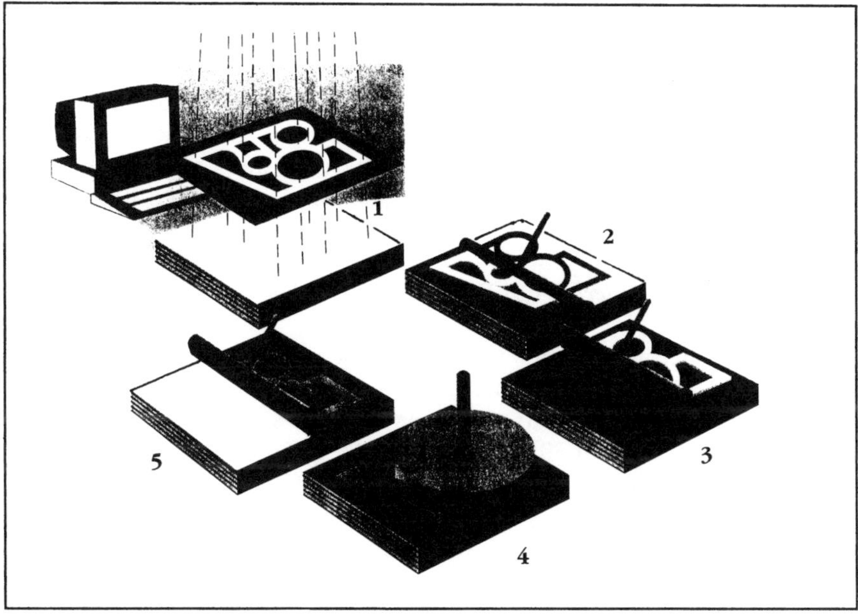

Fig. 9 Illustration of solid ground curing process.

3) A wiper removes and collects unsolidified resin.
4) Wax filling in all cavities, to replace the removed resin.
5) Cooling the wax, producing a solid volume.
6) Milling of the layer to guarantee the exact layer thickness. The carriage then drops down slightly in preparation for the next layer and returns to the exposure station.

After all layers have been formed, the workpiece is removed from the carriage and placed in an oven to melt the wax and release the parts trapped inside. Because the wax is also water-soluble, final cleaning is usually done with a warm water bath. Brushes may also be used to help remove wax from narrow passages.

4.3. Layer-Additive Laser Sintering

Sintering is the process whereby compacted powder material fuses into solid under the influence of heat. The principal governing variables in sintering are temperature, time, and the atmosphere. Traditional sintering is done in a controlled atmosphere furnace where the powder material is heated to a temperature just below its melting point, but sufficiently high to allow bonding of the individual particles. However, the physical processes in laser sintering are substantially different from traditional bulk sintering. In laser sintering, the powder material is actually heated by laser energy to the melting threshold and then fusion occurs in a very short duration.

Similar to stereolithography process, laser sintering also works in a layer-additive fashion. In laser sintering, the object is formed by converting loose powder to solid shape using the heat of an infrared laser beam. For each layer to be built, a thin layer of heat-fusible powdered thermoplastic material is spread on the build surface. Then, the pattern of the corresponding cross section is drawn by the laser on the powder surface. With amorphous materials, the laser heat causes powder particles to melt and fuse to one another at their points of contact, forming a solid mass. After one layer is formed, a layer of fresh powder is spread over the previous layer and the process repeated until all layers have been formed. The finished part is embedded within a cake of loose powder, which must be removed.

4.3.1. ***Selective Laser Sintering (SLS):*** *TM Corporation – Austin, Texas, U.S.A.*

Selective Laser Sintering process was invented by Carl Deckard and Michael Feygin and marketed by DTM Corporation. Currently, DTM manufactures and sells the "Sinterstation 2000" system to its customers to build SLS parts. The Sinterstation 2000 has a work envelope of 305 mm diameter by 381 mm high (12″Dia. × 15″H). Common materials used in the machine are Polycarbonate, nylon, investment casting wax.

The SLS process of building a part also begins with an STL file. Then the STL file is software sliced into a set of parallel two-dimensional cross sections. At the same time, the machine needs to be prepared by loading powder cartridges and heating the build chamber to its operating temperature and purged with nitrogen.

After the machine is prepared, the control computer obtains the first few cross sections and stores them in a buffer. The following sequence then takes place for each layer of the object to be built as illustrated in Figure 10.

After all layers are built, the part piston is raised to the scanning surface. The part is then removed from the powder bed and taken to a "rough breakout station"

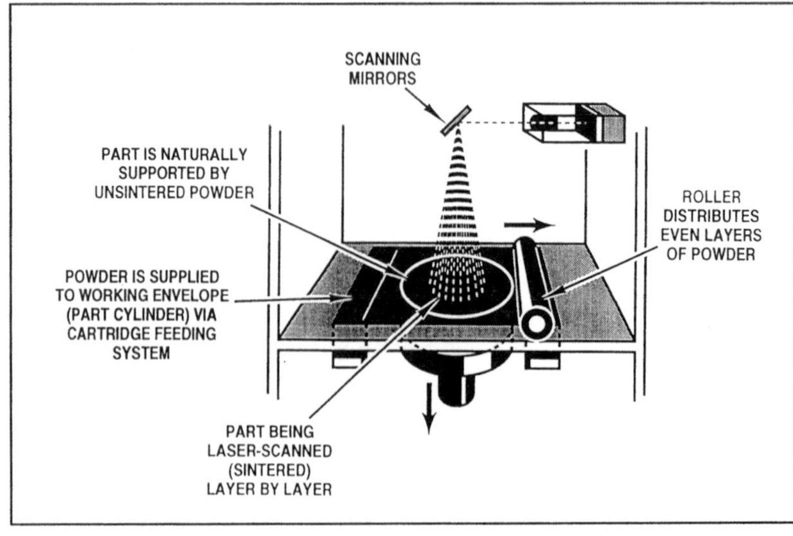

Fig. 10 Illustration of selective laser sintering process.

where the powder is dusted off using brushes, low-pressure air and/or other tools. Depending upon the specific requirements, some secondary operations may be performed.

4.4. Layer-Additive Freeform Extrusion

In the basic extrusion process, liquid or pliable material is placed into a chamber and is forced through a die opening by a ram. This is similar to forcing the paste through the opening of a toothpaste tube. In traditional applications, extrusion is usually used to form objects that have a constant cross-section shape. However, in the layer-additive freeform extrusion, the extrusion nozzle is mounted on a programmable linear slide which can be programmed to move in X and Y directions. Additional motion axes can be incorporated in the nozzle and/or build platform. In such a way, material can be extruded layer by layer to form complex 3-dimensional objects. Figure 11 illustrates the freeform extrusion process.

4.4.1. ***Fused Deposition Modeling (FDM):*** *Stratasys, Inc. – Eden Prairie, Minnesota, U.S.A.*

Invented by Scott Crump and marketed by Stratasys, Fused Deposition Modeling is the first system available using freeform extrusion principle. Stratasys currently offers a number of system configurations, 3D Modeler, FDM 1000, FDM 1500 and FDM 1600. 3D Modeler has a 305 mm (12 inches) cube work envelope while other FDM machines have a 254 mm (10 inches) cube work envelope. Different system models are configured to extrude different type of materials.

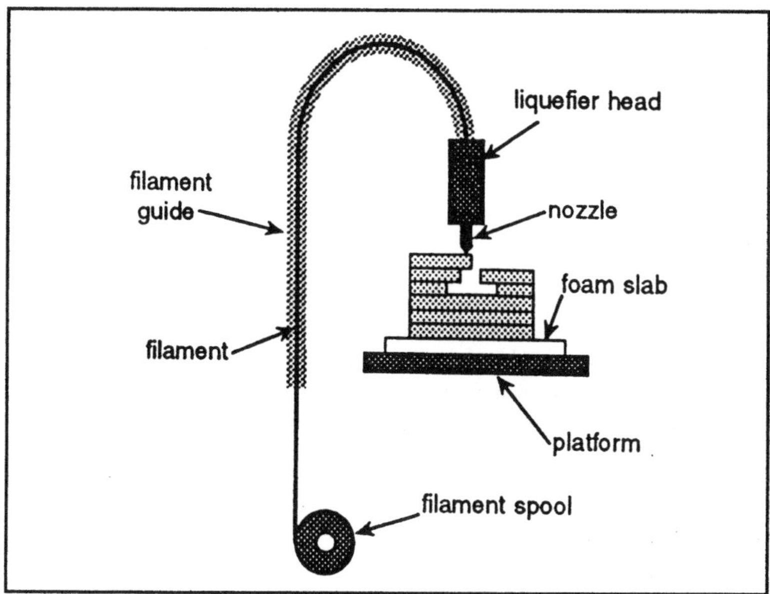

Fig. 11 Illustration of freeform extrusion process.

Materials currently available for use with the FDM process are investment casting wax, machinable wax, nylon, and ABS. Materials are supplied in the form of a filament on a spool.

Like other rapid prototyping processes, FDM builds objects by sequentially stacking thin layers. In FDM, each layer is generated by pumping molten thermoplastic through the orifice of a nozzle, while the nozzle traces out the contours of the layer in the horizontal plane. When the material exits the orifice, it is squeezed between the nozzle tip and the previous layer, and sheared by the nozzle's motion. Since the surrounding temperature in the work envelope is maintained below the material's melting temperature, the extruded molten material quickly solidifies to form a solid lamination. The completed part achieves its final mechanical properties upon cooling.

The FDM process of building an object also begins with an STL file. Then, support structures may need to be added to the part data file. Other build parameters also need to be specified. Before starting the build process, the machine must be preheated to the specified operating temperatures for the material to be used. The operator must check that the spool contains enough filament for the part to be built. When necesary, a new spool must be loaded onto a spindle on the back of the building chamber. The operator also needs to place a removable styrofoam slab on the platform surface as a base for the part to build on. Finally, the machine starts and the operator needs to verify the platform and extrustion head positions and makes adjustments as needed. Then the following sequence occurs for each layer of the object:

1) The platform descends by the thickness of the layer to be built, the nozzle moves to the start position of the cross-section, and the pump begins to force molten material through the orifice.
2) The nozzle moves in the horizontal plane at a controlled speed, laying down a line of material in the shape of the cross-section. The width and thickness of this line depends on both the speed of the nozzle and the feed rate of the pump. As the nozzle traces curves and goes around corners, its speed changes. The flow rate of material changes as well to maintain control over line dimensions. At the end of the line (or line segment), the platform may briefly descend a short distance, or the nozzle may move sideways, to help detach the nozzle from the deposited material.

Additional lines of material must be deposited to fill any area wider than one line width. After all layers are built, the head is moved out of the way and the platform is raised to its starting height. The part is then removed from the building chamber and allowed to cool. Any support structures are cut off manually and the part surfaces may then be smoothed to minimized layer stair steps or irregularities.

4.5. Layer-Additive Droplet Deposition

The droplet deposition method of layer-additive process is a result of research at the Massachusetts Institute of Technology (MIT) by a team let by Emanuel Sachs and Michael Cima. In this process, a powdered material is laid in layers and a stream of binder is deposited on it in successive patterns representing the cross sections of the desired object. The binder is deposited through standard ink-jet nozzles, as are used in 2-dimensional ink-jet printing. When the binder strikes the surface of the powder and penetrates a short distance into it, the powder and binder form a composite structure in which the powder particles are trapped in a matrix of the binder material. This "green part" is then later "fired" to cure the binder and strengthen the object. This process, called three-dimensional or 3-D printing is potentially applicable to all manner of powder/adhesive combinations, including plastics, metals, ceramics and glasses. Figure 12 illustrates the 3-D printing process developed by MIT.

*4.5.1. **Direct Shell Production Casting (DSPC):** Soligen, Inc. – Northridge, California, U.S.A.*

Direct Shell Production Casting (DSPC) is Soligen's proprietary technology for creating ceramic casting molds for metal parts and tooling directly from a computer-aided design (CAD) database. The DSPC process builds investment casting shells by selectively binding layers of ceramic material. It is based on the MIT's 3D Printing technology, which is exclusively licensed to Soligen for metal casting. Figure 13 illustrates Soligen's Direct Shell Production Casting machine.

Similar to other rapid prototyping processes, the DSPC process first requires a STL file of the CAD model. Also, structures to guide the molten metal into the casting shell must be incorporated. Soligen's software will allow the user to position multiple copies of a shell in the build-volume and connect them into a casting tree with gates and runners. Once the STL model of the casting shell is completed, it is sent to the DSPC machine. The DSPC machine then builds the shells by selectively spraying a colloidal silica binder onto layers of alumina powder. First, a layer of the ceramic powder is deposited and smoothed by a roller. The layer thickness is

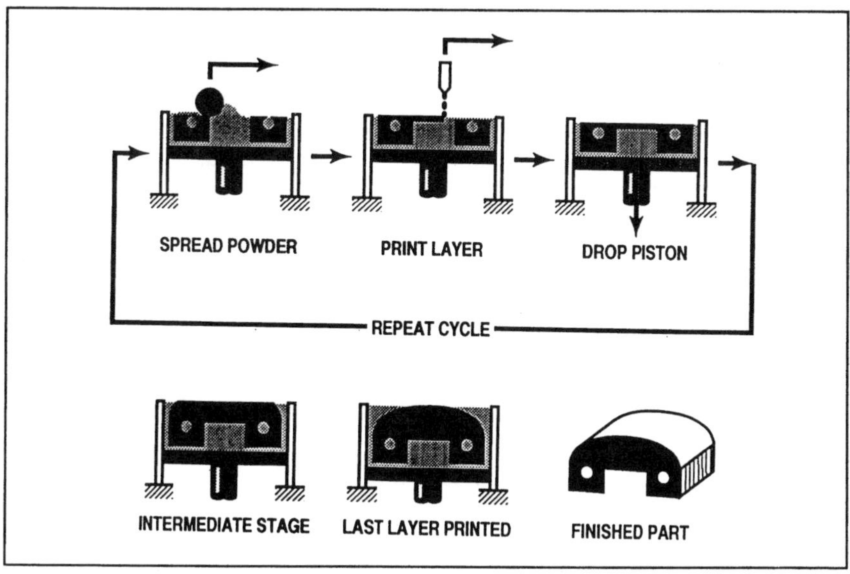

Fig. 12 3D printing process developed by MIT researchers.

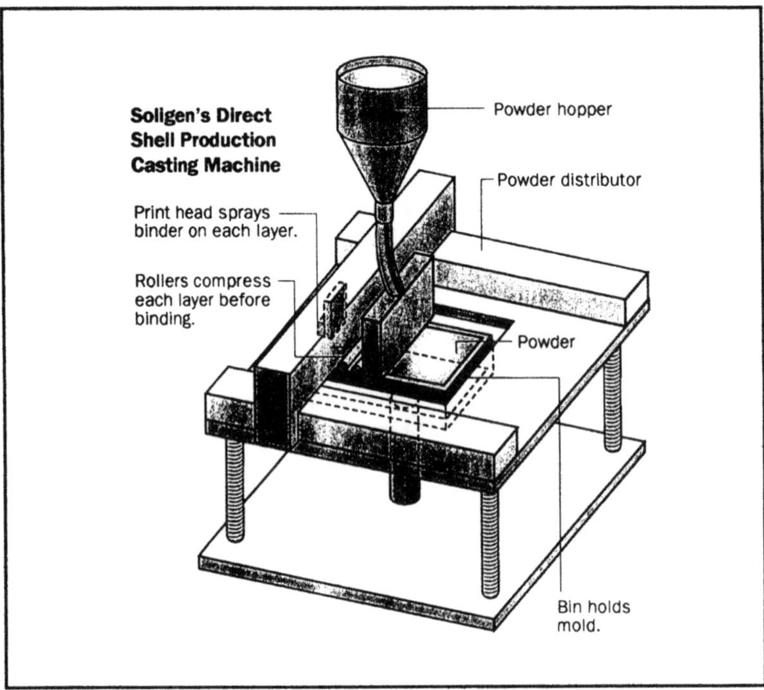

Fig. 13 Soligen's direct shell production casting machine.

approximately 0.178 mm (0.007 inches). Next, the DSPC print head, in a raster pattern, "prints" the current cross-section of the casting shell.

The binder flows through the print head continuously, but is selectively diverted by an electrical charge either onto the layer, or back into the binder reservoir. Keeping the material constantly flowing through the jets prevents clogging which might occur if the flow were actually turned off and on. Once all the layers have been printed, the cake of ceramic powder containing the casting shell is removed from the machine and taken to a "break-out table" where excess powder is removed from the exterior of the shell. The casting shell is then fired in a two-phase kiln cycle, first at approximately 100° C to remove any moisture left in the shell, and then at 700 – 800°C to fully harden the powder and binder. Once this is done, the shell is ready to be poured with molten metal. The maximum build-volume is 203 mm × 305 mm × 203 mm (8″ × 12″ × 8″). Figure 14 is a schematic drawing demonstrates the production steps of the DSPC system.

*4.5.2. **Drop-On-Demand Ink-Jet Plotting:** Sanders Prototype, Inc. – Wilton, New Hampshire, U.S.A.*

Sanders Prototype manufactures and sells the Drop-On-Demand Ink-Jet Plotting rapid prototyping system. Its current offering is the Model Maker MM-6B machine which has a cube of 152.4 mm × 152.44 mm × 152.4 mm (6″ × 6″ × 6″) work volume. The system is based on ink-jet printing technology similar to the 3D printing technology developed at MIT.

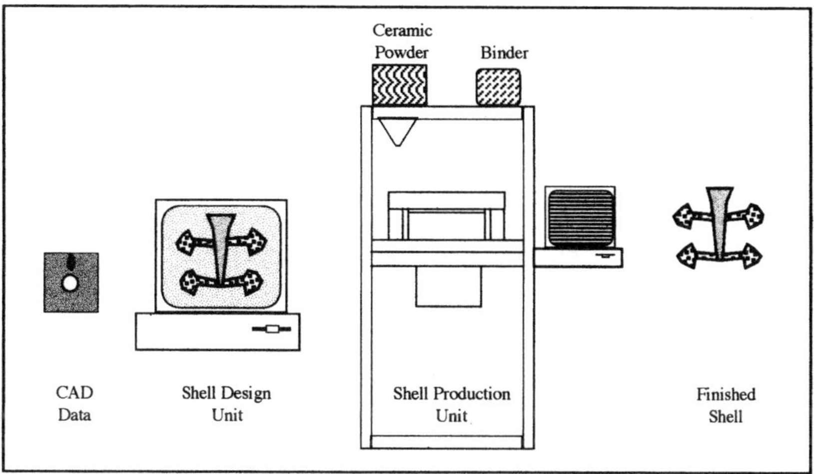

Fig. 14 Schematic drawing of production steps and components of the DSPC system.

The process uses dual ink-jet print heads, one to deposit thermoplastic building material, the other to deposit supporting wax. As with most other commercially available rapid prototyping systems, an object is built on a platform which lowers by one layer thickness after each layer is deposited. The liquid build material cools as it is ejected from the jet and solidifies upon impact. The wax is laid down to provide a flat, stable surface for deposition of build material in subsequent layers in areas which would otherwise be unsupported. Each layer is approximately 0.0762 mm (0.003 inches) thick as it is deposited. After each layer is printed, a cutter planes approximately 0.0254 mm (0.001 inch) of material off the layer's top to provide a smooth, even surface for the next layer. The material is deposited at a rate of approximately 254 mm (10 inches) per second. Figure 15 is the schematic diagram of the Model Maker MM-6B.

The Model Maker MM-6B machine is designed for unattended operation. After several layers are built, the print-heads are moved to a monitoring area on the machine to determine if the jets have become clogged. At this point, if material is flowing freely through the jets, the build process is resumed. If, however, the jets are blocked, they are purged, then the cutter removes any layers deposited since the last check of the jets, and restarts the building process from that point.

Once the build process is complete, the object is removed from the build chamber. Wax supports can be easily removed using wax solvent. Figure 16 shows the functionality of the Model Maker MM-6B.

4.6. Layer Laminating and Laser Trimming

In this process, a laser cuts a pattern representing a layer of the desired object in a sheet material with adhesive coating. A new sheet of material is then laid over the previous one and bonded to it. The laser precisely cuts another layer and the process is repeated.

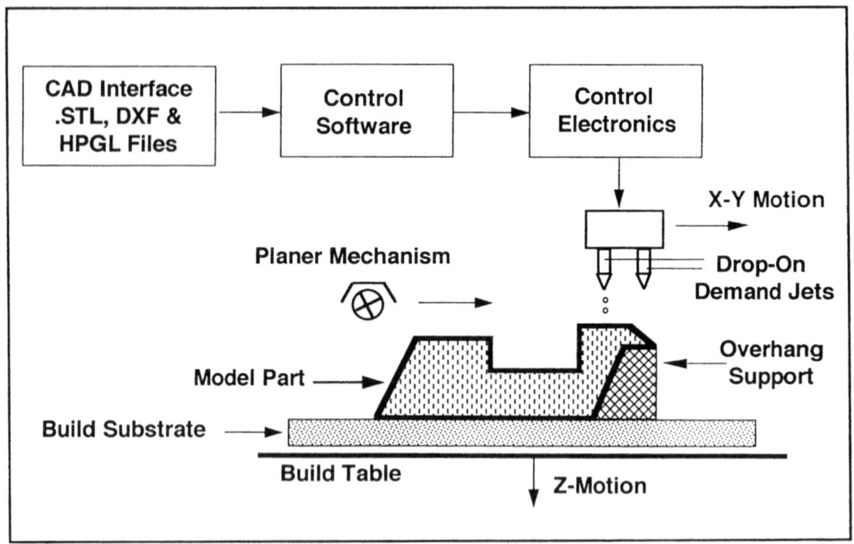

Fig. 15 Schematic diagram of the model maker from sanders prototype inc.

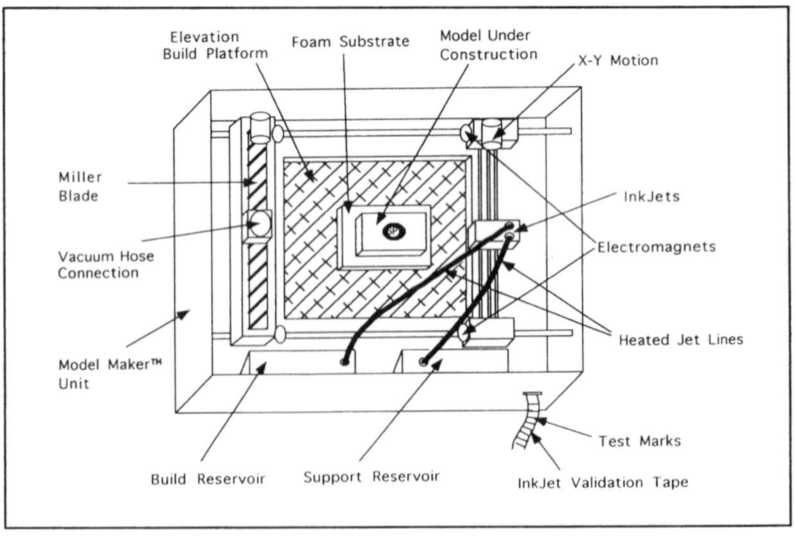

Fig. 16 Model maker MM-6B functionality.

*4.6.1. **Laminated Object Manufacturing (LOM):** Helisys, Inc – Torrance, California, U.S.A.*

The Laminated Object Manufacturing (LOM) technology developed and marketed by Helisys, Inc. builds parts by laminating and laser trimming the material delivered in sheet form. (Figure 17 is a schematic of Helisys' LOM rapid prototyping system.)

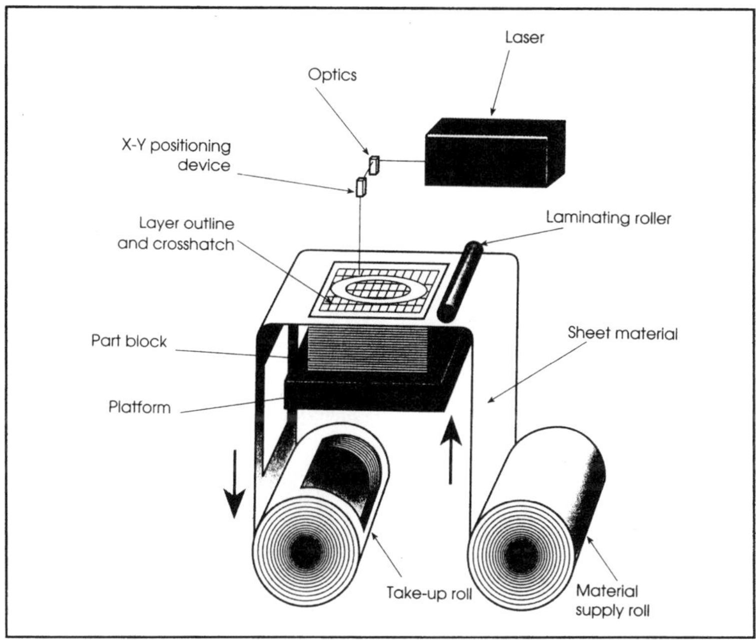

Fig. 17 Schematic of helisys' LOM rapid prototyping system.

The sheets of material are laminated into a solid block using a thermal adhesive coating. Currently the most popular material used in LOM machines is polyethylene-coated paper.

The LOM process starts with slicing of STL representation of 3D objects into thin cross-sections. The cross-sectional information is then fed into the LOM control computer, which guides the laser beam trimming around the periphery of the layer. The laser cuts to the depth of one layer of the material. The material which surrounds the cross-section is usually cross hatched by the laser beam into squares, to facilitate separation at the completion of the process. Next the platform moves down. The sheet material advances by an increment exceeding the length of a cross-section onto the rewinding roll. (See Figure 18.) As the platform moves up, the heated roller moves across the stack while pressing the sheet against the stack and bonding it to the previous layer. The laser then cuts a new cross-section. The process continues until all of the cross-sections have been deposited and cut. Figure 18 illustrates the LOM process.

5. FUTURE DEVELOPMENT TRENDS

The entire rapid prototyping industry is marked by dynamic and aggressive developments in three basic areas: 1) Material, 2) Computer software and 3) Process control & automation. New developments in these areas will result in significant improvements of currently available systems and realization of new processes.

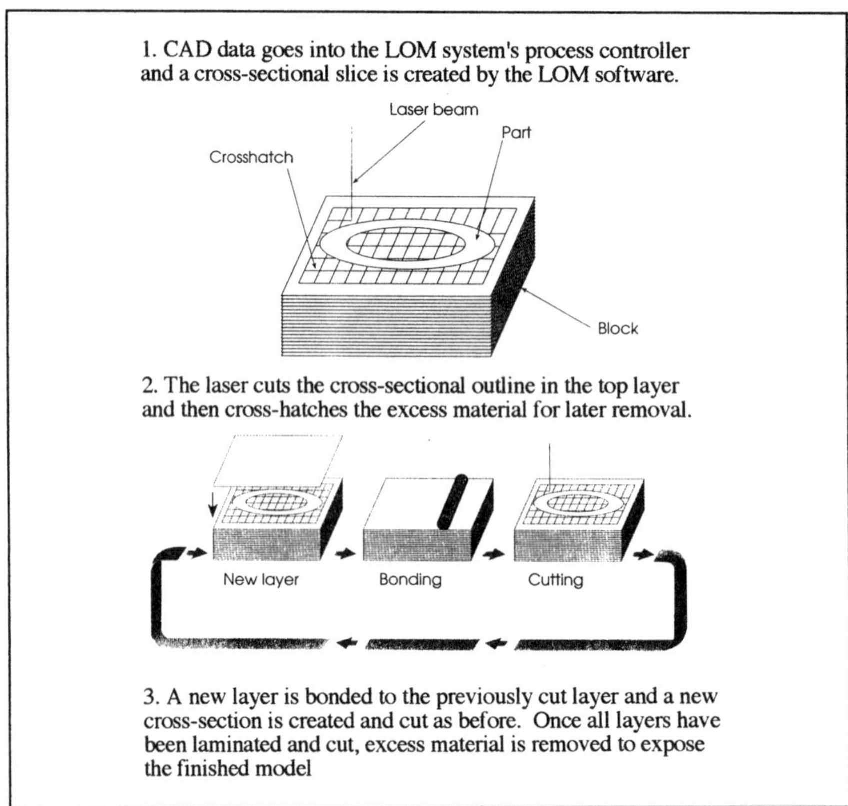

Fig. 18 Illustration of laminated object manufacturing process.

5.1. Material Development

When a specific product or component is designed, the material which it will be made of is as important as its geometry. Any prototype built to simulate a finished part in specific tests must be made of a material with properties close to the final production material. In today's rapid prototyping systems, the most commonly used materials are liquid photopolymer systems. Unfortunately, all of these systems experience shrinkage during the phase transformation from liquid to solid. This results in stress and subsequently in strain deformations. The most serious problem is that the resulted prototypes or parts can not meet the engineering requirements. Other material systems used in the rapid prototyping processes also exhibit some shrinkage or deformation due to phase transformation and/or residual stress built into the layering process. Therefore, one of the major material developments is to come up with new materials which have reduced shrinkage and distortion.

Other areas of material developments are focused on ceramics, metals and advanced composites and materials have other mechanical, thermal, electrical and optical properties than plastics and wax. Material safety and ease of handling are two other areas that have received strong user and manufacturer emphasis.

5.2. Computer Software

All of the current rapid prototyping processes begin with a three-dimensional CAD model or some type of three-dimensional computer database, such as data from 3D digitizer or computer tomography scan. In majority of the RP processes, these computer representations of a part will then go through some conversion into a certain format that a particular RP system can understand. In many cases that means to convert a 3D CAD model into STL, STereoLithography, or CFL, Cubital Facet List, format which is a tessellated approximation of the intended geometry. Though faceted representation is an easy and quick way to represent a 3D geometry, it is certainly not the most efficient and accurate approach. Rapid prototyping industry is demanding a standard data representation that can describe the true geometry of the part to be built and can interface to all the systems.

Besides the development of interfacing software that links between the CAD and RP systems, software that controls each RP process is also a challenge for the developers of RP equipment. Adopting and fully utilizing advanced software capabilities in the process control software will result in more efficient, intelligent and user friendly software than the ones currently in use.

5.3. Process Control and Automation

A close look at today's RP systems compared with traditional manufacturing and fabrication systems, it is not difficult to discover that these new breed of fabricators are "manipulating" materials at a lower level than traditional systems. They "build" parts by solidifying polymers, melting powder particles, depositing droplets, cutting thin sheets. These type of processes require high degree of precision in the process control to yield satisfactory results. Accurately controlling the amount of energy input, (such as light, heat, forces) and the timing to deliver the energy and the location to apply the energy are crucial to the success of these systems. On top of precise process control, developers of RP systems have built very high degrees of automation into the systems. Most of the RP systems can run unattended for hours or even days.

As revolutionary as today's RP systems are, there are still tremendous developments in process control and automation that need to be done to bring this technology to an even higher level. Many believe that RP technology will march toward the so called "nanotechnology," in which machines that build things, anything, one atom or one molecule at a time. Machines will be capable of automatically rearranging the molecular structures of the input materials in order to generate the programmed object.

In summary, as a commercial technology, rapid prototyping is still in its infancy stage. The various processes are undergoing very rapid refinement and improvement. New processes will arise in the years to come. Many of the future processes will be able to build objects in a way that can not be done with current technologies.

References

M. Burns, *Automated Fabrication*, PTR Prentice Hall, Englewood Cliffs, New Jersey (1993).

Cibatool Resin Handbook, 3D Systems, Inc (1994).

D.E. Coleman and D.C. Montgomery, A Systematic Approach to Planning for a Designed Industrial Experiment, *Technometrics*, **35**(1), 1–27 (1993).

Corporate Focus: Cubital, Ltd., *Rapid Prototyping Report*, **3**(5), CAD/CAM Publishing, Inc., San Diego, CA (1993).
Corporate Focus: Helisys – Laminated Object Manufacturing, *Rapid Prototyping Report* **3**(2), CAD/CAM Publishing, Inc., San Diego, CA (1993).
Corporate Focus: Stratasys, Inc. *Rapid Prototyping Report*, **3**(12), CAD/CAM Publishing, Inc., San Diego, CA (1993).
FDM 1000 Product Specification, Stratasys, Inc. (1994).
B. Hinzmann, *TechMonitoring, Technology Profile: Rapid Prototyping*, SRI International, Menlo Park, CA (1993).
R.K.C. Hsu and S.M. Copley, Producing Three-Dimensional Shapes by Laser Milling. *Journal of Engineering for Industry*, 375–9 (1990).
M. Hyduke and R. Chu, *Rapid Prototyping for the Development of Gas Turbine Engines*. Presented at the American Helicopter Society 49th Annual Forum, St. Louis, MO (1993).
Improving Accuracy, Rapid Prototyping Report, **3**(2), CAD/CAM Publishing, Inc., San Diego, CA (1993).
Interfacing CAD and Rapid Prototyping, *Rapid Prototyping Report*, **2**(1), CAD/CAM Publishing, Inc., San Diego, CA (1992).
P.F. Jacobs (1993) Stereolithography: From Art to Part, *Cutting Tool Engineering* (1993).
— *Rapid Prototyping & Manufacturing*, Society of Manufacturing Engineers, Dearborn, MI.
D. Kochan (ed) *Solid Freeform Fabrication*, Elsevier Science Publishers, Amsterdam, Netherlands (1993).
Material Focus: Rapid Prototyping Materials – Part I, *Rapid Prototyping Report*, **2**(10), CAD/CAM Publishing, Inc., San Diego, CA (1992).
Material Focus: Rapid Prototyping Materials – Part II, *Rapid Prototyping Report*, **2**(11), CAD/CAM Publishing, Inc., San Diego, CA (1992).
Material Focus: Rapid Prototyping Materials – Part III, *Rapid Prototyping Report*, **2**(12), CAD/CAM Publishing, Inc., San Diego, CA (1992).
MM-6B, Model Maker 3D Plotting System Specification, Sanders Prototype, Inc. (1994).
New Products: Ink-Jet Based Rapid Prototyping System, *Rapid Prototyping Report*, **4**(3), CAD/CAM Publishing, Inc., San Diego, CA (1994).
Rapid Prototyping in Japan, *Rapid Prototyping Report*, **4**(2), CAD/CAM Publishing, Inc., San Diego, CA (1994).
Research Notes: Three Dimensional Printing, *Rapid Prototyping Report*, **2**(3), CAD/CAM Publishing, Inc., San Diego, CA (1992).
D.A. Schaub and D.C. Montgomery, Using Experimental Design to Optimize the Stereolithography Process. *Quality Engineering* 9(4), June (1997), 575–586.
Sinterstation 2000 Product Specification, DTM Corporation (1994).
SLA User Guide, Software Release 3.83, 3D Systems, Inc. (1990).
SLA-250/40 Product Specification 3D Systems, Inc (1994).
SOLDIER 5600 Product Specification, Cubital Ltd. (1994).
Soligen: Direct Shell Production Casting System, *Rapid Prototyping Report* **4**(5), CAD/CAM Publishing, Inc., San Diego, CA (1994).
Stereolithography Workstation User Guide for Software Release 1.52, 3D Systems, Inc (1994).
Technology Focus: Fused Deposition Modeling, *Rapid Prototyping Report*, **2**(3), CAD/CAM Publishing, Inc., San Diego, CA (1992).
Technology Focus: Selective Laser Sintering, *Rapid Prototyping Report*, **1**(7), CAD/CAM Publishing, Inc., San Diego, CA (1991).
Technology Focus: Stereolithography, *Rapid Prototyping Report* **2**(9), CAD/CAM Publishing, Inc., San Diego, CA (1992).
P.L. Ulerich, *Tips for Concurrent Engineering*, Machine Design, 54–6 (1992).
Uziel, Yehoram, *An Unconventional Approach to Producing Investment Castings*, Modern Casting (1993).
— *Functional Prototyping — Has the Future Arrived?* FOUNDRY Management & Technology (1993).

CHAPTER 8

Coordinate Measuring Machine (CMM) Technologies and Their Applications in Support of Concurrent Engineering Endeavor: Concepts, Tools and Challenges

YIN-LIN SHEN[a] and F. FRANK CHEN[b]

[a]*Department of Civil, Mechanical and Environmental Engineering, The George Washington University, Academic Center T720A, 801 22nd Street NW, Washington, D.C. 20052, USA;* [b]*Department of Mechanical, Industrial and Manufacturing Engineering, University of Toledo, Toledo, OH 43606, USA*

1. INTRODUCTION

Coordinate measuring machines (CMMs) are used to measure the physical dimensions of manufactured parts. They are rapidly becoming the measurement tool of choice because of their speed, accuracy, and flexibility (Phillips *et al.*, 1994). Toleranced dimensions are a critical attribute of the more than $500B worth of discrete-part products the United States has produced each year (Swyt, 1992). Dimensional inspection of discrete-part products is important because dimensions and tolerances of products play an important role on product functional performance and costs. The basic idea of CMM-based dimensional inspection is that a probe, mounted on the gage head of a three-axis machine tool structure or a horizontal robot arm, is used to measure the coordinates of points on part surfaces. The measurement data is then processed by post-inspection data analysis software to construct ideal geometric features of the corresponding part features and to establish relationships between features (Caskey *et al.*, 1991). The measured results can be used to check the dimensional correctness of the parts and to ascertain the stability of the manufacturing process from which the parts are produced. The advantages of using CMMs to perform dimensional inspection over traditional gaging techniques are greater flexibility, reduced setup time, improved accuracy, reduced operator influence, and improved productivity (Wick and Veilleux, 1987). CMM-based dimensional inspection provides process control information which is critical for quality control (QC) purposes. Computer-controlled CMM allows off-line programming of inspection routines based on part design database (Duffie *et al.*, 1988; Brown and Gyorog, 1990; Menq *et al.*, 1992b). Furthermore, the dimensional measuring interface standard (DMIS) has been developed as a two-way communication interface format between dimensional measuring equipment (DME) and computer-aided design (CAD) systems (CAMI, 1989). DMIS provides the interface support for CMMs to be integrated into computer-integrated manufacturing (CIM) environments. Benefiting from cost economics of electronics and software, CMM prices are steadily dropping relative to capabilities, resulting in increased CMM application opportunities. Powerful control capabilities and sophisticated software allow linking of CMMs with computers, machines, and computer-aided design/computer-aided manufacturing (CAD/CAM) systems for manufacturing integration and automation (Bosch, 1991).

Interests in concurrent engineering and integrated product and process design are rising as design and manufacturing engineers realize the benefits of early collaboration (Beutel, 1993). The concurrent engineering concept has been recently further broadened to encompass a wider scope of simultaneous engineering efforts in manufacturing settings. Chen (1993) proposed a scheme for concurrent consideration of cell physical configuration and required cell control functions to ensure economic implementation of factory cells. Chen and Sagi (1994) investigated a neural-net based methodology for concurrent design, planning, and control of manufacturing cells. Since major metrology equipment such as CMM has gradually become a critical unit of most automated discrete part manufacturing systems, the integration and proper selection of CMM performance functionality will have a profound impact on the success of manufacturing automation.

This chapter reviews CMM technologies and applications, and discusses supports and challenges in concurrent engineering endeavor. Technologies contributing to CMM success are described in Section 2, and CMM applications are presented in Section 3. Concurrent engineering issues in CMM-based dimensional inspection are discussed in Section 4. Technological challenges and research needs in CMM are described in Section 5, and conclusions are drawn in Section 6.

2. CMM TECHNOLOGIES

As numeric control (NC) machines are to manufacturing, coordinate measuring machines are to inspection (Placek, 1988). CMMs are especially suited for inspecting complex parts with many features and tight tolerances. CMMs were developed by modifying precision layout machines. These machines provide digital readouts of the movable machine member for display, computer calculations of geometric measurement routines, or computer-controlled positioning of the machine member. The travelling member, which is a gage head with a mounted probe, contact sensor, or an optical device, is guided along inspection paths. The paths contained in a three-dimensional space formed by three mutually perpendicular axes (X, Y, and Z axes). The machines are equipped with staging tables of certain size, which are made of a dimensionally stable material (usually granite), and lapped to precision geometrical flatness requirements. Machine configurations include cantilever, moving bridge and fixed bridge, column, horizontal arm, and gantry designs (ASNI, 1990; Robison, 1990; CMM combines ..., 1991). In some cases, a rotary table is attached to the staging table to provide the fourth axis of the machine. In recent years, portable CMMs which can be easily carried around shop floors to parts needing dimensional inspection have been developed (Tool company ..., 1992; Portable CMM ..., 1994). Air bearings are often used to provide a nearly frictionless movement along the guideways. The measurement function is accomplished by clamping the workpiece to the staging table and aligning it with the coordinate system of the mutually perpendicular X, Y, and Z guideways (so the relative position and orientation of the workpiece with respect to the machine coordinate system is known). Each slide is equipped with a precision linear-measurement transducer whose position is digitally updated in very small increments (such as 0.001 mm) as the measurement sensing probe is moved from point to point on the workpiece. If a contact sensor is used, the probe is guided to approach the workpiece preferably in a normal direction at a specified measuring speed, and when the probe hits the workpiece a signal is triggered to store the coordinates of the contact point. The coordinates of the contact points then become the measurement data to be used to calculate dimensions, features, and feature relationships (Powaser, 1991; Genest, 1994). A touch trigger probe closes an electronic switch when the probe contacts the part, making it possible to automate the measurement process with direct computer controlled

(DCC) CMMs (Genest, 1994). In fact, the availability of the touch trigger probe, and the development of grating-based transducers with the ability to hold the displacement information at the instant of contact and computer software to process the measurement data, has resulted in a rapid growth in the use of CMMs (Butler, 1991). Since off-line programming does not tie up machines while developing inspection programs, it is highly desirable to have off-line programming capability in automated or computer-controlled CMMs. In addition, geometric product data information stored in the CAD model of the part can and should be used to facilitate inspection program development.

There are several technologies contributing to the success of the CMM applications. This section gives a general overview of CMM technologies, including basic CMM types, measuring capability, manual and computer-controlled CMMs, sensors and probes, measurement programs and software tools, linking CAD systems and CMMs, from QC labs to shop floors, and standards.

2.1. Basic CMM Types

No one CMM configuration is ideal for all dimensional measurement applications. Some of the more commonly used types of CMMs include moving bridge, fixed bridge, cantilever, gantry, column, horizontal arm, and measuring robot (ANSI, 1990; Genest, 1994). Factors in selecting a CMM configuration include operator access, part loading and unloading, machine geometric stability, accuracy requirements, weight of dynamic components, ease of maintenance, part capacity, sensitivity to part weight, measuring volume, and cost.

2.2. Measurement Capability

Accuracy and throughput are important indices of CMM measurement capability. Machining capability can only be verified by measurement instruments (Farago, 1982; New CMM ..., 1993). CMM manufacturers always try to improve accuracy of their products. Many techniques have been developed such as new CMM structures and microprocessor correction. Examples include reducing the spindle mass to increase measuring speed and using ceramic spindle while maintaining maximum mechanical stiffness (Inspection/Testing X, Y, and ..., 1990). Some CMM manufacturers claim that developments, in dynamic, unstressed structures and microprocessor correction of accuracy, are allowing creation of very high accuracy machine with greater environmental tolerance and fewer compromises in machine speed and measuring envelope size. It is true that the microprocessor-based accuracy compensation decreases the cost for higher level of accuracy, as compared to mechanically adjusted error correction.

As CMMs improve in accuracy and repeatability, temperature changes account for a greater portion of measurement uncertainty. It should be noted that thermally-indued errors are still one main error source in machine tool accuracy. In CMM applications, even 1 degree F difference between the part and the CMM can cause 30-millionths error in measurement readings. It is not uncommon to encounter 5 or 10 degree F differences, and it is enough to use up much of the machine's tolerance and give misleading information on process stability and part conformity. Microprocessor-based temperature compensation is now available for shop-floor

CMMs. It attempts to convert the measurement data to 68 degree F standard temperature (Placek, 1988).

Dynamic factors affecting CMM accuracy include measuring speed and acceleration setting, probe approach distance (some part features do not allow optimized probe approach distance), and the amount of extension of the CMM axes or components. An optimization strategy for maximizing CMM productivity has been developed at Ford Motor, and research results showed the trade-off between measurement speed and measurement quality (Jones, 1994a, 1994b).

Throughput is typically a critical productivity indicator on machines used for process verification in a production environment. A CMM's throughput is usually represented by using its point-per-second capability (Placek, 1988; Installing personal..., 1993). Inspection of complex workpieces usually requires dozens to hundreds of short, quick moves. The fastest touch trigger probe CMMs are capable of probing rates in excess of two measurement points-per-second. Some optical CMMs have shown the measuring speed of 10 to 100 times faster than traditional methods. Use of ceramics in the beam and quill has contributed to achieving acceleration rate of 150 mm/sec which enables higher throughput (Page, 1991).

2.3. Manual and Computer-Controlled CMMs

Large automotive manufacturers, such as GM, Ford, and Chrysler, are leading the way in automated inspection. Direct computer-controlled (DCC) CMMs offer major advantages over their manual counterparts in speed, accuracy, flexibility, labor saving, and system integration (Placek, 1992; Desktop manual..., 1993). Much advancesment has been made in this category, so the CMM users are finding it easier to cost-justify and upgrade from manual to automated machines. The inspection time can be cut by factors of at least 10:1, which provides faster feedback for process control. Programmed routines assure that the probes approach from precisely the same angles and distances part-to-part for greater accuracy and repeatability. This is almost impossible with operator setup and control of a manual CMM. DCC CMMs also permit the running of sophisticated, specialized software programs for profiling, scanning, digitizing, and gear measurement (Genest, 1994). Furthermore, DCC CMMs provide automatic information transfer capability to the manufacturing system in that process information (dimensional) can be acquired by CMMs and transmitted to a process control division (Grimes, 1991).

Although the trend is moving toward automated CMMs as their cost comes down, there is still a niche where manual machines make the most sense. This is especially true in a machine shop environment where many different parts are measured. Several factors justify the purchase. They include increased operator or inspection productivity for incoming, in-process, or final inspection, and lower scrap and rework (Manual or ..., 1994). There are machines with a drive system which is disengageable so that it can switch from the automatic mode to the manual mode at any time (Direct computer..., 1991).

2.4. Sensors and Probes

Sensors and probes are devices through which CMMs collect measurement data (Genest, 1994). Hard or fixed probes (ball, tapered plug, and edge probes) have been used for more than twenty years. The major shortcoming of these probes is that they require a certain amount of feel by the machine operator. They may result in inconsistency from one operator to the other. A common solution is the touch trigger probe, which is commonly used in manual and automated CMMs. Available with interchangable tips and extensions, the probe automatically closes an electronic switch when it contacts the part (Butler, 1991; Shams and Butler, 1991). Touch trigger probes have been used on four- and five-axis machine tools utilizing tilt-spindle designs. The probe has become one of the basic building blocks for supporting untended machining in manufacturing cells and systems (Herrin, 1992). They can be used to detect errors in part setup, part fixturing, improper tool use and tool wear, and inspection on the NC (numerical control) machine can catch these common contributors to errors (Five myths ..., 1992).

The most advanced probe system to date is the analog, continuous-scanning probe available on some ultra-precise CMMs. In the scanning mode, the probe follows the part's contours, even when a contour changes direction (Genest, 1994). Continuous scanning makes it possible to collect massive amounts of data from subtly variable geometries. Electro-optical non-contact sensors are being developed and used; they are regarded as the future trends of CMMs (Stovicek, 1991). Non-contact capacitance probes capable of 10 "hits" per second and high accuracy have been developed and are being integrated into standard CMMs (Fast non-touch ..., 1993). The idea is that capacitance probes, or an array of them, can capture surface profile information in-process; this data can be rapidly transferred to a machining process controller for continuous adjustment of machining parameters.

2.5. Measurement Programs and Software Tools

CMMs are capable of collecting large amounts of historical quality control data, and the data is useless without a means of turning that data into meaningful information (Sheehan, 1993). The new automated CMMs feature sophisticated software to sort, organize, and report data. Within the past decade, measurement software has evolved from laborious programming languages to layered packages of modular tools readily accessible by devices such as hierarchical menus and icons (Flexible CMM ..., 1993; Genest, 1994). The modular tools include standard operator interface, programming tools, statistical analysis and graphical tools, mathematical tools, and communication tools. Interactive graphics operator interface has been used to show how to measure the part manually or how to program the CMM for automatic measurement (Desktop manual ..., 1993). Nowadays, it is typical to find CMMs with layered software design. For example, the most basic layer contains a standard operator interface (symbols and procedures for operating a CMM), a second layer offers icon-based programming, and the deepest layer offers a core language for advanced users (Direct computer ..., 1991). This is a critical step in CMM evolvement in that user-friendly software environment is available to CMM operators. It improves productivity and enhances computer integration capability. Powerful measurement programs and software tools make it possible for CMMs to become an integral component in computer-integrated manufacturing systems.

2.6. Linking CAD Systems and CMMs

CAD/CAM systems have been widely used in industry, but manufacturing will not be fully integrated until the loop back to CAD/CAM is closed by CMMs (Zink, 1989). It is advantageous to write inspection procedures using the same database created by the CAD system. However, efforts are needed in linking CAD/CAM, flexible manufacturing system (FMS), and flexible inspection system (FIS). There is potential for major gains in productivity and quality by making computer-controlled machining and inspection key elements in CIM systems. It provides early detection and correction of manufacturing processes. The CMM can operate fully automated and untended or with relatively little attention from production line personnel (rather than quality control technicians). Using the CAD database, the inspection procedure can be created before the first part is actually made, shortening system start-up time, saving inspection programming time and costs, and insuring that parts conform to design tolerances (Sheehan, 1993).

The CMM crowd has largely solved the postprocessor problem that is still plaguing the NC crowd, thank to DMIS. DMIS allows developers of off-line CMM programming systems to write only one software program for converting their inspection routines into a format that any CMM will accept. DMIS defines that format. Likewise, CMM builders have to write only one software program to convert inspection routines from the DMIS format into the language their CMMs require – unless DMIS is the "native language" of the CMM, which is what the originators of DMIS hoped for all along (Albert, 1994). DMIS is a fully functioning interface between CAD and CMM, and provides a means of sharing information between the design and inspection process (Schreiber, 1990a). It enhances the ability to design quality into products.

DMIS Version 2.0 became an ANSI (American National Standards Institute) standard in February 1990. The standard enables users to pick a CMM suited for their needs without worrying about proprietary interfaces with CAD systems. Before DMIS, the interface between CAD systems and CMMs may need special efforts and maintenance to keep pace with new CMM technology and new CAD developments. DMIS is bi-directional. When the first part is created and goes to the CMM for measuring, inspection procedures flow from the CAD system to the CMM, and results flow back to the CAD system. There are two versions of DMIS; interactive and batch (Owen, 1991a). The batch version defines and implements a subset of DMIS that defines CMM commands to move and measure, report data, define part geometry, and execute tool change. For interactive version, it defines the communications protocol and error processing needed to run CMMs in a real-time, interactive mode.

2.7. From QC Lab to Shop Floors

CMMs traditionally are used for post-process inspection (Grimes, 1991). In recent years, the CMMs are being moved out of the quality control or metrology laboratories onto the production floor (Fix and Franck, 1989; Zink, 1989; Owen, 1991b; CMM combines ..., 1991). The need to perform process control, instead of product control, and to prevent defects is driving CMMs to the shop floor and into production roles. Caterpillar has put CMMs on the line with machine tools and under the control of the operators (Grimes, 1991). The need to verify the process with early gaging of critical characteristics means that CMMs must be located on the manufacture line, rather than off-line. In this case, the CMM is used for on-line process control. This capability is especially needed for small- to medium-batch manufacturing, since dedicated gaging systems may not be available, and CMMs provide an ideal solution for such situations. The CMM enables operators to inspect all parts for a fraction of the cost of dedicated air gages. It also provides the data needed to implement statistical methods of process verification. Shop-floor CMMs serve process control in two ways. First, they provide fast, accurate inspection feedback to machine tool operators for quick intervention in process correction. Secondly, they provide comprehensive data collection on parts produced for statistical process control (SPC) and management analysis (Bosch, 1991). Shop-hardened CMMs are developed for greater reliability and tolerance of production shop-floor environmental factors (Owen, 1991a). Shop-floor CMMs need to have an environmental enclosure to buffer temperature variations in the plant and keep contaminants out (Grimes, 1991). The enclosure is not air conditioned, so parts can be inspected at ambient temperatures, with measurement data automatically adjusted by the CMM's temperature compensation system.

It is reported that at GM and Ford plants, portable CMMs shuttle from job to job, and at a Chrysler plant CMMs sit next to the jobs they will measure. On today's factory floor, both portable CMMs and fixed-base (floor-mounted) CMMs are widely used. If the parts to be measured are small and can be carried, and where extreme accuracy is needed, fixed-base CMMs are recommended. This is mainly due to the fact that the portable CMM is exposed to varying temperatures and has difficulty in maintaining high accuracy. However, the portable CMMs have high flexibility and they have been increasingly used (Tool company..., 1992; Portable CMM, 1994). Several companies are using portable CMMs to maneuver from workstation to workstation while on the shop floor to obtain accurate measurement data in difficult areas without the problems of transporting large parts to the CMM and without the required setup times (Simon, 1991). The machine selection issue is important and needs to be considered in pursuit of automated dimensional inspection. The moving of CMMs to the shop floor symbolizes the beginning of an integrated quality effort.

2.8. Standards

The first CMM standard in the U.S. did not exist until 1985 when ANSI B89.1.12–1985 "Methods for Performance Evaluation of Coordinate Measuring Machines" was approved. This represents a critical point in CMM technology evolution. It clarifies the performance evaluation of CMMs and facilitates performance comparisons between machines. The standard was revised in 1990, and it contains four major sections: general machine specification, machine environmental specification and responses, machine performance, and subsystem performance (ANSI, 1990). Two industry standards commonly referenced by U.S. manufacturers of commercial CMMs are ASNI B89.1.12–1990, and the German standard VDI/VDE 2617 (Phillips, 1993).

3. CMM APPLICATIONS

CMMs represent a flexible and versatile dimensional measurement technology. Applications are summarized in this section.

3.1. Machine Setup Verification

CMMs are used to inspect the first part of a production run to verify the machine setup. Once the setup is verified, it then measures parts on a random basis (Wick and Veilleux, 1987). In this case, the CMM is placed at the end of the production line or in an inspection area.

3.2. Process Control

A CMM is an integral process control tool in a manufacturing cell or system; data about manufacturing processes can be collected on a corporate-wide basis to analyze processes statistically, to monitor tooling, and to do process control (Grimes, 1991). Dimensional data can be used to detect drift in a manufacturing process before a scrap is manufactured, and can also be used to institute corrective action (Genest, 1994). This is due to the fact that CMMs generate variable data instead of attribute data (pass/fail), which makes it possible to facilitate process control function and provide real-time process feedback for operators (Beard, 1990; Installing personal ..., 1993).

3.3. Flexible Gaging

CMMs offer flexible programming capability and can be used to accommodate engineering or design changes efficiently in the product development cycle.

3.4. Reserve Engineering

CMMs can be used to digitize parts, and download the resulting digitized data to CAD systems. Part models or drawings can be created in the CAD systems (Sheehan, 1993).

3.5. Manufacturing Capability Qualification

CMMs can be used to perform capability studies to show that manufacturing processes are dimensionally accurate (Genest, 1994). In addition, CMMs can be used to verify manufacturing equipment specifications in purchasing and testing equipment.

3.6. Gage Qualification

CMMs can be used to certify and recertify hand-held measurement tools and fixed gaging to make sure they can measure within specified levels of confidence (Genest, 1994).

3.7. Inspection of Contoured Parts

CMMs equipped with three-dimensional (3D) analog probes and specialized software packages with powerful analytical capability make them ideal for use in measuring complex contoured surfaces of parts. An analog probe is capable of gathering data at up to 50 points per second at spacing of less than 0.001 inch (0.025 mm) if desired (Sheehan, 1993; CMM cuts ..., 1994).

3.8. Quality Assurance Certification

The need to establish a quality assurance registration, according to ISO 9000 standards and increasingly stringent buyer specifications, have motivated manufacturers to purchase CMMs (Page, 1991). Part suppliers need to install CMMs and use them with a documented program of statistical process control (Genest, 1994).

4. CONCURRENT ENGINEERING ISSUES

The task of measuring must be integrated with process planning to achieve the quality and productivity objectives of the manufacturing organization (Bosch, 1992). Most computer-aided process planning (CAPP) research and development activities have concentrated on the manufacture of discrete-part products, and little attention has been devoted to process planning for dimensional inspection applications until recent years (Brown and Gyorog, 1990; Menq *et al.*, 1992b). Research work in CAD-directed inspection developed supports for using the geometric data in a CAD database to facilitate automated dimensional inspection (Hopp, 1982; Duffie *et al.*, 1988). However, additional information is needed to support automated dimensional inspection, and dimensional inspection planning presents a critical link between the islands of automation of CAD and CMM. The following technical and analytical supports are anticipated from the dimensional inspection planning system: definition of inspection scope, decomposition of tolerances to measurable surfaces, selection of appropriate inspection machine, selection of probes, design of sampling strategies, selection of fixtures, determination of inspection sequence and inspection path, verification of collision-free inspection, and generation of DMIS commands and inspection program. Expert systems make it possible to process the planning knowledge in the form of rules and facts, and can be applied to construct the process planning system (Alting and Zhang, 1989; Chang, 1990; Chang *et al.*, 1991; Warnecke and Muthsam, 1992). The advantages of such a planning system include: (1) the dimensional inspection plans are accurate and consistent, (2) the inspection productivity can be increased, (3) the inspection accuracy can be assured, and (4) inspection expertise can be captured in computer systems. Technical and analytical supports are being developed and used to facilitate dimensional inspection in concurrent engineering endeavors.

Process planning of CMM-based dimensional inspection has an impact on suggesting product definition information needed to support concurrent product and process (inspection) design. This is an effort devoted toward developing the complete product definition in computer-integrated manufacturing. Efforts in this area can be witnessed by efforts devoted in STEP (Standards for the Exchange of Product model data) and PDES (Product Data Exchange using STEP) (Carver and Bloom, 1991).

A framework of the inspection planning system for CMM-based dimensional measurement can be described by the following seven modules (Shen, 1993).

4.1. Define Inspection Scope

This module performs the following tasks: (1) identifies features which need to be inspected, (2) identifies datum reference frames and their associated features, (3) identifies non-dimensional features, and (4) identifies inspection accuracy requirements.

4.2. Decompose Inspection Features

This module performs the following tasks: (1) decomposes inspection features into surfaces, (2) decomposes datum reference features into surfaces, (3) groups inspection features into corresponding datum reference frames, and (4) determines inspection sequence (surface level).

4.3. Determine Inspection Machine, Probe, Part Setup, and Fixture

This module performs the following tasks: (1) selects inspection machine, (2) selects probes and probe configuraion, (3) determines parts setups, and (4) determines fixturing and clamping devices.

4.4. Formulate Inspection Strategy

This module perfoms the following tasks: (1) determines sampling strategy, (2) determines inspection sequence (point level), and (3) selects data analysis algorithms.

4.5. Determine Inspection Process Parameters

This module determines inspection process parameters such as machine operation mode (manual, automated, or interactive), probe travel speed, probe measuring speed, probe offset distance (before and after a specific sample), and probe qualification requirements.

4.6. Generate Inspection Process Plan and DMIS Inspection Program

Using the output from the aforementioned modules, knowledge bases, and inference procedures, the inspection process plan can be generated. The dimesional inspection process planning information is then used to generate an inspection program in compliance with the DMIS format.

4.7. Verify Dimensional Inspection Plan and DMIS Inspection Program

Verification of the plan is achieved by visually simulating the planned inspection operation in a CAD environment.

4.8. Recent Research and Development

Tasks specified in the above seven modules need research and development efforts that will provide analytical and technical support. Research and development in recent years is described here in order to assess the current status of concurrent engineering supports via CMM-based dimesional inspection. Hopp and Lau (1982, 1983) developed a control system architecture for a CAD-directed inspection project at the National Institute of Standards and Technology (NIST), and a conceptual framework was proposed. EIMaraghy and Gu (1987) developed an inspection planning system based on expert systems. This is one of the first significant efforts directed toward automated dimensional inspection planning. Examples were given about the inspection knowledge representation and planning logic. The system is mainly applied to simple cylindrical parts. Menq *et al.* (1992 a) developed an optimal match scheme that aligns the measurement data with the design data in CAD-directed dimensional inspection of sculptured surfaces. An algorithm with computational efficiency and robustness against singularity was developed. Menq *et al.* (1992b) also developed an intelligent planning environment for automated dimensional inspection using CMMs. It contains three levels of automation technology, including facility, information, and decision automation. The planner is a knowledge based system which utilizes artificial intelligence to automate the decision making in inspection planning. Merat and Radack (1992) developed an automated inspection planning system for CMMs in a feature-based design environment. The inspection planner uses inspection code fragments (ICFs) associated with various individual features, and pieces them together to form a complete inspection plan. The system is compatible with ANSI Y 14.5 and DMIS. It demonstrated promise for fast generation of reasonably efficient inspection plans suitable for small lot size manufacturing applications.

Pham *et al.* (1991) developed a knowledge-based software package (a preprocessor) for inspection using CMMs. The package contains two main modules: function definition and program synthesis. They are used to preprocess CAD system output information into a neutral-data-file format in support of off-line programming of CMMs. The research incorporated expert system methodology in its development. Lim and Menq (1994) studied probe accessibility and inspection path generation in CMM applications. An algorithm was developed to determine all feasible probe orientations for the inspection task using a touch-trigger probe, and the feasible probe orientations ensure that there is no collision with the probe or the probe stylus when the probe tip is in contact with the workpiece. The alogirthm handles internal holes and complex surfaces. In addition, a heuristic method was developed to determine feasible probe orientations that can pick the optimal probe orientation and create a collision-free inspection path. Lu *et al.* (1994) studied the optimum collision free inspection path to improve the CMM throughput. A modified 3D ray tracing technique is applied to an octree database of a CMM configuration space to detect obstacles between any two target points. The developed algorithm is capable of eliminating the dynamically undersirable characteristics of octree based algorithms and saves searching time in congested work spaces by finding paths around colliding objects. Spyridi and Requicha (1993) developed a planner capable of generating automatically the instructions needed to drive a CMM, so as to inspect a set of specified surface features of a part, and to determine if the part satisfies its tolerance specifications. The upper-level planner deals with part setups, probes, and the machine to be used. The lower-level planner deals with sampling (measurement) point selection and collision-free path generation.

Brown (1993) reported a feature-based tolerancing capability that complements a geometric solid model with an explicit representation of conventional and geometric tolerances. The capability is focused on supporting an intelligent inspection process definition system. The benefits of the model include: (1) advancing complete product definition initiatives (STEP-Standard for Exchange of Product model data), (2) supplying computer-integrated manufacturing applications (including inspection planning and CMM part program generation) with product definition information, and (3) assisting in the solution of measurement performance issues. There are commercial software packages which can be used to generate inspection programs in DMIS format by taking into account the product geometry database and simulating the inspection sequence to check for collisions and near misses (Beard, 1990; CMM inspects ..., 1994).

5. CHALLENGES AND RESEARCH NEEDS

Coordinate measuring machines have been widely used by industry as dimensional measurement tools. Advances in supporting technologies have improved CMM technology and measurement processes in recent years. However, there are still critical challenges in CMM technology. They are accuracy in measurement results, sampling strategies, computational metrology, and geometric dimensioning and tolerancing (GD&T).

5.1. Accuracy and Measurement Uncertainty

Although CMMs provide fast and accurate dimensional measurements, accuracy is a critical aspect of all CMM applications (Swyt, 1992; Shen, 1993). The accuracy issue is particularly important since many CMMs are being used on the shop floor for process control. Errors in the CMM measurement process include machine errors (including geometric errors, machine dynamics errors, thermal errors, and load-induced errors), probe errors, and algorithm errors (to be discussed later) (Hocken *et al.*, 1993; Phillips *et al.*, 1993). The challenge needs to be addressed from two perspectives: accuracy enhancement and measurement uncertainty evaluation. Accuracy enhancement attempts to improve the accuracy of the measurement process by improving machine design, precision manufacturing, probe design, and error compensation. It is aimed to reduce effects of error sources and thus enhance the accuracy of the CMM measurement process. Measurement uncertainty evaluation is another critical factor for CMM technology, since it is typically required that measurement uncertainty be less than the tolerance by a fixed fraction of the tolerance (Kalpakjian, 1992; Phillips *et al.*, 1994).

CMM accuracy enhancement is one of the goals of the research work in the precision engineering community. Typically three methods are used to evaluate performance of CMMs.

(1) Parametric calibration: based on measuring various error terms independently; for a three-axis machine, twenty-one errors are measured at various locations in the working volume (Hocken *et al.*, 1977; Zhang *et al.*, 1985).

(2) Measuring volumetric errors with kinematic reference standards, such as ball, disk, etc. (Kunzmann, 1993).

(3) Measuring standard artifacts (Kunzmann, 1990), has the advantage of the close replication of real measurements; but it takes special efforts to maintain and certify the artifact, and it is difficult to determine error sources and their magnitudes.

Existing standards in many countries use parametric checking, kinematic reference standards, and artifacts or a combination of these methods. Error compensation attempts to model the measurement process and measure or predict the errors. The errors are then eliminated or reduced by means such as software compensation. In 1985, an error compensation method for CMMs was designed, and a factor of ten improvement in accuracy was obtained at NIST (National Institute of Standards and Technology) (Zhang *et al.*, 1985). Today nearly every measuring machine being produced uses the Microprocessor Enhanced Accuracy (MEA) techniques developed at NIST (Bosch, 1994). However it should be noted that the effectiveness of error compensation depends on whether error modeling and prediction can be done correctly and efficiently under various working conditions.

Measurement uncertainty of a CMM represents the capability of a measurement instrument. Once a CMM has been calibrated, it is necessary to periodically check to see if the calibration is still valid (Phillips, 1993). Such an interim test should include as many components of the measurement system as possible. Ball bars are typically used in such processes (ANSI, 1992). It is obvious that equipment and techniques which can rapidly assess the performance of CMMs are needed. NIST has developed a novel interim testing artifact which allows the frequent testing of CMMs to ensure that they measure parts correctly (Phillips *et al.*, 1994). The interim testing artifact is a critical technological innovation since its application increases confidence in CMM performance between CMM calibrations.

Kunzmann *et al.* (1993) at the PTB (Physikalisch-Technische Bundesanstalt) in Germany proposed a virtual CMM method to calibrate CMMs in a sense that measurements are performed virtually by computer simulations by taking into account the CMM's error (geometric, probe, and software) (Schwenke *et al.*, 1994). The concept seems to have a significant impact on CMM technology. It should be noted that it is essential that all errors be quantified correctly and combined properly to produce an assessment of measurement uncertainties in all CMM tasks. Ni *et al.* (1992) developed and implemented a multi-degree-of-freedom measuring system for CMM geometric errors. Precision measurement of straightness, pitch, yaw, and roll errors of the moving axes of a CMM can be simultaneouly performed. The sytem is based on the principles of laser alignment and autocollimator. Chen *et al.* (1993) developed an error compensation system to enhance the time-variant volumetric accuracy of a three-axis machining center. A methodology has been developed to a synthesize both the geometric and thermal errors of machines into a volumetric error model. Sensing, metrology, and computer control techniques have been applied.

The errors introduced by probes are very significant, often exceed the errors from other sources, and they remain largely uncompensated (Bulter, 1991). Due to the part feature characteristics, a touch trigger probe with a long stylus may be needed for deep features such as a deep hole. In such cases, the long stylus probe usually produces larger errors as compared to probes with shorter styli. Other factors affecting probing accuracy include measuring speed, probe approach direction, probe configuration, and probing force setting, etc. Innovative ideas and vigorous research efforts are needed to effectively enhance accuracy and assess measurement uncertainties of CMM applications.

5.2. Sampling Strategy

Sampling strategy can be defined as the number of points (samples) and the spatial distribution of these points on the part surface from which the feature is being inspected. Sampling strategy needs to be synthesized before the dimensional inspection and it definitely has significant impact on measurement results. Usually a relatively small number of samples are measured to evaluate a part feature in practical CMM applications since (1) the number of samples represents time and cost, and (2) the consequences of such "under-sample" approaches are not known. The origin of the sampling issue is that the measurement is performed in the presence of CMM errors and form errors of parts in order to achieve the goal of verifying tolerances specified on the part features (Hocken *et al.*, 1993). It is obvious that the sampling strategy depends on part design tolerances, manufacturing capabilities, and measurement capabilities. However, form errors associated with different manufacturing processes are usually unknown, and CMM errors at each measurement point are usually not well quantified. The most obvious way to examine the effects of sampling in the presence of form and machine errors is to simply acquire a large number of parts, measure them using different sampling strategies, and compare the results. Another common approach is to use computer experiments or simulations. Hocken *et al.* (1993) studied the computer simulation approach using features such as lines, planes, circles, and cylinders. In the computer simulation, systematic and random errors are generated and added to the perfect feature. The feature is then sampled according to different strategies and with different data densities. From this research, it is found that the current inspection techniques, used daily in manufacturing, drastically under-sample geometric features in the presence of unknown part form and measuring machine errors. The finding leads to two corollaries:

(1) much higher sampling densities than currently used must be incorporated (inspection time then increased) and a new generation of machines capable of high-speed scanning of part surfaces (if inspection time not to be drastically increased) are needed, and
(2) decision regarding how to measure a part and the choice of algorithm(s) for data-analysis are often counter-intuitive. This leads to the conclusion that intelligent decision systems or procedures need to be used to control the inspection and analysis process.

Hocken *et al.* (1993) also suggested that a great deal of work still needs to be done. The problems include partial features (to measure the radius of a sphere or toroid when only a small section of the surface has been produced) and interrupted features (such as cylinders with key slots).

It should be noted that modeling of manufacturing error pattern of part surfaces (with which the part form errors are associated) is one critical aspect in suggesting effective sampling strategies in coordinate metrology. In this regard, knowledge of form error patterns associated with various manufacturing processes needs to be captured, and modeling methodologies capable of characterizing part form errors need to be developed in order to effectively handle the sampling strategy issue in coordinate metrology. Research work to date uses Fourier series and random noise to characterize part surfaces, so computer simulation can be carried out to ascertain the effectiveness of various sampling strategy designs with form errors included (Cakey *et al.*, 1991; Chang and Lin, 1993). Menq *et al.* (1992) developed a statistical model to determine the minimum number of required measurement points so that the sampled points could closely represent the entire population based on manufacturing process capability and tolerance specification in surface profile inspection applications.

From the concurrent engineering viewpoint, sampling strategy is truly an issue that design, manufacturing, and metrology engineers need to consider and collaborate on in the product development cycle.

5.3. Computational Metrology

Data analysis software is needed to compute the substitute geometry based on coordinate data measured by CMMs. This is a major source of error in a measurement system. Factors affecting software performance include the choice of analysis method, the quality of the software, and characteristics of the specific measurement task (Hocken *et al.*, 1993; Hopp, 1993). Data analysis software packages are typically proprietary software provided by CMM manufacturers, and consistent variability among data analysis software packages has been reported (Porta and Waldele, 1986; Waldele *et al.*, 1993; NIST, 1994). In fact, the data analysis software issue is not new. It has been reported in a Government-Industry Data Exchange Program (GIDEP) alert (GIDEP, 1988), and known as methods of divergence (Schreiber, 1990b). Since most CMMs perform least-squares fits, in some cases the algorithms selected would not give the correct feature size (Hocken *et al.*, 1993). This is certainly an issue closely related to the concurrent engineering endeavor, since design engineers and metrology engineers need to understand the consequences of selecting certain fitting algorithms in dimensional inspection tasks. Certification of CMM software needs to be addressed. CMM users are becoming aware of the importance of software certification in CMM applications, and more efforts are needed in this area (Charlton and Walker, 1993). Theories, tools, and standards in computational metrology are beginning to emerge such that improved utility of data analysis software is anticipated (Hopp, 1993).

5.4. Geometric Dimensioning and Tolerancing

Geometric dimensioning and tolerancing (GT&T) is the basis engineers use to present design intentions, to design manufacturing process plans and production plans, and to design dimensional inspection plans. GD&T is used to define dimensions of part features and their relationships, and is used throughout the production cycle. However, the dimensioning and tolerancing issue has been identified as one of the technological challenges facing the manufacturing community, and more research is needed (Tipnis, 1990). The issue can be called "specification ambiguity," i.e., ambiguity in the definition of what is to be measured

(Voelcker and Srinivasan, 1993). The main problem is that current dimensioning and tolerancing standards are ambiguous in representing dimensions and tolerances, particularly when coordinate measurement instrumentation is used to verify dimensions and tolerances. A new American national standard, Y14.5.1M-1993 "Mathematical Definition of Dimensioning and Tolerancing Principles" has been approved (ANSI, 1993), which provides a mathematical basis for dimensioning and tolerancing specification. In addition, a major revision to the GD&T standard, ANSI Y14.5, is currently being finalized (approved in 1994) (Krulikoski, 1994). Nowadays, more CMM users and manufacturers understand GD&T, and CAD vendors are beginning to integrate Y14.5 into their products (Charlton and Walker, 1993). It is obvious that tolerancing and metrology are closely related and that advances in these areas are needed for CMM technology to fully support the concurrent engineering endeavor.

6. CONCLUSIONS

The evolution of several technologies contributing to CMM success have made CMMs a key measurement tool in modern manufacturing settings. CMMs are versatile in that they can be applied in various situations and provide fast and accurate dimensional measurement functions. This chapter discusses supports needed to incorporate CMMs in concurrent engineering endeavors. A framework of a dimensional inspection planning system is described and a general survey of research and technological advancement within this framework has been given. Challenges and research needs essential for the support of concurrent engineering endeavor are identified. It should be noted that these challenges are interrelated. Vigorous and systematic research efforts are needed to develop sound technical and analytical supports for concurrent engineering endeavors.

References

M. Albert, Don't let DMIS die!, *Modern Machine Shop*, **66**(4), p. 8 (1994).

L. Alting and H. Zhang, Computer-aided process planning: the state-of-the-art survey, *International Journal of Production Research,* **27**(4), pp. 553–585 (1989).

ANSI/ASME B89.1.12M-1990 *Methods for Perfonmance Evaluation of coordinate Measuring Machines*, ASME, New York (1990).

ANSI/ASME Y14.5.1-1993 *Mathematical Definition of Dimensioning and Tolerancing*, ASME, New York (1993).

T. Beard, An integrated approach to quality, *Modern Machine Shop*, **62**(3), pp. 52–61 (1990).

D. Beutel, The practioner's view of measurement system application, *Proceedings of the International Fourm on Dimensional Tolerancing and Metrology*, pp. 119–130 (1993).

J. Bosch, Trends in measurement, *Automation* (3/91), pp. 54–55 (1991).

J. Bosch, The changing roles of coordinate measuring machines, *Industrial Engineering*, (11/92), pp. 46–48 (1992).

J. Bosch, On-line compensation of measuring machines, *Proceedings of the First S.M. Wu symposium on Manufacturing Science*, Evanston, IL (May 1994), pp. 227–229 (1994).

C. Brown and D. Gyorog, Generative inspection process planner for integrated production. *Proceedings of the ASME 1990 Winter Annual Meeting*, pp. 151–162 (1990).

C. Brown, Feature-based tolerencing for intelligent inspection process definition, *Proceedings of the International Fourm on Dimensional Tolerancing and Metrology*, pp. 249–258 (1993).

C. Bulter, An investigation into the perfomance of probes on cordinate measuring machines, *Industrial Metrology*, **2**, pp. 59–70 (1991).

CAMI: Computer-Integrated Manufacturing-International, Inc. *Dimensional Mesuring Interface Specification*, Version 2.1, R-89-DMB-01 (1989).

G.P. Carver and H.M. Bloom, *Concurrent Engineering Through Product Data Standards*, NISTIR 4573, National Institute of Standards and Technolgy, Gaithersburg, MD (1991).
G. Caskey *et al.*, Samling techniques for coordinate measuring machines, *Proceedings of the NSF Design and Manufacturing Systems Conference*, Austin, Texas, pp. 779–786 (1991).
H. Chang and T. Lin, Evaluation of circularity tolerance using Monte Carlo simulation for coordinate measuring machine, *International Journal of Production Research*, **31**(9), pp. 2079–2086 (1992).
T. Chang, *Expert Process Planning for Manufacturing*, Addision-Westly Publishing, Massachusetts (1990).
T. Chang, R. Wysk and H. Wang, *Computer-Aided Manufacturing*, Prentice Hall, New Jersey (1991).
T. Charlton and R. Walker, Session 8: Questions and discussion, *Proceedings of the International Fourm on Dimesional Tolerancing and Metrology*, pp. 261–267 (1993).
F.F. Chen Concurrent cell design and cell control system configuration, in *Concurrent Engineering: Contemporary Issues and Modern Design Tool*, (eds H.R. Parsaei and W.G. Sullivan), Chapman & Hall, London, pp. 231–247 (1993).
F.F. Chen and S.R. Sagi, A neural-net based methodology for concurrent design, planning, and control of manufacturing cells, *Proceedings of the Third Industrial Engineering Research Conference*, Atlanta, GA, USA, pp. 520–525 (1994).
J. Chen, J. Yuan, J. Ni and S. Wu, Real-time compensation for time-variant. volumetric errors on a matching center, *Journal of Engineering for Industry*, **115**(4), pp. 472–479 (1993).
CMM combines laboratory precision with production capability, *Modern Machine Shop*, **63** (9/90), pp. 150–152 (1991).
CMM cuts airfoil inspection time for Georgia manufacturer, *Modern Machine Shop*, **67**(11), p. 140 (1994).
CMM inspects directly from CAD model, *American Machinist*, **138**(11), p. 54 (1994).
Desktop manual to desktop DCC CMM, makes for easy transition, *Modern Machine Shop*, **65**(10), pp. 112–116 (1991).
Direct computer control added to desktop CMM *Modern Machine shop*, **63**(4), pp. 140–146 (1991).
N. Duffi, S. Feng and J. Kann, CAD-directed inspection, error analysis, and manufacturing process compensation using trucubic solid database, *Annals of the CIRP*, **37**(1), pp. 149–152 (1988).
H. ElMaraghy and P. Gu, Expert system for inspection planning. *Annals of the CIRP*, **36**(1), pp. 85–89 (1987).
F.T. Farago, *Handbook of Dimensional Measurement*, 2nd edn, Industrial Press, New York (1982).
Fast non-touch probes for CMM, *American Machinist*, **137**(5), pp. 12 (1993).
Five myths about touch probes on NC machines, *Modern Machine Shop*, **64**(4), pp. 62–65 (1992).
S. Fix and G. Franck, Process control: emerging role for CMM, *Automation* (4/89), pp. 30–32 (1989).
Flexible CMM makes quick saving for die-caster, *Modern Machine Shop*, **65**(4), pp. 127–128 (1993).
D. Genest, Coordinate measuring machines (CMMs), *Quality in Manufacturing*, **5**(2), pp. 21–23 (1994).
GIDEP: Government and Industry Data Exchange Program (GIDEP) *Alert No. X1A–88–01* (1988).
C. Grimes, Shop floor CMM is key to process control, *Modern Machine Shop*, **63**(5), pp. 86–88 (1991).
G. Herrin, Five axis probing, *Modern Machine Shop*, **65**(8), pp. 66–68 (1992).
R. Hocken, Three dimensional metrology, *Annals of the CIRP*, **26**(2), pp. 403–408 (1977).
R. Hocken, J. Raja and U. Babu, Sampling issues in coordinate metrology, *Proceedings of the International Fourm on Dimensional Tolerancing and Metrology*, pp. 97–111 (1993).
T. Hopp, CAD-directed inspection, *Annals of the CIRP*, **31**(1) (1982).
T. Hopp, A hierachical model-based control system for inspection, *ASTM STP 862*, (ed L.B. Gardner), American Society for Testing and Materials, Philadelphia, PA, pp. 169–187 (1983).
Inspection/Testing/Quality Control *Modern Machine Shop*, **63**(8), p. 470 (1990).
Installing personal flexible gages gets manufacturing department's attention, *Modern Machine Shop*, **65**(4), pp. 120–122 (1993).
S. Jones and A. Ulsoy. An optimization strategy for maximizing coordinate measuring machine productivity: part I: quantifying the effects of the operating speed on measurement quality, *Proceedings of the First S.M. Wu Symposium on Manufacturing Science*, Evanston, IL (May 1994), pp. 211–218 (1994).
S. Jones and A. Ulsoy, An optimization strategy for maximizing coordinate measuring machine productivity: part II: problem formulation, solution, and exprimental results, *Proceeings of the First S.M. Wu Symposium on Manufacturing Science*, Evanston, IL (May 1994), pp. 219–226 (1994).
S. Kalpakjian, *Manufacturing Engineering and Technology*, 2nd edn, Addison Wesley Publishing, Massachusetts (1992).
A. Krulikowski, GD&T challenges the fast draw, *Manufacturing Engineering* (92/94), pp. 2–31 (1994).
H. Kunzmann Performance of CMMs, *Annals of the CIRP*, **32**(1), pp. 633–640 (1983).
H. Kunzmann, A uniform concept for calibration, acceptance test, and periodic inspection of coordinated measuring machines using reference objects, *Annals of the CIRP*, **39**(1), pp. 561–564 (1990).
H. Kunzmann, *et al.*, Concept for the traceability of measurements with coordinate measuring machines, *Seventh International Precision Engineering Seminar*, Kobe, Japan, pp. 1–14 (1993).

C. Lim and C. Menq, CMM feature accessibility and path generation, *Internationl Journal of Production Research*, **32**(3), pp. 597–618 (1994).

E. Lu, J. Ni and S. Wu, An algorithm for the generation of an optimum CMM inspection path, *Journal of Dynamic Systems, Measurement and Control*, **116**(4), pp. 396–404 (1994).

Manual or automatic CMM–which is best? *Modern Machine Shop*, **67**(11), p. 160 (1994).

C. Menq, H. Yau and G. Lai, Automated precision measurement of surface profile in CAD-directed inspection, *IEEE Transactions on Robotics and Automation*, **8**(2), pp. 268–278 (1992).

C. Menq, H. Yau and C. Wong, An intelligent planning environment for automated dimensional inspection using coordiate measuring machines, *ASME Journal of Engineering for Industry*, **114**(2), pp. 222–230 (1992).

F. Merat and G. Radack, Automatic inspection planning within a feature-based CAD system, *Journal of Robotics and Computer–Integrated Manufacturing*, **9**(1), pp. 61–69 (1992).

New CMM handles typical job shop workpices, *Modern Machine Shop* **65**(4), pp. 138–140 (1993).

J. Ni, P. Huang and S. Wu, A multi-degree freedom measuring system for CMM geometric errors, *Journal of Engineering for Industry*, **114**(3), pp. 369–362 (1992).

NIST *Manufacturing Engineering Laboratory*, National Institute of Standards and Technology, Gaithersburg, Maryland (1994).

J. Owen, CMMs for process control, *Manufacturing Engineering*, **(8/91)**, pp. 39–41 (1991).

J. Owen, CMMs on the shop floor, *Manufacturing Engineering*, **(4/91)**, pp. 66–70 (1991).

M. Page, Over-croweded in the CMM market, *Modern Machine Shop*, **63**(1), pp. 50–52 (1991).

D. Pham, K. Martin and L. Khoo, A knowledge-based preprocessor generator for coordinate measuring machines, *International Journal of Production Research*, **29**(4), pp. 677–694 (1991).

S. Phillips, A comparison of national standards for the performance evaluation of coordinate measuring machines in terms of length-based dimensional quantities, *SME Precision Metrology of CMMs Workshop*, Nashville, TN (1993).

S. Phillips, B. Borchardt and G. Caskey, *Measurement Uncertainty Considerations for Coordinate Measuring Machines*, NISTIR 5170, National Institute of Standards and Technology, Gaithersburg, MD (1993).

S. Phillips, *et al.*, A novel CMM interim testing artifact, *Proceedings of the 1994 Measurement Science Conference* (1994).

C. Placek, The state-of-the-CMM-art, *Quality* (8/88), p. Q-5 (1988).

C. Placek, CNC/DCC models boost benchtop CMM sales, *Quality* (6/92), pp. 47–50 (1992).

C. Porta and F. Waldele, *Field Testing of Three Coordinate Measuring Machine Evaluation Algorithms*, Physikalisch Technische Bundesanstalt (1986).

Portable CMM arm puts muscle in design of trimming dies, *American Machinist*, **138** (10/94), p. 20 (1994).

D. Powaser, Linking CAD with CMM, *Automation* (4/91), pp. 52–53 (1991).

G. Robison, Ultra-precise CMMs, *Tooling & Production* (2/90), pp. 92–94 (1990).

R. Schreiber, CMMS: Traits, trends, triumphs, *Manufacturing Engineering* (4/90), pp. 31–37 (1990).

R. Schreiber, The methods divergence dilemma, *Manufacturing Engineering* (5/90), p. 10 (1990).

H. Schwenke, E. Trapet, F. Waldele and U. Weigand, Experience with the error assessment of coordinate measurement by simulation, *Proceedings of the Third International Conference on Ultraprecision in Manufacturing* (eds M. Weck and H. Kunzmann), Aachen, German, pp. 370–373 (1994).

I. Shams and C. Butler, Performance of an opto-electronic probe used with coordinate measuring machines, *SPIE*, **1589**, pp. 120–124 (1991).

K. Sheehan, Measurement verifies machine performance, *Modern Machine Shop*, **66** (10), pp. 76–83 (1993).

Y. Shen, Automated dimensional inspection using coordinate measuring machines, *Proceedings of the 1993 US/Taiwan Joint Automation and Productivity for Small to Medium Scale Manufacturing Industry Workshop*, Taipei, Taiwan, pp. 10:1–12 (1993).

L. Simon, CMMs go portable, *Tooling & Production* (10/91), pp. 42–43 (1991).

A. Spyridi and A.A.G. Requicha, Automatic planning for dimensional inspection, *Proceedings of the International Forum on Dimensional Tolerancing and Metrology*, pp. 219–228 (1993).

D. Stovicek, Precision vision, *Automation* (5/91), pp. 22–23 (1991).

D.A. Swyt, *Challenges to NIST in Dimensional Metrology: The Impact of Tightening Tolerances in the U.S. Discrete-Part Manufacturing Industry*, NISTIR 4757, National Institute of Standards and Technology, Gaithersburg, MD (1992).

V. Tipnis (ed) *Research Needs and Technological Opportunities in Mechanical Tolerancing*, ASME, New York (1990).

Tool company solves fixture inspection problem with portable CMM *Modern Machine Shop*, **64** (1), pp. 128–129 (1992).

H. Voelcker and V. Srinivasan, Preface, *Proceedings of the International Forum on Dimensional Tolerancing and Metrology*, pp. iii–v (1993).

F. Waldele, B. Buittner, K. Busch, R. Drieschner and R. Elligsen, Testing of coordinate measuring machine software, *Precision Engineering*, **15**(2), pp. 121–123 (1993).

H. Warnecke and H. Muthsam, Knowledge-based systems for process planning in *Intelligent Design and Manufacturing* (ed A.Kusiak), John Wiley & Sons, New York, pp. 377–396 (1992).

C. Wick and R. Veilleux (eds) *Tool and Manufacturing Engineering Handbook, Vol.4: Quality Control and Assembly*, SME, Michigan (1987).

G. Zhang, R. Veale, T. Charlton, B. Borchardt and R. Hocken, Error compensation of coordinate measurement machines, *Annals of the CIRP*, **34**(1), pp. 445–448 (1985).

J. Zink, Closing the CIM loop with CMM, *Automation* (1/89), pp. 49–50 (1989).

LIST OF ABBREVIATIONS AND ACRONYMS

ANSI	American National Standards Institute
CAD	computer-aided design
CAM	computer-aided manufacturing
CAMI	computer-Integrated Manufacturing-International, Inc.
CAPP	computer-aided process planning
CIM	computer-integrated manufacturing
CMM	coordinate measuring machine
DCC	direct computer control
DME	dimensional measuring equipment
DMIS	dimensional measuring interface standard
FIS	flexible inspection system
FMS	flexible manufacturing system
GD&T	geometric dimensioning and tolerancing
ICF	inspection code fragment
NC	numeric control
NIST	National Institute of Standards and Technology
PDES	Product Data Exchange using STEP
PTB	Physikalisch-Technische Bundesanstalt
QC	quality control
SPC	statistical process control
STEP	Standards for the Exchange of Product model data

CHAPTER 9

Intelligent Support Systems for Concurrent Product and Process Design

D. T. P HAM and S. S. DIMOV

Intelligent Systems Research Laboratory, Cardiff School of Engineering, University of Wales Cardiff, Cardiff CF2 1XH, United Kingdom

1. INTRODUCTION

The demand for higher quality, on-time delivery, and lower cost products with shorter design and manufacturing lead time for the dynamic global market is forcing companies to introduce new product and process design strategies. The main aim of these new design strategies is to reduce significantly the time from design concept to manufacture by applying a concurrent rather than sequential approach to the various product and process design activities. The full realisation of this concurrent design approach is a challenging task that requires an indepth understanding of the designer's decision making logic and comprehensive models describing the information flow between concurrent activities in product and process design. The concurrent design methodology raises two major issues that need to be addressed.

The first of these relates to the chosen design representation scheme which should be capable of capturing the designer's intents and of being used for various downstream applications [Case *et al.*, 1994; Krause *et al.*, 1993]. It is widely recognised that a feature-based design model provides the ability to represent the product using geometrical and functional information and allows different applications to access the model and have their own "interpretation" of the data. Features, such as geometrical primitives, can be used to build the model and simultaneously can play the role of a communication medium between product and process design.

The second issue relates to the provision of intelligent system support for concurrent product and process design. This involves the identification of the required information to support the different design activities and the determination of adequate techniques for information selection and provision. The designer is, through his decisions, responsible for 70–80% of the product cost [Suh, 1988] but during the design process there is no access to existing knowledge about the manufacturing environment which could, to a large extent, facilitate and direct his decision making activities. The decision logic employed in concurrent design is primarily based on simultaneously matching the product's functional requirements to the component's technological requirements and the technological requirements to the manufacturing capabilities [Krause *et al.*, 1993; Sohlenius, 1992; Peters *et al.*, 1990]. This chapter is mainly concerned with reasoning about component technological requirements and manufacturing system capabilities, and more specifically, with the development of decision support systems for selecting and providing manufacturing information (MI) to help the designer in taking downstream manufacturing concerns into careful consideration. Through the application of artificial intelligence (AI) techniques, it is possible and realistic to acquire, structure and represent knowledge about previous manufacturing cases and past process planning decisions and use the acquired knowledge for intelligent design support. In this chapter, emphasis will be given to the use of AI techniques in the development of design decision support systems.

The next section briefly reviews some of the prototype systems for concurrent design support.

2. EXISTING PROTOTYPE SYSTEMS FOR CONCURRENT DESIGN SUPPORT

Recently, the development of design support systems has attracted the attention of researchers and is considered a key to the successful implementation of the concurrent design methodology [Krause *et al.*, 1993]. These systems are expected to minimise unnecessary iterations during design.

In the ESOP project [Lenau and Alting, 1989], a design support system for selection of surface treatment processes was developed. Taking as an input the product design characteristics and applying its stored knowledge about surface treatment, the system suggests suitable processes together with advantages and disadvantages for each process. The system was built using an expert system development environment, where knowledge is represented as frames and rules.

Another system, called First-Cut [Cutkosky and Tenenbaum, 1990], was constructed to support detail design and high-level process planning. These are tasks that reside at the interface between design and manufacturing. The features of First-Cut are:

- Manufacturing engineering is done as the design takes shape, instead of after it has been completed;
- The design/manufacturing process takes place in a highly interactive, distributed environment consisting of teams of human and computer experts.

During the design process, manufacturing modes that capture process characteristics at an abstract level allow the designer to be kept continually in touch with manufacturing processes. Working in these "manufacturing modes" is similar to designing by features, except that features are promptly translated into process plans. In First-Cut, design for manufacturability is foremost a question of assuring that any part designed on the system can actually be fabricated with available machines, tools and fixtures.

An opportunistic approach to process planning within a concurrent engineering environment has been described in [Herman *et al.*, 1993]. The paper presents a Computer-Aided Process Planning (CAPP) system, called XTURN, and a decision support environment, IDEEA, on which the system is based. XTURN enables the designer to explore alternative process plans at an early design stage. The designer controls process selection strategies, while XTURN determines the interactions between the various choices, identifies those that are inconsistent, and offers advice on resolving any inconsistencies. The system is implemented within the IDEEA environment which integrates various computer tools such as solid modelling packages, finite element analysis, database storage packages, and database storage schemes with multiple knowledge representations and reasoning paradigms.

A blackboard approach has been used in a system called CIMROT [Domazet and Lu, 1993] to integrate product and process design. On this system for concurrent design and process planning of rotational parts, the designer can perform a manufacturability analysis. The analysis is based on a generated process plan and an actual process data inventory. CIMROT can detect such inconsistencies as the unavailability of the required tools or machines, or the use of an extremely costly manufacturing operation and automatically sends warning messages to the designer. The system process planner module XROT has been implemented using the same environment (IDEEA) as the XTURN system.

Abdalla and Knight [Abdalla and Knight, 1994] presented an approach for concurrent product and process design of mechanical parts. This approach enables designers to ensure that the product can be made within the existing manufacturing environment. The proposed concurrent design environment is composed of integrated expert and CAD systems. The expert system contains extensive information about product features and manufacturing facilities which allows the designer to check the process limits and product feasibility and also to detect design inconsistencies.

As the above examples have indicated, work in the area of concurrent product and process design is varied in terms of application domains and the modelling techniques adopted. It is fair to state that, although impressive achievements have been demonstrated, a comprehensive and systematic approach is still lacking.

3. ARTIFICIAL INTELLIGENCE TECHNIQUES FOR KNOWLEDGE ACQUISITION

As larger and more complex concurrent design applications are developed, the introduction of new approaches for automatic knowledge acquisition becomes necessary for manufacturing information (MI) processing and for the creation of practical design decision support systems.

Examining the manufacturing data related to the planning, machining, measuring, and assembling procedures specific to a company leads to the conclusion that there are many different sources of knowledge, for example, raw data stored in databases, simulation data, expert decisions, etc. In spite of the diversity and complexity of the available MI, approaches have to be sought to process, represent and utilise it. One possible way is to employ machine learning techniques developed from artificial intelligence research.

Machine learning can be used to process and represent valuable data in a knowledge base (KB). KB models can facilitate some concurrent design activities that need qualitative assessment as well as those requiring exact, quantitative evaluations. The structured MI in a KB can assist designers in creating more optimal products that fully utilise the available manufacturing capabilities. Machine learning can help to systematise and make more explicit the knowledge that is implicitly embedded in manufacturing data and process planning decisions.

In the following sections, two machine learning algorithms used for autonomous processing of design and manufacturing data and for constructing the KB are described.

3.1. Rules-3 Inductive Learning Algorithm

RULES-3 is a simple algorithm for extracting a set of classification rules for a collection of objects each belonging to one of a number of given classes [Pham and Aksoy, 1995]. The objects together with their associated classes constitute the set of training examples from which the algorithm has to induce general rules. An object must be described in terms of a fixed set of attributes, each with its own set of possible values. For example "Heat_ treatment"and "Material"might be attributes with possible values {Yes, No} and {Steel_1045, Steel_3135, Aluminium} respectively.

In RULES-3 an attribute-value pair constitutes a condition. If the number of attributes is n_a, a rule may contain between one and n_a conditions, each of which must be a different attribute-value pair. Only conjunction of conditions is permitted in a rule and therefore the attributes must all be different if the rule comprises more than one condition.

For each example that has not been classified, the rule forming procedure may require at most n_a iterations, the first iteration producing rules with one condition and the second iteration resulting in rules with two conditions, etc. An array called the "array of attributes and values" is constructed, the elements of which are attribute-value pairs associated with those examples. The total number of elements in the array is equal to the number of attributes in the example. In the first iteration, each element of the array is examined to decide whether it can form a rule with that element as the condition. For the whole set of examples, if a given element applies to only one class, then it is a candidate for forming a rule. If it pertains to more than one class, it is passed over and the next element is examined. When all elements of the array have been looked at and if no rule can be formed, the second iteration starts. This involves inspecting elements of the array in pairs to ascertain whether they apply to only one class in the whole set of examples. As before, for those pairs of elements that pertain to unique classes, candidate rules are obtained. If no rule can

Step 1.	Define ranges for attributes that have numerical values and assign labels to those ranges
Step 2.	Set the minimum number of conditions (n_{comin}) for each rule
Step 3.	Select an unclassified example and set $n_{co} = n_{comin} - 1$.
Step 4.	IF $n_{co} < n_a$ THEN $n_{co} = n_{co} + 1$; ELSE use the example as a rule and go to Step 9.
Step 5.	Take all values or labels contained in the example
Step 6.	Form objects which are combinations of n_{co} values or labels taken from the values or labels obtained in Step 5.
Step 7.	IF at least one of the objects belongs to a unique class THEN form rules with those objects; ELSE go to Step 4.
Step 8.	Select the rule which classifies the highest number of examples.
Step 9.	Mark examples covered by the selected rule as classified.
Step 10.	IF there are no more unclassified examples THEN STOP; ELSE go to Step 3.

Fig. 1 Induction procedure in RULES-3. n_{co}-number of conditions, n_a-number of attributes.

be formed the procedure continues until the number of iterations for the chosen example is equal to n_a. If after iteration n_a, the algorithm still cannot form a rule, the example itself is taken as the rule for this particular case. At the end of an iteration, if there is more than one candidate rule, the rule that classifies the largest number of previously unclassified examples is added to the set of extracted rules. The examples, covered by that rule are marked as classified. A new unclassified example is taken from the training set and the procedure repeated. This continues until all examples are correctly classified.

Fig. 1 summarises the induction procedure of RULES-3.

3.2. ART Algorithm

The Adaptive Resonance Theory (ART) [Carpenter and Grossberg, 1988] competitive learning algorithm is an example of an unsupervised learning algorithm. The algorithm is associated with a neural network of the same name. During training, only input patterns are presented to the neural network which automatically adapts the weights of its connections to cluster the input patterns into groups with similar features. There are different versions of the ART network. Fig. 2 shows the ART-1 version for dealing with binary inputs. Later versions, such as ART-2, can also handle continuous-valued inputs.

As illustrated in Fig. 2, an ART-1 network has two layers, an input layer and an output layer. The two layers are fully interconnected, the connections are in both the

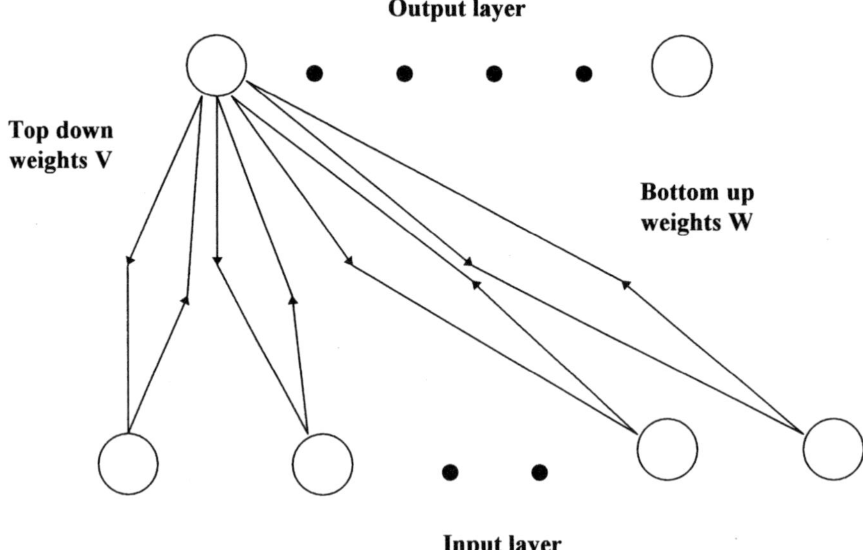

Fig. 2 An ART-1 network.

forward (or bottom-up) direction and the feedback (or top-down) direction. The vector $\mathbf{W}_i$ of weights of the bottom-up connections to an output neuron i forms an exemplar of the class it represents. All the $\mathbf{W}_i$ vectors constitute the long-term memory of the network. They are employed to select the winning neuron, the latter being the neuron whose $\mathbf{W}_i$ vector is most similar to the current input pattern. The vector $\mathbf{V}_i$ of the weights of the top-down connections from an output neuron i is used for "vigilance" testing, that is, determining whether an input pattern is sufficiently close to a stored exemplar. The vigilance vectors $\mathbf{V}_i$ form the short-term memory of the network. $\mathbf{V}_i$ and $\mathbf{W}_i$ are related in that $\mathbf{W}_i$ is a normalised copy of $\mathbf{V}_i$, viz.

$$\mathbf{W}_i = \frac{\mathbf{V}_i}{\epsilon + \Sigma \mathbf{V}_{ji}} \tag{1}$$

where ϵ is a small constant and $\mathbf{V}_{ji}$, the jth component of $\mathbf{V}_i$ (i.e. the weight of the connection from output neuron i to input neuron j).

Training in an ART-1 network occurs continuously when the network is in use and involves the following steps:-

(i) initialise the exemplar and vigilance vectors $\mathbf{W}_i$ and $\mathbf{V}_i$ for all output neurons, setting all the components of each $\mathbf{V}_i$ to 1 and computing $\mathbf{W}_i$ according to Eq. (1). An output neuron with all its vigilance weights set to 1 is known as an "uncommitted" neuron in the sense that it is not assigned to represent any pattern classes;
(ii) present a new input pattern $\mathbf{x}$;
(iii) enable all output neurons so that they can participate in the competition for activation;

(iv) find the winning output neuron among the competing neurons, i.e. the neuron for which $\mathbf{x} \cdot \mathbf{W}_i$ is largest; a winning neuron can be an uncommitted neuron as is the case at the beginning of training or if there are no better output neurons;

(v) test whether the input pattern $\mathbf{x}$ is sufficiently similar to the vigilance vector $\mathbf{V}_i$ of the winning neuron. Similarity is measured by the fraction r of bits in $\mathbf{x}$ that are also in $\mathbf{V}_i$, viz.

$$r = \frac{\mathbf{x} \cdot \mathbf{V}_i}{\Sigma x_i} \tag{2}$$

$\mathbf{x}$ is deemed to be sufficiently similar to $\mathbf{V}_i$ if r is at least equal to the "vigilance threshold" $\rho\,(0 < \rho \leq 1)$;

(vi) go to step (vii) if $r \geq \rho$ (i.e. there is "resonance"); else disable the winning neuron temporarily from further competition and go to step (iv) repeating this procedure until there are no further enabled neurons;

(vii) adjust the vigilance vector $\mathbf{V}_i$ of the most recent winning neuron by logically ANDing it with $\mathbf{x}$, thus deleting bits in $\mathbf{V}_i$ that are not also in $\mathbf{x}$; compute the bottom-up exemplar vector $\mathbf{W}_i$ using the new $\mathbf{V}_i$ according to Eq. (1); activate the winning output neuron;

(viii) go to step (ii).

The above training procedure ensures that if the same sequence of training patterns is repeatedly presented to the network, its long-term and short-term memories are unchanged (i.e. the network is "stable"). Also, provided there are sufficient output neurons to represent all the different classes, new patterns can always be learnt, as a new pattern can be assigned to an uncommitted output neuron if it does not match previously stored exemplars well (i.e. the network is "plastic").

4. CONCEPT FOR DESIGN DECISION SUPPORT

Two levels of abstraction should be considered in providing design decision support. At the higher level, there is a need to reason about the design model globally and the support system should address the problems relating to the retrieval and utilisation of manufacturing solutions for similar components and ease the building of the component manufacturing models. Furthermore, the support system should provide an estimate of the component manufacturability and identify problematic form features. Manufacturability can be defined as ease of machining a component or feature according to the specified technological requirement. In this chapter, only manufacturability from a machining view point will be considered (that is, the feasibility of machining a component or feature in a particular production environment with a defined tool database and machining facility).

In the proposed concept, the manufacturing model is seen as an integrated object-oriented database that incorporates a component design model (feature-based model) with all related information, such as the workpiece model, the manufacturing process, the required manufacturing cells, fixture configurations and cutting tools for the machining (Fig. 3). The Pro/Engineer CAD system is employed in the research as a feature-based modeller and a concurrent design environment for building manufacturing models. The design support system should help at this high level of

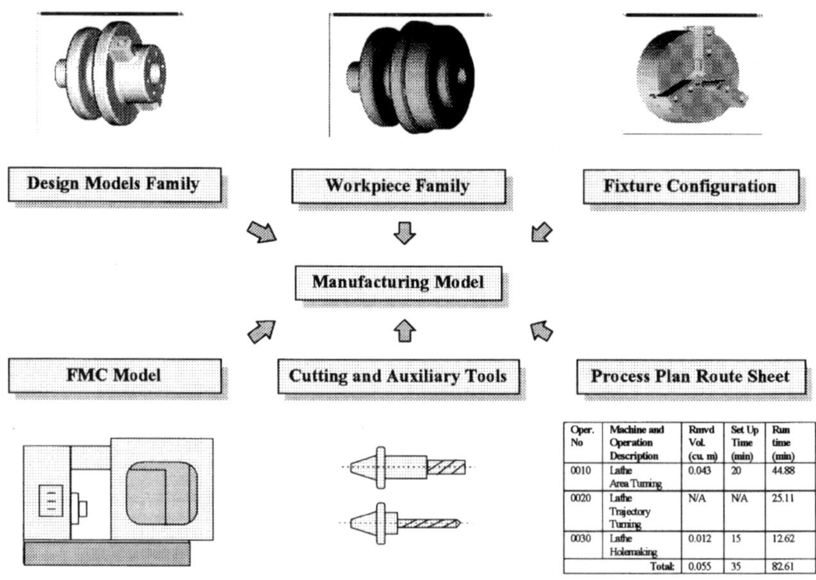

Oper. No	Machine and Operation Description	Rmvd Vol. (cu. m)	Set Up Time (min)	Run time (min)
0010	Lathe Area Turning	0.043	20	44.88
0020	Lathe Trajectory Turning	N/A	N/A	25.11
0030	Lathe Holemaking	0.012	15	12.62
	Total:	0.055	35	82.61

Fig. 3 Manufacturing model (integrated object-oriented data base).

abstraction to identify the facilities required and retrieve manufacturing solutions for similar components.

At the lower level of abstraction, design decision support on a feature by feature basis should take place. Form features are considered the main building blocks of the components and are employed as the communication medium between the design process and decision support procedures. The geometrical and technological requirements specified by the designer for each feature are used as input information for reasoning about manufacturability and cost.

For both levels of abstraction, the support systems have a similar general architecture as shown in Fig. 4. The implementation of the approach described in this chapter for design decision support requires the use of a modeller that supports two solid model representation schemes: constructive solid geometry (CSG) and boundary representation (Brep). The CSG scheme describes the component in terms of the geometrical primitives used (form features defined by the user). In this research, the term form features refers to recognisable shapes which cannot be further decomposed otherwise they will reduce to meaningless geometrical primitives such as lines, points and surfaces [ElMaraghy, 1993]. The form features can be generated by employing two groups of commands (feature commands in Pro/Engineer terms). The first group defines features whose shape is fixed and whose size only can vary (e.g. rounds, chamfers, etc.). The second group provides more freedom to the designer and allows features such as protrusions, cuts, slots, holes, etc., to be generated for which both the size and shape can vary. These commands can generate an infinite set of form features that make the identification of the existing mapping between the features and the manufacturing considerations relating to them (recognition of manufacturing features) very difficult. In spite of the close association

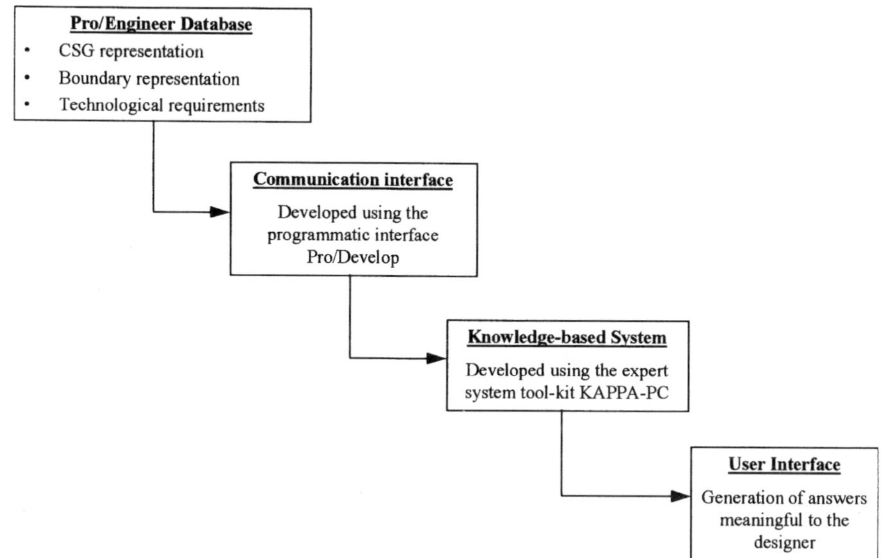

Fig. 4 Main components of the design support systems.

between machined volumes and form features which makes the CSG component data very attractive, this representation scheme does not provide sufficient information for mapping form features to manufacturing features. The component description has to be enhanced by using two representation schemes simultaneously. The Brep scheme which contains detailed information about faces, edges and vertices of the components provides the needed complementary data to the CSG model. This data is essential for reasoning about the facilities requirements of each form feature (identification of the existing relationship between form and manufacturing features). An example of a feature model that can be extracted from the Pro/Engineer database is shown in Fig. 5. Note that in addition to the available information from the CSG and Brep schemes, technological requirements (roughness, dimensional and geometrical tolerances) are also included in the feature model. Reference [Case *et al.*, 1994] gives a comprehensive classification of the tolerance information defined in British Standard BS 308 [Parker, 1984]. The modeller employed (Pro/Engineer) provides a convenient, interactive means for the input, modification and representation of the tolerance information. The information is stored in the database as technological requirements attached to features.

The second module of the overall system architecture plays the role of an interface between the Pro/Engineer database and the knowledge-based (KB) system. This interface allows the required information for each design decision support procedure to be extracted and represented in object-oriented form. For example, when there is a need to reason about a given form feature, a feature model as shown in Fig. 5 will be extracted from the database. The information link between the concurrent design and KB environments has been developed by employing the Pro/Engineer module Pro/Develop. Pro/Develop consists of C functions that provide direct access to the

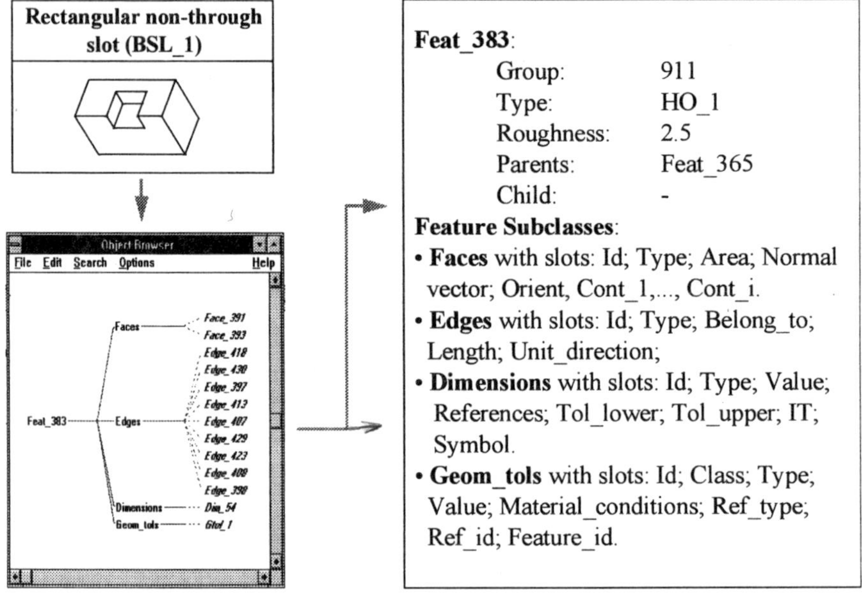

Fig. 5 An example of a feature model.

Pro/Engineer database. Some of the procedures for extraction of data from the database are easy to realise and implement but there are tasks, for example, the determination of the access direction for machining a given form feature and reasoning about a feature relationship, that require development of complex reasoning mechanisms.

Existing CAD systems lack sophisticated data management facilities and are also unable to handle the large volume of information required for provision of effective design decision support. To enhance the concurrent design environment capabilities, some complementary features should be added by linking the CAD system with a specialised environment which utilises the strengths and capabilities of both a database management system and an expert system.

The next module that each design decision support system should include is an environment with such features as object-oriented representation of the extracted data (facts) from the design database and representation techniques that allow for convenient storing and handling of previous manufacturing experience and process planning knowledge. For the development of the prototype systems described in this research, the KAPPA-PC expert system tool-kit has been employed.

Finally, for the successful implementation of a design decision support system, it is necessary to develop a user interface for efficient bi-directional communication between the system and the designer. Through the user interface, the use of the support system should become a natural part of the designer's working practice. Therefore before developing the user interface, there is a need to establish the designer's requirements and specify what information should be exchanged between different concurrent design activities, for example what information will be useful for the designer when he is making decisions about technological requirements on a feature by feature basis. For each abstraction level, an analysis of the designer's requirements has been conducted and the relevant manufacturing information has been identified.

5. DESIGN DECISION SUPPORT AT THE FEATURE LEVEL

The design decision support system provides the designer with relevant manufacturing information to facilitate his decision making activities on a feature by feature basis. The overall architecture of this support system is shown in Fig. 6. The system architecture for the level of abstraction considered differs from the general architecture (see Fig. 4) in some specific modules. In particular, the KAPPA-PC expert system tool-kit has been employed to create the following three modules:

- a feature recognition module;
- an inductive learning module;
- a knowledge-based system for storing and handling procedural and declarative manufacturing knowledge for a given production environment.

The other three modules depicted in Fig. 6 are generic programs with similar functions as those outlined in the support systems general architecture. The following sections give a detailed description of the specific modules.

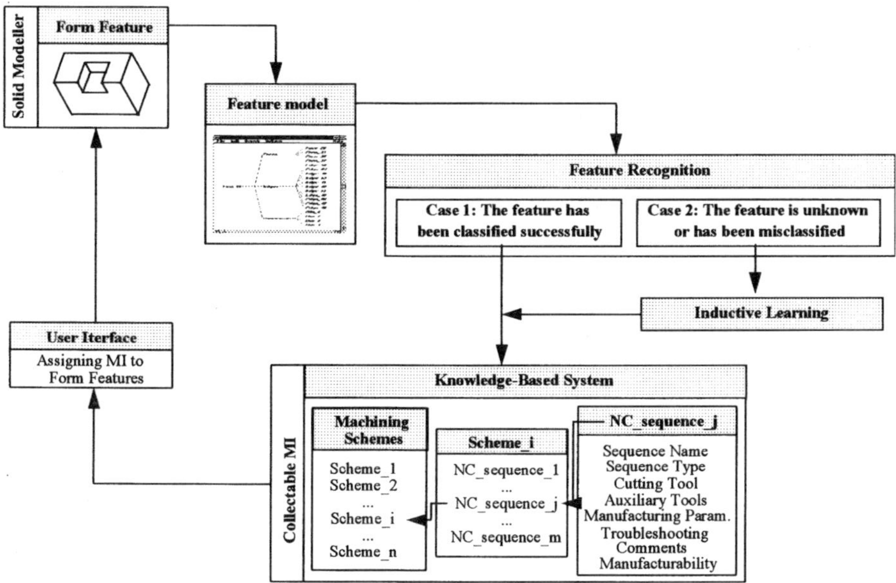

Fig. 6 Overall architecture of the support system at the feature at the feature level.

5.1. Feature Recognition and Inductive Learning Modules

A feature-based model of a part can be created by applying two different approaches [De Martino *et al.*, 1994; Case *et al.*, 1994; Krause *et al.*, 1993; Salomons *et al.*, 1993]: design by features and feature recognition. Both approaches, however, have deficiencies. With the first approach, design by features, the following deficiencies can be listed:

- the user needs to define features prior to object design;
- predefined features are oriented towards a specific application and cannot be shared between different domains;
- the feature dimensioning schemes (feature parametric models) are predetermined which greatly restricts the freedom of the designer who can only modify but not add new parameters.

Some of the deficiencies of the feature recognition approach are as follows:

- it is computationally a very expensive procedure;
- it does not retain functional information relating to features (e.g. relations between features);
- it provides only recognition of known form features.

In this research a hybrid approach is adopted. The designer has the freedom to create his own features by means of a solid modeller and specify for them different dimensional and technological parameters. When there is a need to reason about a given feature (e.g. to identify its machining schemes, assess its manufacturability, etc.), a feature model (see Fig. 5) is extracted from the solid modeller database. This model is then analysed by a feature recognition module (pattern classifier) and similarities with already known feature classes are identified.

The most interesting characteristic of this hybrid approach is its capability for autonomous formation of feature recognition rules. A stand-by inductive learning system is activated in case the designer is not satisfied with the outcome of the recognition process. This machine learning system allows the rule base to be updated and data about a new feature (salient topological and geometrical parameters) to be incorporated into the knowledge base. The next time the designer needs to reason about a similar feature, the feature recognition module will recognise it correctly. In this way, relevant manufacturing considerations associated with the class to which the feature belongs will be activated and brought to the designer's attention.

The inductive learning algorithm employed is RULES-3 Plus, an enhanced version of RULES-3 which has been described in Section 3.1. This algorithm, which has the added benefits of a more efficient rule searching procedure and a simple metric for assessing the generality and accuracy of candidate rules [Pham and Dimov, 1995] allows classification rules to be extracted from a collection of feature models belonging to different feature classes. The feature classes considered in this research are shown in Table 1, but the proposed approach is not limited to this set only. If there is a need, other feature classes can be appended to the set and manufacturing considerations for them can be retrieved and conveyed to the designer. The feature recognition module is an open system which can easily incorporate new types of form features created by the designer.

TABLE 1
Feature taxonomy.

No	Feature Group	Feature Class	Code
1	Protrusions	Circular	PR_1
2		Rectangular	PR_2
3		Hexagonal	PR_3
4	Steps	–	ST
5	Pockets	Rectangular	PO_1
6		Hexagonal	PO_2
7		Octagonal	PO_3
8		Obround	PO_4
9		Partial Round	PO_5
10	Blind Holes	–	BH
11	Holes	Circular	HO_1
12		Rectangular	HO_2
13		Hexagonal	HO_3
14		Octagonal	HO_4
15		Obround	HO_5
16		Partial Round	HO_6
17	Non-through Slots	Rectangular	BSL_1
18		Partial Round	BSL_2
19		V-shaped	BSL_3
20	Through Slots	Rectangular	SL_1
21		Partial Round	SL_2
22		V-shaped	SL_3
23	Corner Notches	Straight	CN_1
24		External Radius	CN_2

RULES-3 Plus induces feature recognition rules from a collection of known form features with their associated classes. To prepare the feature data for the induction process, each feature model should be preprocessed and its characteristic vector extracted. This vector describes the form features in terms of a fixed set of attributes, each with its own set of possible values. The feature attributes used in this research are shown in Table 2.

The proposed approach was tested using a collection of 104 feature models belonging to 24 different classes. These models were preprocessed and a characteristic vector extracted for each of them. An example illustrating the extraction of a characteristic vector is given in Fig. 7. The set of evaluated vectors thus formed constitutes the training set for induction of feature recognition rules (Fig. 8).

The set of feature recognition rules generated by RULES-3 Plus for the given training set is shown in Table 3. A postprocessor is used to transform the feature recognition rules into KAPPA-PC Application Language (KAL) form for handling in KAPPA-PC's environment. The following is an example of a rule in KAL (rule 8) for recognising a rectangular through-slot feature (SL_1):

```
MakeRule (rule_7, [],
instance:orient_pos #= 0 And instance: angle_60_120 #= 2,
Append To List (instance: Class, SL_1#_#0.315309));
```

TABLE 2
Feature attributes.

No	Attribute name	Attribute	Values
1	Pro/Engineer feature type	pro_type	911–928
2	Number of faces	n_faces	0–25
3	Feature interference	interf	0 or 1
4	Number of cylindrical faces	cyl_faces	0–25
5	Number of planar faces	pl_faces	0–25
	Number of faces with:		
6	one contour	cont_1	0–25
7	two contours	cont_2	0–25
8	more than two contours	cont_3	0–25
	Number of faces with:		
9	positive orientation	orient_pos	0–25
10	negative orientation	orient_neg	0–25
	Number of angular relationships between faces which fall into the range:		
11	0° to 60°	angle_0_60	0–25
12	60° to 120°	angle_60_120	0–25
13	120° to 180°	angle_120_180	0–25

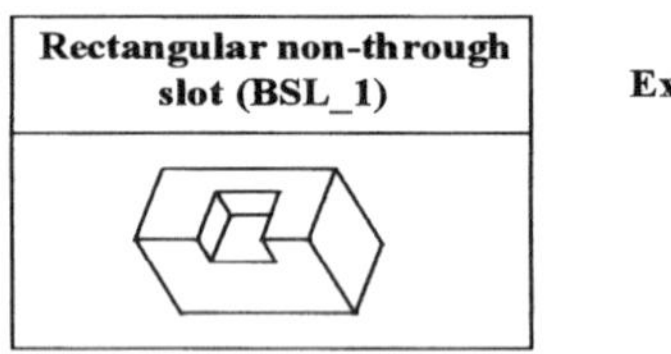

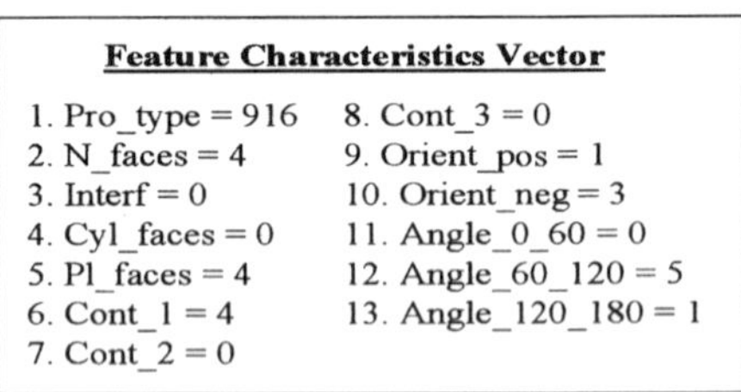

Fig. 7 Extraction of a feature characteristics vector.

When the designer creates a new form feature, the feature recognition process is activated. Forward chaining is initiated to recognise this new feature based on its evaluated vector extracted from the solid modeller database. As a result of the reasoning process, there are three possible outcomes:

1. The mapping between the new feature and the known feature classes is one-to-one. In other words, the feature has been recognised and belongs to only one class.
2. The mapping is one to many. This means that the new feature has common attributes with more than one feature class. For example, the recognition process will detect that the feature shown in Fig. 9 has common characteristics with two feature classes: Rectangular Non-through Slot (BSL_1) and Partial Round Non-through Slot (BSL_2). In this case, the designer will have two possibilities:

   - accept one of the proposed classes or
   - create a new class and then activate the inductive learning process which will update the rule base.

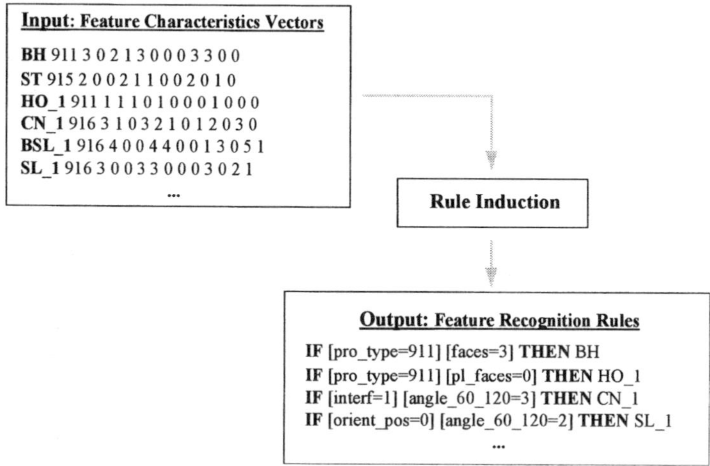

Fig. 8 Induction of feature recognition rules.

TABLE 3
Feature recognition rules generated by RULES-3 Plus.

No	Rule Description	H-measure
Rule 1	[orient_pos = 5]⇒PR_2	0.238889
Rule 2	[pro_type = 911] [faces = 3]⇒BH	0.238889
Rule 3	[faces = 2] [cont_1 = 1]⇒ST	0.238889
Rule 4	[pro_type = 911] [pl_faces = 0]⇒HO_1	0.365
Rule 5	[interf = 1] [angle_60_120 = 3]⇒CN_1	0.298021
Rule 6	[angle_60_120 = 5]⇒BSL_1	0.400854
Rule 7	[orient_pos = 0] [angle_60_120 = 2]⇒ SL_1	0.315309
Rule 8	[faces = 9]⇒PO_3	0.315309
Rule 9	[pl_faces = 5] [cont_1 = 5] [orient_pos = 1]⇒PO_1	0.400854
Rule 10	[angle_60_120 = 6]⇒PO_4	0.365
Rule 11	[cyl_faces = 1] [angle_60_120 = 2]⇒ PO_5	0.315309
Rule 12	[faces = 7] [orient_pos = 1]⇒ PO_2	0.315309
Rule 13	[cont_1 = 6]⇒ HO_3	0.315309
Rule 14	[faces = 8]⇒ HO_4	0.365
Rule 15	[cyl_faces = 1] [angle_60_120 = 1]⇒ HO_6	0.315309
Rule 16	[faces = 4] [cyl_faces = 2]⇒ HO_5	0.365
Rule 17	[pl_faces = 4] [orient_pos = 0]HO_2	0.365
Rule 18	[pl_faces = 1] [orient_pos = 1]⇒ BSL_2	0.222957
Rule 19	[cyl_faces = 0] [orient_pos = 1] [angle_60_120 = 2]⇒ BSL_3	0.222957
Rule 20	[orient_pos = 2]⇒ CN_2	0.315309
Rule 21	[pro_type = 916] [pl_faces = 0]⇒ SL_2	0.222957
Rule 22	[pl_faces = 2] [angle_60_120 = o]⇒ SL_3	0.222957
Rule 23	[pro_type = 915] [pl_faces = 0]⇒ SL_2	0.222957
Rule 24	[orient_pos = 7]⇒ PR_3	0.176885
Rule 25	[orient_pos = 3]⇒ PR_1	0.176885
Rule 26	[cyl_faces = 1] [orient_neg = 3]⇒ BSL_2	0.222957
Rule 27	[interf = 0] [angle_60_120 = 3]⇒ BSL_3	0.222957
Rule 28	[pl_faces = 2] [cont_1 = 2] [orient_pos = 0]⇒ SL_3	0.315309
Rule 29	[orient_pos = 1] [angle_60_120 = 1]⇒ CN_1	0.210733

(Note: The H-measure [Pham and Dimov, 1995] is a metric for assessing the generality and accuracy of a rule)

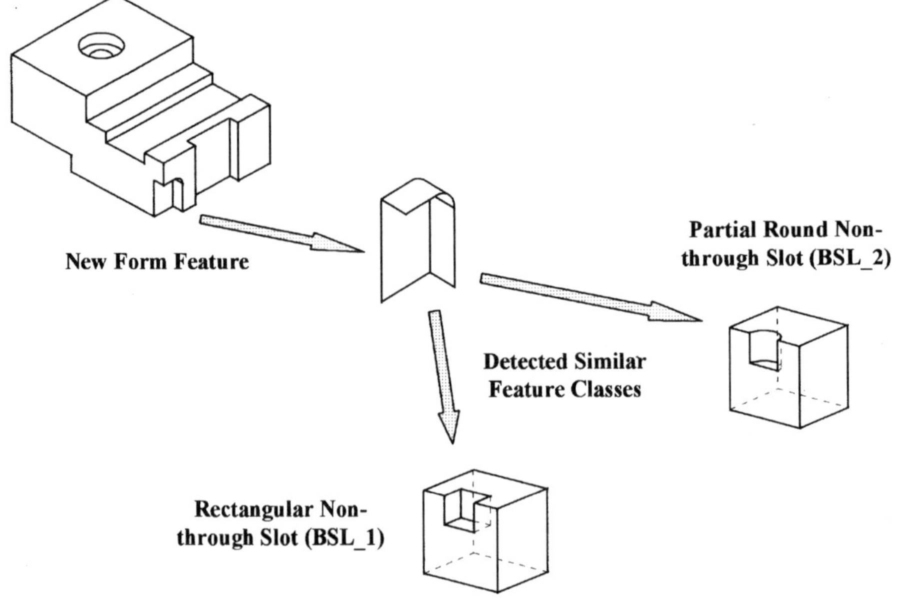

Fig. 9 An example of feature recognition.

3. The new feature does not belong to any known feature class. A new class should be created and then the learning mode activated to update the rule base.

For further reasoning, only the manufacturing knowledge associated with the class to which a given feature belongs will be taken into consideration.

5.2. Knowledge-Based System

At the design stage, the knowledge-based system allows the designer to assess the feasibility for a given feature to be machined and the specified technological requirement achieved utilising the available manufacturing facilities. The overall structure of the system is shown in Fig. 10.

The most important issues to be addressed when the system is built are the selection of an adequate model for knowledge representation and the use of an efficient inference mechanism. In this research, two schemes are used to encapsulate knowledge: object-oriented and rule-based representation schemes with standard reasoning strategies for conducting the inference. The design decision support system embeds in the knowledge base:

- declarative knowledge about cutting and auxiliary tools, manufacturing constraints, processing capabilities of machining facilities, test and measurement facilities, etc.;
- procedural knowledge in the form of a feature transition graph expressed as sets of rules.

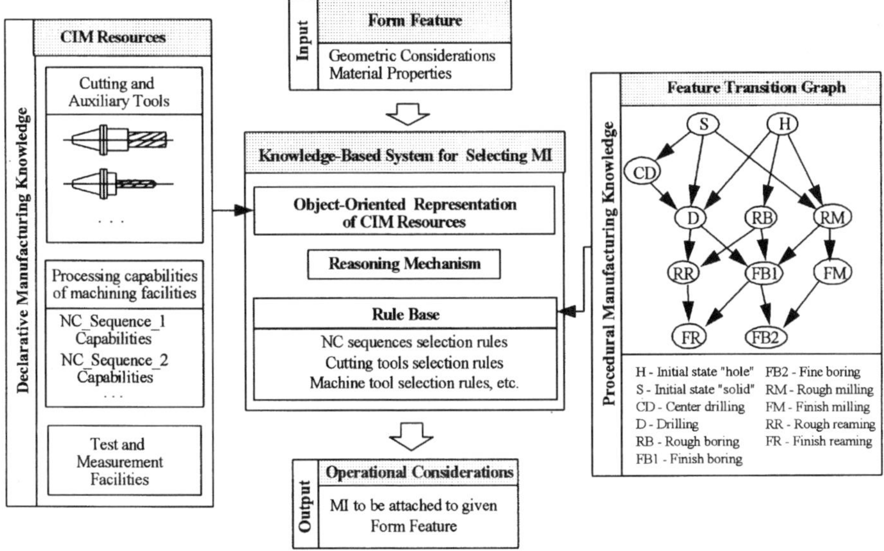

Fig. 10 Knowledge-based selection procedure.

The declarative knowledge deals with manufacturing process data while the procedural knowledge controls the inference process. To facilitate the latter, both types of knowledge are partitioned and associated with different feature classes. The procedural knowledge is divided into rule sets while the declarative knowledge is grouped into slots and methods attached to the different feature classes (Fig. 11). When there is a need to reason about the manufacturability of a given feature only the knowledge associated with the class to which the feature belongs will be used.

For example, the information relating to the capabilities of the processes that can be applied for machining a through-hole feature are represented as methods attached to class HOLES (see Fig. 11). As these methods encapsulate two types of data and also have two different functions, the methods can be divided informally into two groups. The first group embodies the constraints relating to the machining process (pre-conditions for including a given operation in a machining scheme). This group of methods can be considered as process selection rules which guide the search by assessing if a particular operation is pertinent to the machining of a given feature. The acquisition of these selection rules can be conducted either manually or using inductive learning algorithms. In both cases, historic data about expert decisions is processed to form the rules. The second group of methods encapsulates data about the machining accuracy for the available manufacturing facilities. These methods allow the search process to detect the stage when the specified technological requirements are achieved and to terminate the exploration of a given path in the feature transition graph.

The declarative knowledge is mostly domain-dependent. Therefore, the implementation of the proposed approach for design decision support requires manufacturing knowledge about the available production facilities and process

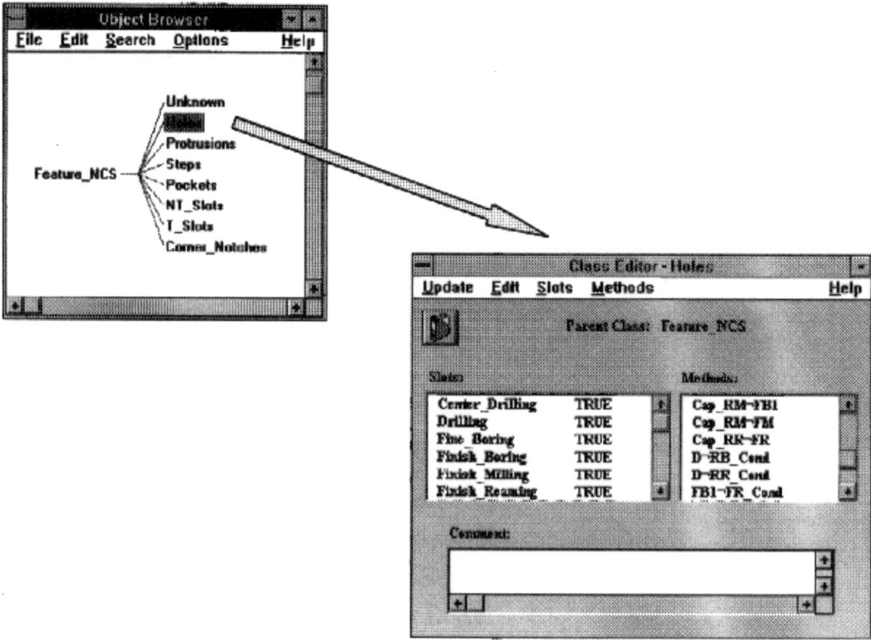

Fig. 11 Declarative knowledge for class HOLES grouped in slots and methods.

planning practice to be gathered and the acquired knowledge to be represented in a suitable (object-oriented) form.

The reasoning strategy in the provision of design support at the feature level is modelled as a generative process. This process gradually forms feature machining schemes (an ordered set of machining operations) by choosing between alternatives in the feature transition graph. In other words, the generative process is a formal search task for technological solutions that can ensure the specified technological requirements. The search is performed in the space of all technological solutions which contains manufacturing constraints to limit and guide the search. Each feature class has its own specific transition graph which is formally represented as a set of rules (Fig. 12). The feature transition graph (procedural knowledge) and machining process selection rules (declarative knowledge) associated with each multi-route node form the knowledge base for the search process. In addition, monitoring of the feasibility of the machining scheme is carried out concurrently with other reasoning operations.

Communication between the rules which embed the procedural knowledge and the declarative knowledge is realised by messages that the rules send to the appropriate methods when the reasoning process is activated. For example, the rule CD¬D has the following structure:

```
MakeRule(CD¬D,[],
    {
    Holes:Center_Drilling #= TRUE And
```

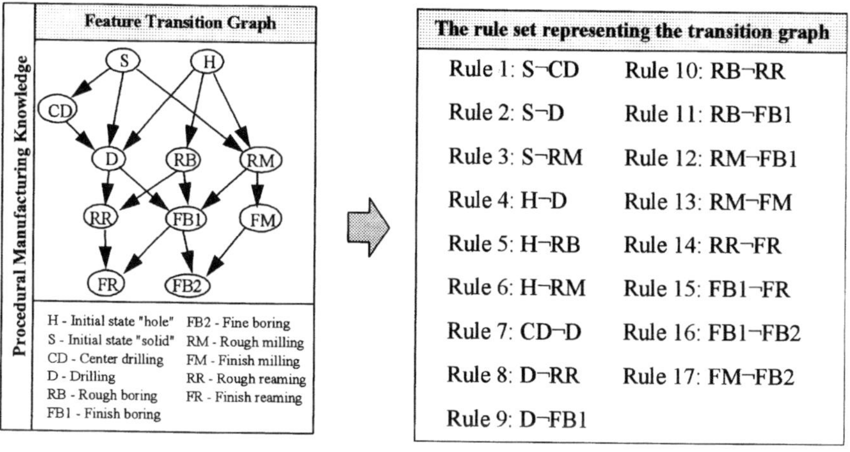

Fig. 12 The rule set associated with feature class HOLES.

```
        (Not(Method?(Holes,CD¬D_Cond)) Or
        SendMessage(Holes,CD¬D_Cond));
        },
        {
        Add_NC_Sequence(Holes,Center_Drilling,Drilling);
        If(Not(Method?(Holes,Cap_CD¬D)) Or
        SendMessage(Holes,Cap_CD¬D))
            Then Solution()
            Else Assert(Holes:Drilling);
         });
SetRulePriority(CD¬D, 1);
AppendToList(Feature_NCS:Holes, CD¬D);
```

where: CD¬D_Cond is a method representing the pre-conditions for including drilling in a machining scheme after the centre drilling operation; Cap_ CD¬D is a method representing the achievable technological requirements when drilling follows the centre drilling operation.

The proposed rule structure allows the system to reason about features even before the declarative knowledge is acquired. The knowledge base can be gradually refined to represent adequately a specific production environment. To facilitate the setting-up of the knowledge base, autonomous acquisition of the procedural knowledge (transition graphs for different feature classes) is conducted by applying a pseudo-induction approach. This approach allows a collection of feature machining schemes (process planner decisions) to be processed incrementally and a context-free non-recursive grammar to be formed. This grammar represents the existing relations between a set of processes for machining a given feature class. The grammar rules are used as an initial structure for generating the procedural knowledge. Fig. 13 illustrates the application of this pseudo-induction approach on a set of schemes for machining holes.

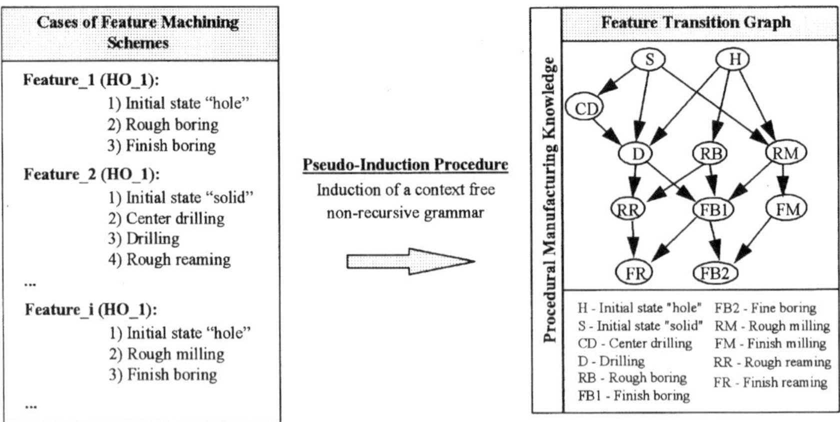

Fig. 13 Pseudo-induction approach for generating the procedural knowledge.

To illustrate the reasoning process of the knowledge-based system for design decision support at the feature level, it is assumed that the designer decides to place a through-hole feature at some location on the model. Then, after creating the feature by means of the solid modeller, he specifies the feature technological requirements as they are shown in Fig. 14. It is also assumed that the designer is not sure if the

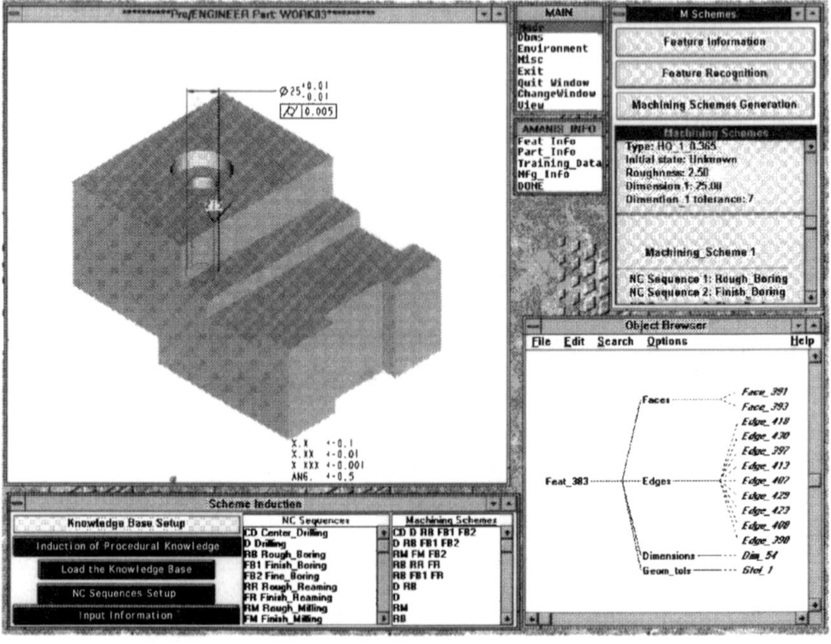

Fig. 14 Design decision support system at the feature level.

specified requirements can be achieved in the available manufacturing facilities and, therefore, he initiates the reasoning process by choosing the feat_info option in the system menu. This option activates a sequence of procedures, in particular:

1. Extraction of feature data from the solid modeller;
2. Feature recognition. Selection of pertinent rules and feature classes for the inference process;
3. Feature manufacturability assessment and determination of feature facilities requirements, such as machining time, cutting and auxiliary tools, etc.;
4. Analysis of the influence of the specified technological requirements on the facilities requirements of the feature.

The most significant benefit of the proposed approach for design decision support at the feature level is the possibility of ensuring the best match between the product functional carriers (form features) and available manufacturing facilities. The support system allows the designer to conduct, simultaneously with the design process, an initial process design by attaching to form features manufacturing considerations for downstream applications. As a result, at an early stage, with the provision of design decision support at the feature level, better integration between product and process design can be achieved and, as a consequence, market requirements can be met at a lower cost.

6. DESIGN DECISION SUPPORT AT THE COMPONENT LEVEL

At the highest level of abstraction, when there is a need to reason about a design model globally, procedures for concurrent product and process design should provide information about component manufacturability and facilitate the building of component manufacturing models. In this research, as already mentioned, the manufacturing model is seen as an integrated object-oriented database (see Fig. 3) that incorporates a component design model with all related information, such as the workpiece model, the complete manufacturing plan, and the flexible manufacturing cell and fixture configuration models used for machining the component.

The manufacturability assessment is conducted in a similar fashion as at the feature level. The only difference is that the set of features belonging to a component are processed and analysed consecutively. The support information that the procedure for manufacturability assessment can provide includes:

- a list of problematic features (features whose technological requirements cannot be achieved with the available manufacturing facilities);
- part machining times and costs.

A detailed description of this procedure has already been given in the previous section. Therefore, this section only covers the procedure for utilising the MI about components for which the manufacturing models have been already created.

The building of manufacturing models involves expenses not only for component design, but also for process design which specifies the component facility requirements. These expenses can be minimised by utilising the available MI about components already machined in the given manufacturing environment. The category of MI that relates to the available manufacturing facility can be represented by case bases of fixture configurations, Flexible Manufacturing Cell (FMC) models and manufacturing models. As shown in Fig. 15, the management, control and use of these case bases can be considered a practical means of conveying useful MI and experience for use in the downstream concurrent design activities. The proposed approach is based on the assumption that geometrical similarities in design models are an important consideration in the selection of manufacturing cases which may be reused. The first step in defining the required selection procedure is the identification of salient design characteristics from a manufacturing point of view that can be employed to classify design models and also to cluster the available MI about them. The selection procedure can support the building of new manufacturing models through a variant design process that accesses prior design and manufacturing experience. Such a support procedure can facilitate the reuse of:

- component design models;
- workpiece design models;
- fixture configurations;
- FMC models with the associated cutting and auxiliary tools;
- complete manufacturing models.

The selection procedure will help at this high level of abstraction to identify component facility requirements and search for the existence of similar manufacturing cases.

The overall architecture of the support system developed is given in Fig. 16. At the level of abstraction considered, it is presupposed that a component definition in terms of form features is already available. The feature-based design model contains information on salient design characteristics such as overall shape, relative location of major form features, the presence of certain features, etc. A parser is created to scan the feature-based model and extract a set of salient design characteristics, as

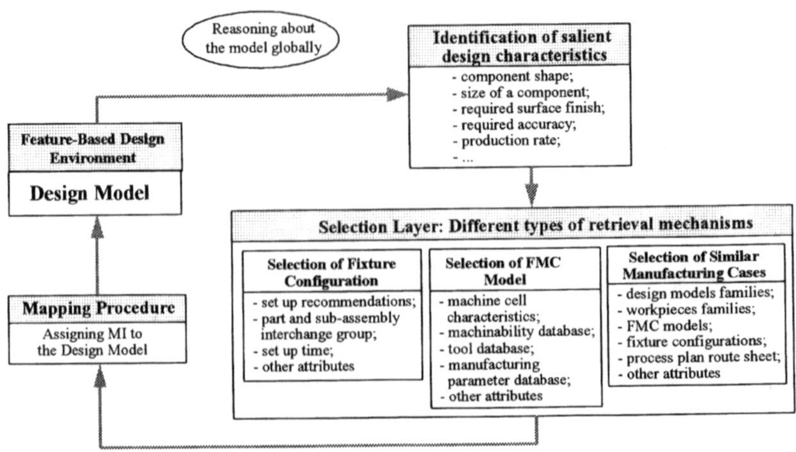

Fig. 15 An approach for selecting manufacturing information for a design model.

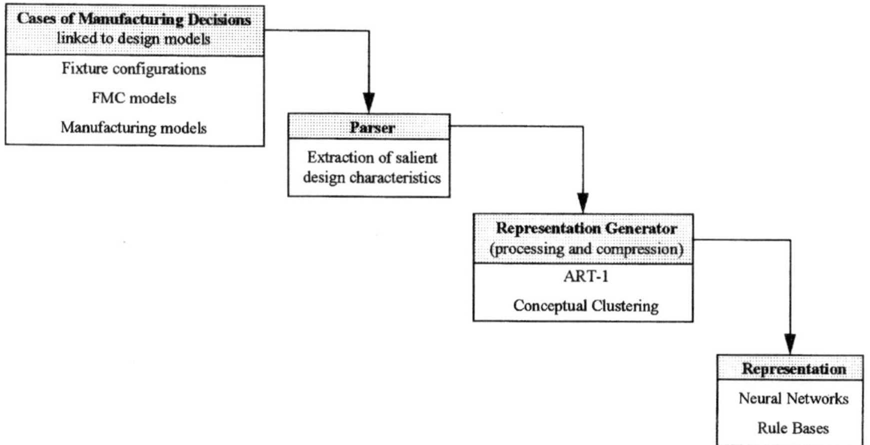

Fig. 16 Overall architecture of a support system at the component level.

shown in Fig. 17. The parser plays the role of an interface between the Pro/Engineer and KAPPA-PC environments.

To exploit similarities among design models through classification into component families, conceptual clustering [Michalski and Stepp, 1983] and ART based approaches [Escobedo *et al.*, 1993] can be employed to generate a rule base or a neural network representation of the manufacturing cases. The generated representation schemes allow cases to be discriminated according to specific design characteristics and the degree of discrimination to be varied.

In this research, the ART-1 algorithm described in Section 3 is used to process and store the manufacturing cases. As already mentioned, ART-1 is a self-organising clustering algorithm that is able to retain previously learned cases and also learn new facts. Thus, the proposed procedure for manufacturing information processing and selection supports two modes. The first is a learning mode which allows design models to be classified on the basis of the extracted salient design characteristics and consequently for the manufacturing models associated with them to be structured and attached to a given cluster of design models. An example of the information that can be attached and consequently retrieved in response to queries from the designer is shown in Fig. 18.

The second mode allows the designer to query and consult the created neural database and, in this way, to identify clusters of manufacturing models relevant to the component being designed. An important feature of this consultation mode is the ability to vary the degree of cluster discrimination by specifying interactively the required level of similarity between the component and manufacturing models stored in the database.

The formal representation of the manufacturing facility and the identification of the salient design characteristics are both critical for the successful implementation of the proposed approach for MI selection. The chosen representation schemes must explicitly capture, on the one hand, the manufacturing capabilities of a given CIM

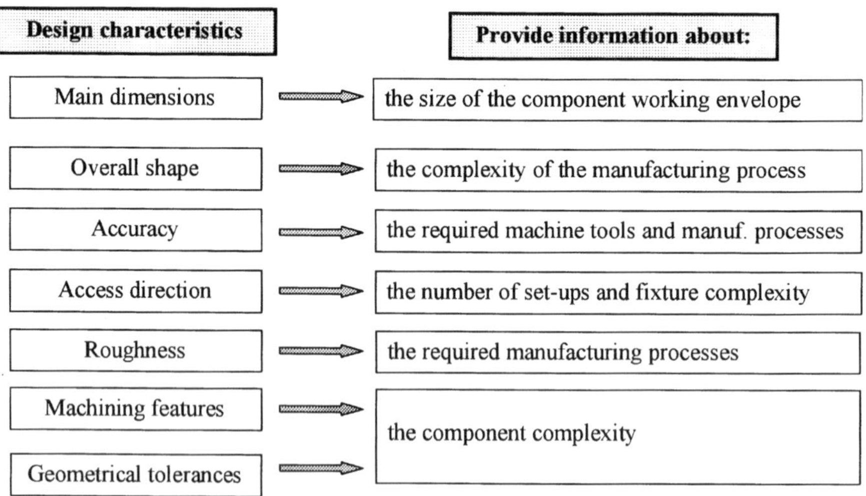

Fig. 17 Salient design characteristics.

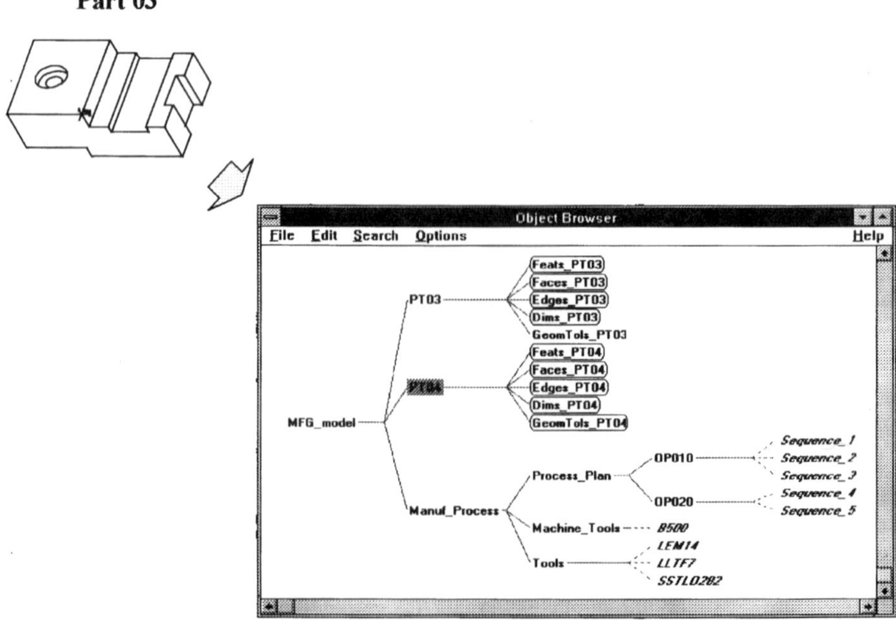

Fig. 18 Information that can be retrieved about component manufacturing models.

environment together with the manufacturing cases tested on the shop floor and, on the other hand, manufacturing similarities of design models.

7. CONCLUSION

This chapter has presented an approach and a prototype system for supporting concurrent product and process design activities. The key issues in this approach are the development of:

- a framework for structuring manufacturing information and maximising the information-carrying capacity of design models;
- a procedure for intelligently mapping pertinent manufacturing considerations to form features (design decision support at the feature level);
- a procedure for utilising the available manufacturing information about components already machined within the considered manufacturing environment (decision support at the component level).

The proposed approach for concurrent product and process design provides a more natural way for conveying manufacturing information to the designer. The distinguishing feature of the proposed framework for design decision support is the application of a variety of AI techniques for knowledge acquisition and deductive reasoning, in particular:

- inductive learning for autonomous formation of feature recognition rules;
- object-oriented and rule-based systems for representing the manufacturing knowledge about the available production facilities and the existing process planning practice;
- self-organising neural networks for acquiring, structuring and representing knowledge about previous manufacturing cases.

The ultimate goal of the research is to assist the designer in taking downstream manufacturing concerns into careful consideration. The concurrent design modelling technology proposed in this chapter should be able to support the representation and exploration of process and product alternatives in order to reduce costly iterations and increase product flexibility.

Acknowledgements

The research described in this chapter was performed within the EC-funded BRITE/EURAM project 5139 "Advanced Manufacturing Information System for the Designer" carried out in collaboration with the Technical University of Clausthal (Germany) and the National Technical University of Athens (Greece).

References

H.S. Abdalla and J. Knight, An expert system for concurrent product and process design of mechanical parts. *Proc. IMechE, Part B: Journal of Engineering Manufacture*, **208**, 167–172 (1994).

G.A. Carpenter and S. Grossberg, The ART of adaptive pattern recognition by a self-organising network. *Computer*, 77–88 March (1988).

K. Case, J.X. Gao and N.N.Z. Gindy, The implementation of a feature-based component representation for CAD/CAM integration. *Proc. IMechE, Part B: Journal of Engineering Manufacture*, **208**, 71–80 (1994).

P. Clark and T. Niblett, The CN2 induction algorithm. *Machine Learning* 3, Kluwer, The Netherlands, 261–283 (1989).

M.R. Czutkosky and J.M. Tenenbaum, A methodology and computational framework for concurrent product and process design. *Mech. Mach. Theory*, **25** (3), 365–381 (1990).

T. De Martino, B. Falcidieno, F. Giannini, S. Hassinger and J. Ovtcharova, Feature-based modelling by integrating design and recognition approaches. *CAD J.*, **26**(8), 646–653 (1994).

D.S. Domazet and S.C.-Y. Lu, Concurrent design and process planning of rotational parts. *Annals of the CIRP*, **42** (1), 181–184 (1993).

H.A. ElMaraghy, Evaluation and future perspectives of CAPP. *Annals of the CIRP*, **42**(2), 739–751 (1993).

R. Escobedo, S.D.G. Smith and T.P. Caudell, A neural information retrieval system. *Int. J. Adv. Manuf. Technol.*, **8**, 269–274 (1993).

A. Herman, M. Lawley, S.C.-Y. Lu and D. Mattox, An opportunistic approach to process planning within a concurrent engineering environment. *Annals of the CIRP*, **42**(1), 545–548 (1993).

F.L. Krause, F. Kimura and T. Kjellberg *et al.*, Product modelling. *Annals of the CIRP*, **42**(2), 695–706 (1993).

C. Lee, Generating classification rules from databases. *Proc. 9th Conf. on Application of AI in Eng.*, Malvern, Pennsylvania, USA, 205–212 (1994).

T. Lenau and L. Alting, Intelligent support systems for product design. *Annals of the CIRP*, **38**(1), 153–156 (1989).

R.S. Michalski, A theory and methodology of inductive learning. *Readings Machine Learning*, Eds: J.W. Shavlik and T.G. Dietterich, Morgan Kaufmann, San Mateo, CA, 70–95 (1990).

R.S. Michalski and R.E. Stepp, Learning from observation: Conceptual clustering, in *Machine Learning: An Artificial Intelligence Approach, Vol. 1* (R.S. Michalski, J.G. Carbonell and T.M. Mitchell, Morgan Kaufmann), San Mateo, CA, 331–363 (1983).

M. Parker, (ed.) *Manual of British Standards in Engineering Drawing and Design*, 2nd edition, British Standards Institute, London (1984).

J. Peters *et al.*, Design: an integrated approach. Keynote Paper, *Annals of the CIRP*, **39**(2), 599–607 (1990).

D.T. Pham and M.S. Aksoy, A new algorithm for inductive learning. *J. Systems Eng.*, **5**(2), 115–122 (1995).

D.T. Pham and S.S. Dimov, RULES-3 Plus inductive learning algorithm, Technical Report, Intelligent Systems Lab., Cardiff School of Engineering, University of Wales Cardiff (1995).

O.W. Salomons, F.J.A.M. Van Houten and H.J.J. Kals, Review of research in feature-based design. *J. Manuf. Systems*, **12**(2), 113–132 (1993).

G. Sohlenius, Concurrent engineering, *Annals of the CIRP*, **41** (2), 645–655 (1992).

N.P. Suh, Basic concepts in design for producibility, Keynote Paper, *Annals of the CIRP*, **37** (2), 1–9 (1988).

CHAPTER 10

Process Control

NONG YE

Department of Mechanical Engineering, University of Illinois at Chicago, 2039 Engineering Research Facility, 842 West Taylor Street, Chicago, Illinois 60607–7022, USA

1. INTRODUCTION

Manufacturing systems carry out the manufacturing process that transforms the design of a product into the actual product. Features of the product design are converted into performance requirements for the manufacturing process which in turn are implemented by manufacturing systems through a proper control.

Process control is chained with other functions of the manufacturing enterprise. Process control is designed according to performance requirements for manufacturing systems that support the product design. During the process control, the state data of manufacturing process is collected and maintained. The state data provide information essential to quality engineering. The state data are also important to production management in that they draw a dynamic, interrelated picture of production operations which in turn lays the basis of monitoring production operations to meet production objectives. Since the computer technology has been widely used for process control, the computer networking of the process control function with other functions of the manufacturing enterprise can be established.

This chapter presents control principles used in two types of manufacturing environments: continuous process and discrete-part manufacturing. Hardware involved in process control is described. Human roles in designing and maintaining process control are discussed. In this chapter we use the term "process" to refer to systems being controlled, such as chemical plants for continuous process, machining tools and assembly equipment for discrete-part manufacturing.

2. CONTINUOUS PROCESS CONTROL

The control of a continuous process usually aims at regulating the process input to maintain the process output at a desired level or to optimize process performance. The process output is affected by various kinds of disturbances (e.g., changes of raw materials, changing environment, friction, etc.). The disturbances cause deviations of the process output from its desired value. Therefore, the control input must be adjusted to compensate for the disturbances. To achieve the goal of process control, the process output is measured using sensors, and actuators are driven by the process input to generate control actions affecting the process output. The disturbances are often not measurable, and their effects on the process output cannot be predicted. Several commonly used control schemes are presented in this section. A detailed discussion of continuous process control and its applications can be found in a reference (Balchen and Mumme, 1988).

2.1. Feedback Control

The feedback control determines the process input based on the deviation (or error) of the process output from the desired output (set point). The feedback control is often referred to as the close-loop control. Figure 1 presents a block diagram of the feedback control. The feedback control is generally built on three types of control responses, namely proportional control, integral control and derivative control. Three control responses differ in how the process input is determined from the error of the process output.

2.1.1. Proportional Control

The proportional control can be described by the following function,

$$u(t) = K_p e(t) \tag{1}$$

where $u(t)$ is the process input to generate a control action, K_p is the proportional control gain, $e(t)$ is the error of the process output and is calculated by subtracting

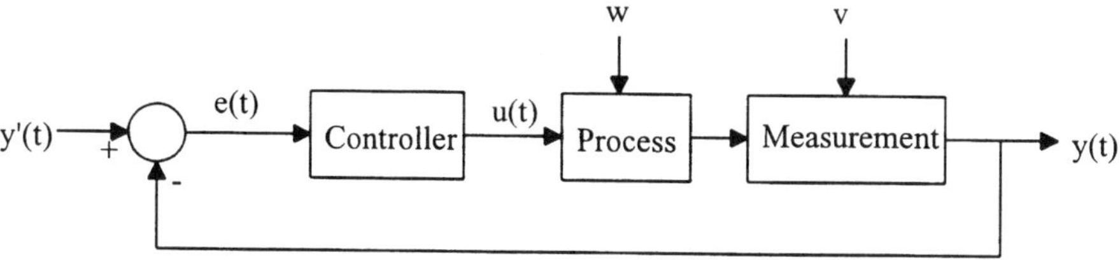

u(t): process input.
y(t): process output.
y'(t) desired process output, or set point.
e(t): y'(t) - y(t), difference between the desired value and the actual value of process output.
w: disturbances.
v: measurement noise.

Fig. 1 Feedback control.

the actual process output $y(t)$ from the desired process output $y'(t)$, and t is the time. When $e(t)$ is present, a control action proportional to the magnitude of $e(t)$ is used to drive the process output to its set point. This control action reduces the error. The larger the error, the larger the control input to the process. When $e(t)$ is zero, no control action is necessary.

The proportional control gain determines the responsiveness of the controller to the error of the process output. A low gain makes the controller take longer to drive the process output to its set point than a high gain. A mathematical model of the process dynamics showing the relationship between the process input and the process output can help in calculating the proportional control gain. The control gain also needs to be fine tuned through trial-and-error experimentation.

Eq. (1) describes the proportional control using analog control hardware. If digital control hardware is used, the process output is sampled at discrete points in time, and the process input is adjusted in every sampling cycle. The discrete form of the proportional control is given by:

$$u_n = K_p e^n \tag{2}$$

where e_n is the error of the process output in the nth sample period, and u_n is the process input produced in the same period.

The disadvantage of the proportional control is that when the output error is reduced to a low level after a series of control actions, the control input may be too small to drive the control device which is subject to long-term sustained disturbances such as friction. This results in a low, but constant level of the output error and the control input no matter how long the controller tries.

2.1.2. *Integral Control*

The integral control overcomes the effects of sustained disturbances by using the error accumulated over time rather than the error at a specific instance. The integral control is described as follows,

$$u(t) = K_i \int_0^t e(x)dx \tag{3}$$

where K_i is the integral control gain. The discrete form of the function is

$$u_n = K_i \sum_{j=1}^{n} Te_j \tag{4}$$

where T is the sample interval. That is, the error of the process output is integrated over time to obtain an accumulated error term for generating the control action. As the integral error term continues to build up over time from the small error, the control input continues to increase until the point when the control input is large enough to drive the process output to its set point. However, when the process output reaches its set point, the control input continues to drive the process output because the integral error term is not equal to zero. This results in an overshoot and a negative error. When the negative error eventually reduces the integral error term to 0, there is still a negative error of the process output. When the integral error term obtains a negative value, the control input drives the process output changing in an opposite direction, returning back to the set point, and passing the set point. The oscillations of the process output can be magnified by quickly changing disturbances. The quicker the change of the disturbances, the quicker the change of the error, the more severe the oscillations of the process output, and the less stable the process. In the worst case, the process output changes without a bound, and the process becomes unstable.

2.1.3. Derivative Control

The derivative control provides the capacity of responding to the quickly changing error. The control function is described below.

$$u(t) = K_d \frac{de}{dt}, \tag{5}$$

the discrete form is:

$$u_n = K_d \left(\frac{e_n - e_{n-1}}{T} \right), \tag{6}$$

where K_d is a derivative control gain. The disadvantage of the derivative control is that no control input is generated for a constant error.

2.1.4. PID Control

The proportional control provides a basic control model. The integral control can be added to the proportional control to form the PI control. The derivative control can be added to the proportional control to form the PD control. The most commonly used control is the PID control which combines the proportional control, the integral control, and the derivative control as follows.

$$u(t) = K_p e(t) + K_i \int_0^t e(t)dt + K_d \frac{de}{dt} \tag{7}$$

The PID control uses the proportional term to force the error to 0, the integral term to decrease the long-term error, the derivative term to speed up the response to the quickly changing error. The stability constraint derived from the process dynamics can be used to determine a range for the control parameters, K_p, K_i, and K_d. The control parameters can then be fine tuned within their range through trial-and-error experimentation.

The discrete form of the PID control is similar to Eq. (7) except that the continuous variable, time t, is replaced by a discrete variable to indicate discrete sampling points in time.

2.2. Feedforward Control

If the disturbances driving the process output away from its desired level cannot be measured or predicted, the process control has to rely on continuously monitoring process variables, responding to changes in process variables, thereby restraining the growth of the error, as seen in the feedback control. However, if the disturbances are measurable or predictable and their effects on the process are known, a simple feedforward control can be built based on an accurate model of the process and the disturbances. The process input is precisely calculated to compensate for the effects of the disturbances before the disturbances occur, thus keeping the process output at the desired level. Since the feedforward control receives no feedback concerning the operation of the process, it is often referred to as open-loop control. Figure 2 shows a block diagram of the feedforward control.

The feedforward control does not need to deal with the stability problem, an advantage over the feedback control. However, perfect knowledge of the disturbances and their effects on the process is hardly available. Even with perfect knowledge, it sometimes is impossible to develop a mathematical model to precisely describe the dynamics of the process and the disturbances. Therefore, the feedback control is often combined with the feedback control, as shown in Figure 3.

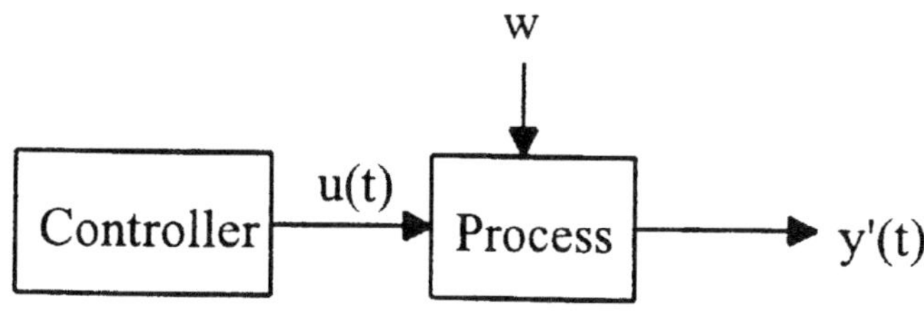

Fig. 2 Feedforward control.

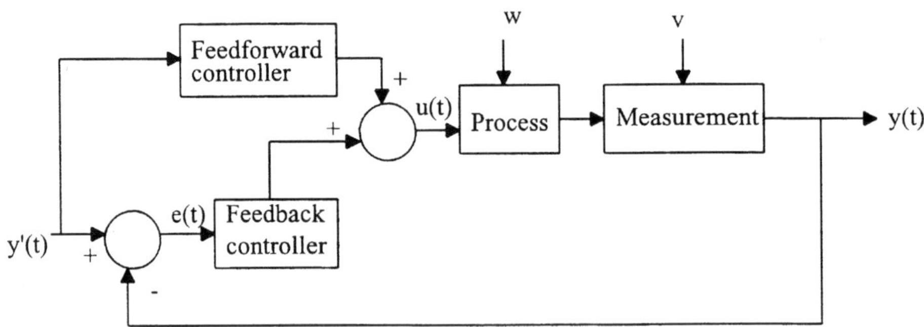

Fig. 3 A combination of feedforward control and feedback control.

2.3. Cascade Control

If a process involves many stages of operation, the control input signal has to pass through a long path to affect the process output. During the long time lapse between the control input and its effect on the process output, the disturbances may quickly develop and change the state of the process. As a result, the control input always lags behind the present state of the process. This in turn causes the failure of maintaining the process output at its set point. Therefore, the time delay violates the essence of the feedback control which is to control the growth of the error by immediately responding to changes in the process.

The cascade control reduces the length of the control path by breaking down the process into subprocesses, each with a measurable output. Inner control loops are used to control subprocesses. Inner control loops use intermediate process variables as the feedback. Figure 4 shows a block diagram of the cascade control. Since inner control loops quickly suppress the disturbances in subprocesses, the effects of the disturbances on the process output are minimized. This results in improved control performance.

2.4. Supervisory Control

In the control schemes discussed above, the human monitors the process and sets the desired process output. However, the human may have difficulty to monitor many processes in a plant. A computer system can be programmed to set the desired outputs for those processes according to the operation objective of the plant. The objective of the plant may be established based on performance criteria such as product quality, energy consumption, material utilization, and production efficiency. This central computer watches over the controllers which are directly connected to the processes, calculates the set points for the processes, and passes those values to the process-level controllers. Hence, the supervisory control uses the central computer to direct and coordinate the activities of the process-level controllers so as to optimize performance criteria of the plant. Figure 5 shows a block diagram of supervisory control.

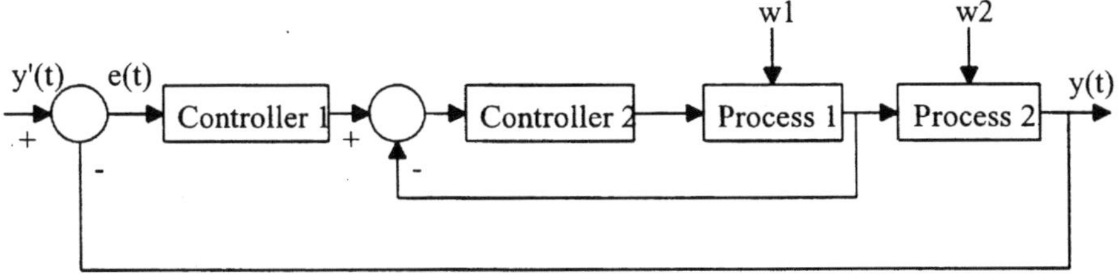

Note: the measurement blocks for both subprocesses are omitted.

Fig. 4 Cascade control.

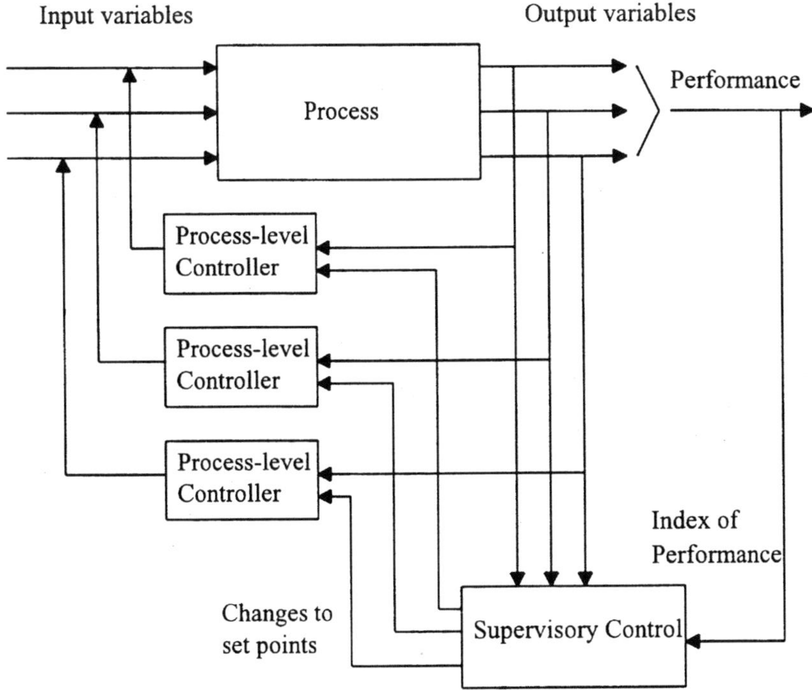

Fig. 5 Supervisory control.

2.5. Control with State Estimation

If the dynamics of the process and disturbances can be described in a mathematical model, the state of the process at any instant can be estimated. Based on the deviation of the estimated state of the process from the desired state of the process, the process input can be determined. Figure 6 shows a block diagram of control with state estimation.

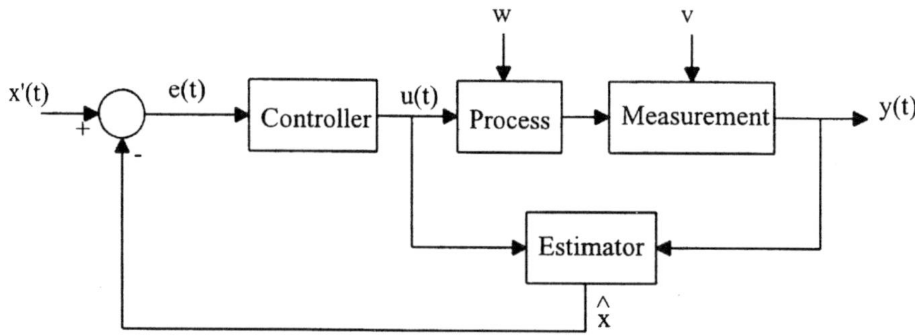

Fig. 6 Control with state estimation.

The best-known state estimator is the Kalman filter (Rouse, 1980). In the Kalman filter, a single-input, single-output, first-order (including one state variable) linear dynamic process is characterized by Rouse (1980) as follows:

$$x(t+1) = \phi x(t) + \psi u(t) + \gamma w(t) \tag{8}$$

$$y(t+1) = hx(t+1) + v(t+1) \tag{9}$$

where u is the process input, x is the state variable, y is the process output, w is a Gaussian disturbance with a zero mean and variance σ_w^2, and v is a Gaussian measurement noise with a zero mean and variance σ_v^2. Eq. (8) specifies that the state of a dynamic process depends on both the process input and the previous state of the process. The process output is an observable or measurable form of the process state, subject to the measurement noise (e.g., inaccuracy of measurement device). From Eq. (8) and Eq. (9), we obtain

$$x(t+T) = \phi^T x(t) + \sum_{i=0}^{T-1} \phi^{T-i-1} \psi u(t+i) \tag{10}$$

$$y(t+T) = hx(t+T) \tag{11}$$

As T increases, the absolute value of ϕ must be less than or equal to one to keep $y(t+T)$ within bound. In a stable process, ϕ^T approaches zero as Y increases. Eventually, the state of a stable process becomes steady in that it does not depend on the initial state of the process.

Knowing the process outputs up to the present time, $y(0), ..., y(t)$, the estimation of a future state is called prediction, denoted by $\hat{x}(t+i|t)$; the estimation of the present state is called filtering, denoted by $\hat{x}(t|t)$; the estimation of a past state is called smoothing, denoted by $\hat{x}(t-i|t)$. The Kalman filter is to obtain $\hat{x}(t+1|t+1)$ based on a tradeoff between the observation of $y(t+1)$ and the state prediction.

Before $y(t+1)$ is observed, the estimate of $x(t+1)$ is determined by the prediction according to Eq. (18):

$$\hat{x}(t+1|t) = \phi\hat{x}(t|t) + \psi u(t). \tag{12}$$

After observing $y(t+1)$, another estimate of $x(t+1)$ is determined according to Eq. (9). This estimate has the value of $\frac{1}{h}y(t+1)$. A tradeoff must be made between the prediction estimate and the observation estimate, depending on the variance of the measurement noise σ_v^2. If σ_v^2 is small, more trust or confidence is placed in the observation estimate. If σ_v^2 is large, more trust or confidence is placed in the prediction estimate. Thus, the Kalman filter estimates the process state as follows.

$$\hat{x}(t+1|t+1) = \hat{x}(t+1|t) + k(t+1)[y(t+1) - h\hat{x}(t+1|t)] \tag{13}$$

where $k(t+1)$ is a filter gain reflecting the tradeoff, and is determined by

$$k(t+1) = \frac{1}{h} \bullet \frac{h^2\sigma_{\tilde{x}}^2(t+1|t)}{[h^2\sigma_{\tilde{x}}^2(t+1 \mid t) + \sigma_v^2]} \tag{14}$$

$$\sigma_{\tilde{x}}^2(t+1|t) = \phi^2\,\sigma_{\tilde{x}}^2(t|t) + \gamma^2\sigma_w^2 \tag{15}$$

$$\sigma_{\tilde{x}}^2(t|t) = [1 - k(t)h]\sigma_{\tilde{x}}^2(t|t-1) \tag{16}$$

where $\sigma_{\tilde{x}}^2$ is the variance of the estimation error (for prediction and filtering). Eq. (14) specifies the relationship between $k(t+1)$ and σ_v^2. If σ_v^2 is small, $k(t+1)$ approaches $\frac{1}{h}$ so that $\hat{x}(t+1 \mid t+1)$ is based on the observation estimate. If σ_v^2 is large, $k(t+1)$ approaches 0 so that $\hat{x}(t+1 \mid t+1)$ is based on the prediction estimate. Eq. (14) also specifies that the variance of the measurement noise is evaluated relatively to the variance of the disturbances which contributes to the variance of the prediction error as described in Eq. (15). Eq. (15) is derived from Eq. (8) to determine the variance of the prediction error from the variance of the filtering error. Eq. (16) is derived from Eq. (13) with some manipulation. Eqs. (12), (13), (14), (15), and (16) give a recursive solution to estimate the state of the process. A computer can be used to calculate the state estimate iteratively.

Many processes may involve multiple input variables, multiple state variables, and multiple output variables. A multiple-input, multiple-output dynamic process can be described as follows.

$$X(t+1) = \Phi X(t) + \Psi U(t) \tag{17}$$

$$Y(t+1) = HX(t+1) \tag{18}$$

Similar equations for the Kalman filter can be derived using the matrix algebra (Rouse, 1980). State estimation is particularly useful when few measurements of the process output variables can be taken conveniently. State estimation allows us to estimate many process state variables from few measured output variables. It should be noted that a precise model of the process dynamics is required to make state estimation.

In practice, a common solution to multiple-variable process control is to separate the process into multiple single-variable processes. A controller is used for each single-variable process. The separation of the process and the development of the mathematical model for the dynamics of each single-variable process must be based on two considerations: 1) minimizing the effects of interactions between single-variables processes; and 2) ensuring the desired relationships between input variables and output variables of the overall process. This control method often yields satisfactory control performance.

2.6. Optimal Control

The optimal control is to adjust the process input according to the optimization of process performance (Rouse, 1980). The optimal control is a kind of feedforward, open-loop control. To use the optimal control, one must have perfect knowledge of the process and disturbances. The steady-state dynamics of the process and disturbances are usually described in a mathematical model. Most control optimization algorithms use this steady-state feature of the process dynamics to generate control solutions. A variety of mathematical techniques, such as differential equations, linear programming, and dynamic programming, have been developed to solve optimal control problems. A typical optimal control solution for a single-input, single-output, first-order linear dynamic process is presented here.

Since the process output is an observable or measurable form of the process state, the calculation of the process output error can be defined directly by the state variable of the process. The measured process output may differ from the true process output due to the measurement noise (e.g., inaccuracy of measurement device). That is, the calculation of the process output error is subject to the measurement noise. Thus, Eq. (9) can be rewritten as follows.

$$e(t+1) = gx(t+1) + v(t+1) \tag{19}$$

Eq. (8) and (19) describe the dynamics of the process and the disturbances. A criterion or index of performance is typically formulated using a quadratic function (Rouse, 1980):

$$J = E\left[\sum_{i=1}^{T} ae^2(t+i) + bu^2(t+i-1)\right] \tag{20}$$

where $E[\,]$ denotes the expected value. The optimal control tries to accomplish two goals: 1) to minimize the squared error of the process output; and 2) to minimize the squared control effort. The ratio a/b reflects the tradeoff between the accuracy of the process output and the control effort. There usually are costs associated with the control effort. More control effort results in less process error but increased control costs. On the other hand, the larger deviation of the process output from its set point may also be costly. The ratio a/b is determined by the process designer. If a/b is greater than 1, cost savings from minimizing the process error are considered more important. If a/b is less than 1, cost savings from minimizing the control effort are considered more important. If a/b is equal to 1, minimizing the process error and minimizing the control effort are considered equally important.

The solution to the optimal control problem is (Rouse, 1980):

$$u(t+i) = c(t+i)x(t+i) \tag{21}$$

where

$$c(t+i) = \frac{-\psi f(t+i+1)\phi}{\psi^2 f(t+i+1)+b} \tag{22}$$

$$f(t+i) = \phi^2 f(t+i+1) + \phi f(t+i+1)\psi c(t+i) + g^2 a \tag{23}$$

$$f(t+T) = 0. \tag{24}$$

Eqs. (21), (22), (23), and (24) give a recursive control solution. The process input or control action can be calculated iteratively. A detailed discussion of optimal state estimation and control can be found in a reference (Medith, 1969).

2.7. Adaptive Control

In some processes, the time-varying environment causes changes of the process dynamics. The parameters of the controller must be adjusted to reflect the changed process dynamics. An adaptive controller can adjust its parameters automatically according to the changing process dynamics. Optimization criteria of the process may be used to adjust the control parameters so that the controller produces the optimal process performance. Figure 7 shows a block diagram of adaptive control.

3. DISCRETE PROCESS CONTROL

Unlike continuous process control to maintain the desired process output or to optimize process performance, discrete process control aims at generating and

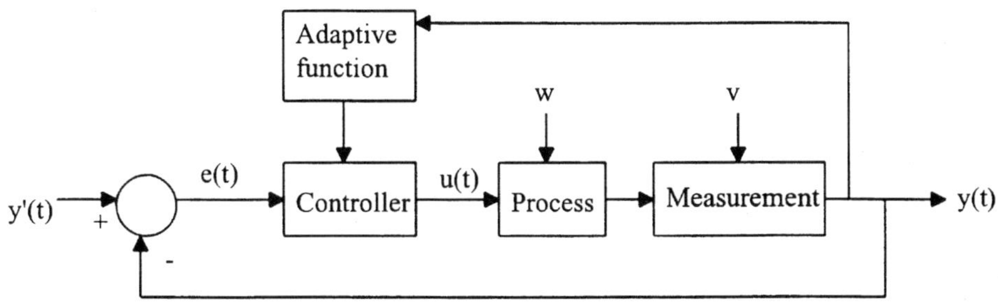

Fig. 7 Adaptive control.

following a sequence of operations. Depending on the nature of operations, the control of discrete-part manufacturing processes may be classified into two types: numerical control and logic control. In numerical control, movements of machining tools are controlled. In logic control, discrete values (e.g., on or off) of process outputs are controlled. A detailed discussion of discrete process control can be found in a reference (Bollinger and Duffie, 1988).

3.1. Numerical Control

The numerical control was originated in the machining tool industry. Before digital computers were developed, metal-cutting machining tools were controlled by the human. A template was usually used to guide the path of a cutting tool. As part shapes became more complex and the market demanded more small-batch productions, the manual control approach was no longer appropriate in terms of both productivity and product quality. When digital computers were introduced at a reasonable cost, the numerical control approach emerged. The numerical control (NC) is to program the preplanned sequence of tool movements. The control unit of the machine tool reads the instructions in the program and controls positions and movements of the machine tool accordingly. When the part shape changes, the sequence of tool movements can be reprogrammed.

The main requirement for numerical control is the positional control of a processing unit (e.g., a cutting tool) in relation to a work part being processed. The relative positions of the processing unit and the work part are described using a coordinate system such as the Cartesian coordinate system for $x, y,$ and z axes plus rotations $a, b,$ and c axes along x, y, z axes respectively. The control of tool movements can be carried out in two levels: 1) point-to-point regardless of the path used to move the process unit from one point to another point; and 2) continuous path with the relative position of the processing unit being continuously controlled in real time to produce a smooth motion. For point-to-point control, the movement on each axe can be performed separately, whereas for continuous path control the movement on one axe must be coordinated with the movements on other axes.

Tool movements can be stored on the computer as a set of points. When generating a series of movement commands, intermediate points are filled in using linear interpolation of the stored points. A motion path can be broken down into a number of segments. Segments can be approximately represented using third-order polynomials which are often referred to as cubic spline functions. The cubic spline functions are continuous at segment boundaries.

In early days of computer applications, computers were expensive. A mainframe computer was used to direct activities of a number of separate NC machines in order to justify the high cost of the computer. This is called direct numerical control (DNC). The computer was connected to the control units of those NC machines. The control program for each NC machine is down-loaded from the computer to the control unit of the NC machine. The computer collects performance data (e.g., part counts, cycle times, machine utilization, etc.) from each NC machine. The performance data provide the basis for a high-level management of the plant.

As the price of computer drops drastically, the cost of computers is no longer a major concern. A small computer is dedicated to each NC machine. The computer directly connects to the NC machine and serves as the machining control unit. This is what is often referred to as computer numerical control (CNC). If CNC machines are connected to a central computer for high-level management, distributed numerical control is formed.

3.2. Logic Control

Many industrial processes concern mainly changes of discrete states (e.g., on or off) of their components. Usually discrete states of components can be described in a binary form. This type of control is called logic control. For instance, the control requirement of a process may be to turn on or off of various output devices according to the condition (on or off) of input devices. Among common input devices are push buttons, limit switches, and relay contacts. Among common output devices are indicator lights and solenoids.

Depending on the requirement for the predetermined sequence of operations, logic control can take two forms: combinational logic control and sequential logic control. In combinational logic control, only the combination of input conditions are used to determine the state of outputs. In sequential logic control, the state of outputs is determined by both the combination of input conditions and the state of the process indicating what has occurred. Therefore, memory functions are needed to support sequential logic control. Sequential logic control can be performed in either a close-loop fashion or an open-loop fashion. In the close-loop control, the feedback on the proper execution of a step is necessary to execute the next step in the sequence of operations. The open-loop control simply executes the predetermined sequence of operations without feedback.

Boolean algebra is the primary tool used to design logic control. Boolean algebra consists of basic operations AND, OR, NOT, NAND, NOR, and XOR (exclusive OR). Table 1 gives the truth table of those operations. Boolean theorems such as $(X+Y)(X+Z)=X+YZ$ can be used to manipulate logical expressions and simplify the logic circuits. The bus driver logic as shown in Table 2 is used to control the information flow over a communication bus shared by a number of components in the process. A bus driver includes an enable input (E) that breaks or establishes the connection from the input (X) to the output (Y). The $\bar{R}-\bar{S}(\overline{RESET}-\overline{SET})$ flip-flop logic as shown in Table 3 is used to construct memory functions. Setting either input $\bar{R}$ or input $\bar{S}$ briefly to 0 changes the content of memory. The memory remains unchanged as long as $\bar{R}$ and $\bar{S}$ are set to 1. The D flip-flop logic as shown in Table 4 is used for data sampling. When the clock input (C) is set to 1, the value of data input (D) is passed to the output X. The outputs of the D flip-flop remain unchanged as long as input C is held at 0. D flip-flops can also be used to construct memory functions.

A set of Boolean equations describing the control relationships between inputs and outputs can be implemented using relay control circuits in which electromechanical relays are wired or computers in which solid-state electronic gates are combined. In relay control circuits, relays are wired to accomplish Boolean equations. In computers, solid-state electronic gates are used to implement basic logic operations and memory functions which in turn provide the basis of programming Boolean equations. Relay control circuits directly connect input devices

TABLE 1
Truth table of logic operations.

		AND	OR	NOT	NAND	NOR	XOR
X	Y	$X \bullet Y$	$X + Y$	$\bar{X}$	$\overline{X \bullet Y}$	$\overline{X + Y}$	$X \oplus Y$
0	0	0	0	1	1	1	0
0	1	0	1	1	1	0	1
1	0	0	1	0	1	0	1
1	1	1	1	0	0	0	0

TABLE 2
Truth table of bus driver logic.

E (enable input)	X	Y
0	0	no connection
0	1	no connection
1	0	0
1	1	1

TABLE 3
Truth table of $\bar{R} - \bar{S}$ flip-flop logic.

$\bar{R}$	$\bar{S}$	X	$\bar{X}$
0	0	1	1
0	1	0	1
1	0	1	0
1	1	X	$\bar{X}$

TABLE 4
Truth table of D flip-flop Logic.

C	D	X	$\bar{X}$
0	0	X	$\bar{X}$
0	1	X	$\bar{X}$
1	0	0	1
1	1	1	0

to output devices. Computers sample logic values from input devices periodically, evaluate Boolean equations, and send logic output signals to output devices. Hence, there are no hard-wired connections between input devices and output devices. The sampling interval must be shorter than the duration of most frequently occurred input/output events.

4. CONTROLLER HARDWARE

4.1. Analog Controller

Traditionally, the PID control was implemented using analog controllers. Figure 8 shows a PID analog circuit (Hunter, 1987). An analog controller is composed of an analog circuit performing the PID control function, a comparator measuring the error of the process output, operator interface and display units. The analog controller allows one to calibrate the control gain and to enter the set point. Information such as the set point, measured output, error of the output, and control input can be displayed. An analog controller is used for one controlled variable. If a complex process requires multiple control loops for multiple variables, a number of analog controllers must be used. This may induce substantial hardware cost.

4.2. Relay Control Circuits

Logic control was implemented using relay control circuits in the past. Relay control circuits are rungs of input devices wired with output devices. Typical input devices are electromechanical relays consisting of coil and switching contacts. When the coil is energized, a core in the magnetic field moves. When the coil is de-energized, the core is pulled back by a spring to its original position. When the core moves, the switching contacts open and close. The normally open contacts stay in an open position when the coil is de-energized. The normally closed contacts stay in a closed position when the relay coil is de-energized. Other input devices include push-button switches and limit switches. Typical output devices are solenoids, switches, and lamps.

Ladder diagrams were a conventional tool to design relay control circuits. In a ladder diagram, input devices in each rung are placed on the left side, and output devices in the same rung are placed on the right side. Each rung is a separate circuit. Two long vertical lines represent the electrical power. Figure 9 shows a ladder diagram for logic operations. Since Boolean equations describing the control sequence are hardwired in relay control circuits, changes in control functions are hard to make.

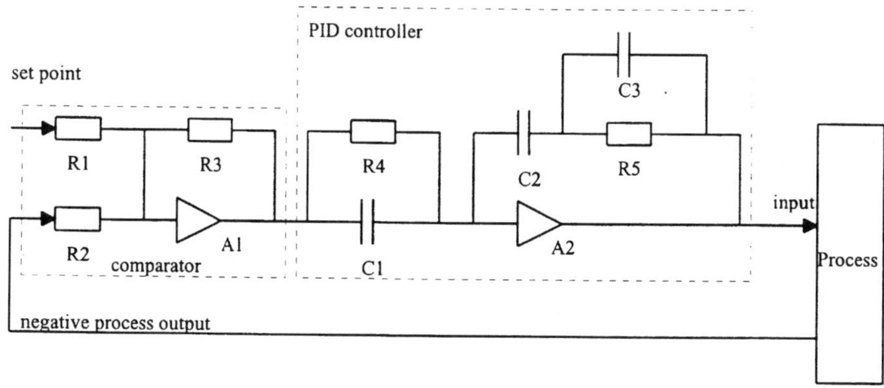

Fig. 8 A PID analog controller using resistors, capacitors and a signal amplifier.

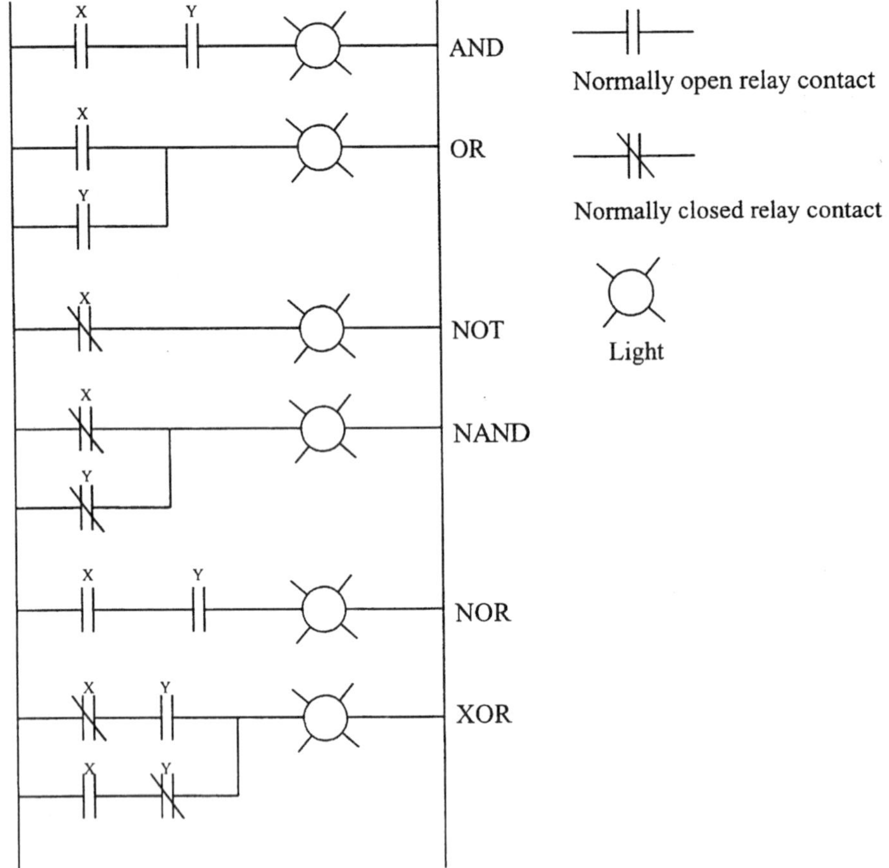

Fig. 9 Ladder logic diagram for various logic operations.

4.3. Programmable Logic Control

Computers, which are designed specially for logic control and have convenient interfaces with input and output devices, are called programmable logic controllers (PLCs). They were developed to replace relay control circuits. However, modern PLCs can accept both analog (e.g., current or voltage) and digital input signals, and generate both analog and digital output signals. The control sequence is programmed using a series of computer instructions to evaluate Boolean equations. Therefore, changes in the control sequence can easily be made by changing the control program.

4.4. Computer Digital Control

Digital computers can be used for both continuous process control and discrete process control. For both continuous process and discrete process, the computer operates based on sampled data. At discrete points in time, the computer collects process outputs, evaluates control functions, and sends out process inputs. A series of data sampled over time are often referred to as time series data. For continuous process, the sample period is determined considering the process stability and the growth of disturbances. For discrete process, the sample period must be less than the duration of the shortest input or output event. In addition, the time to accomplish control calculations should be less than the sample period. The calculation of sophisticated control functions may cause a time delay between the sampling of process outputs and the generation of process inputs. To reduce the time delay, more control loops may be included. More calculations may also be carried out in unused time between samples.

Since the computer processes digital signals, analog-to-digital converters (ADCs) and digital-to-analog converters (DACs) are required to connect the computer to the process which receives and produces both analog and digital signals. One common technique of transforming an analog signal to the corresponding digital data is used in the linear ramp encoder. The linear ramp encoder includes a comparator relating the unknown analog voltage to the analog signal produced by a ramp generator. The ramp generator increases its voltage by a fixed amount at a constant rate. A binary counter counts the number of pulses from a digital clock, and thus measures the time passed as the ramp generator is increasing its output voltage. When the produced voltage is equal to the unknown voltage, the binary counter stops counting. A digital value for the unknown analog voltage is then determined by multiplying the passed time to the rate of voltage increase. DACs transform the digital data into a stepwise, continuous analog signal. In each step, a digital value is decoded into an analog voltage which is held for an interval.

Since the computer is faster than input/output devices, the computer usually communicates with input/output devices using the interrupt facility. The interrupt facility of the computer allows computer resources to be used by other programs while input/output devices are not ready to send or receive data. When input/output devices are ready, an interrupt signal is sent to the computer. The interrupt facility can be enabled or disabled. If the interrupt facility is enabled, the computer responds to the interrupt signal by suspending its running program and executing a predetermined interrupt-handling procedure. After the procedure is accomplished, the computer resumes the interrupted program. For example, the data-sampling clock can be connected to an interrupt input, and control functions are included in the interrupt-handling procedure. An operator input such as pushing an emergency stop button can be connected to an interrupt input. Alarm indicators can also be connected to interrupt inputs.

Figure 10 shows common computer interfaces and corresponding software procedures for digital-to-analog conversion (DAC), analog-to-digital conversion (ADC), logic output (DLC), logic input (LDC), as well as enable (CONNECT) and disabled facility of interrupt handling. Each type of interface facility may have multiple channels. Therefore, each procedure must specify the interface channel with which it is associated.

When computers for process-level control are connected to a central computer managing a high-level operation of the plant, a distributed computer network is formed. Some typical structures of communications between computers in the distributed system are shown in Figure 11.

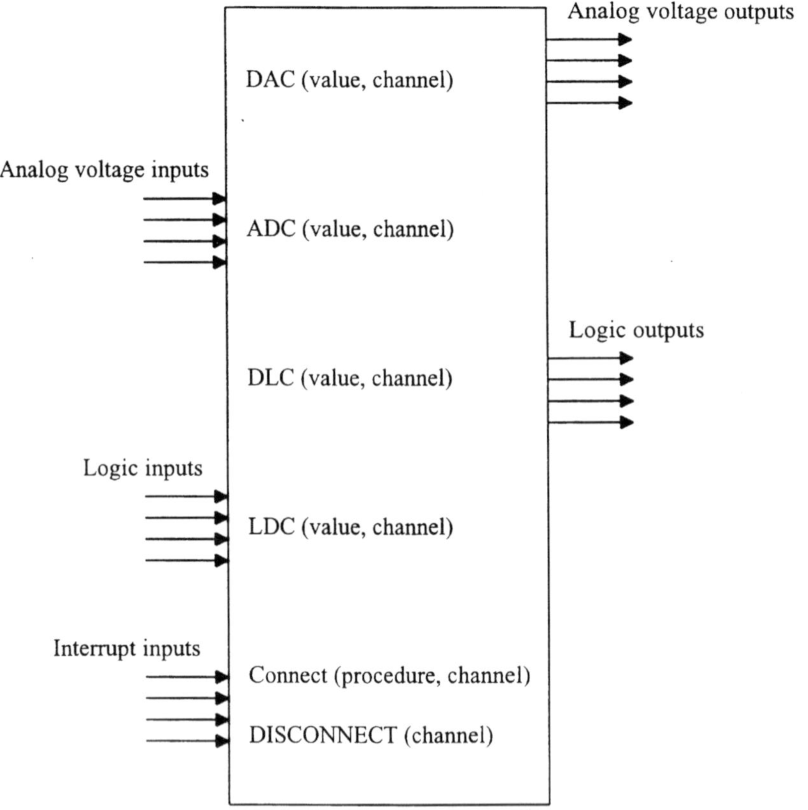

Fig. 10 Computer interfaces with input/output devices and software procedures.

4.5. Sensors and Actuators

Transducers are devices transforming one type of energy to another. Transducers used to measure physical variables are referred to as sensors. Sensors are used as input devices in process control to measure important variables in the process. Sensors may measure position, velocity, acceleration, force, temperature, strain, and so on. Among sensors producing analog outputs are resolvers, thermocouples, tachmeters, and stain gauges. Among sensors producing digital outputs are switches and encoders.

Actuators are devices converting input signals to mechanical motions which in turn change process variables through valves, cylinders, etc. Actuators are used as output devices in process control. Typical actuators are solenoids which use an electrical signal to move a magnetic object. Among other common actuators are linear motion solenoids, rotary motion solenoids, direct current motors, alternating current motors, and stepper motors.

Detailed discussions of sensors and actuators can be found in the references (Soloman, 1994; Parr, 1987).

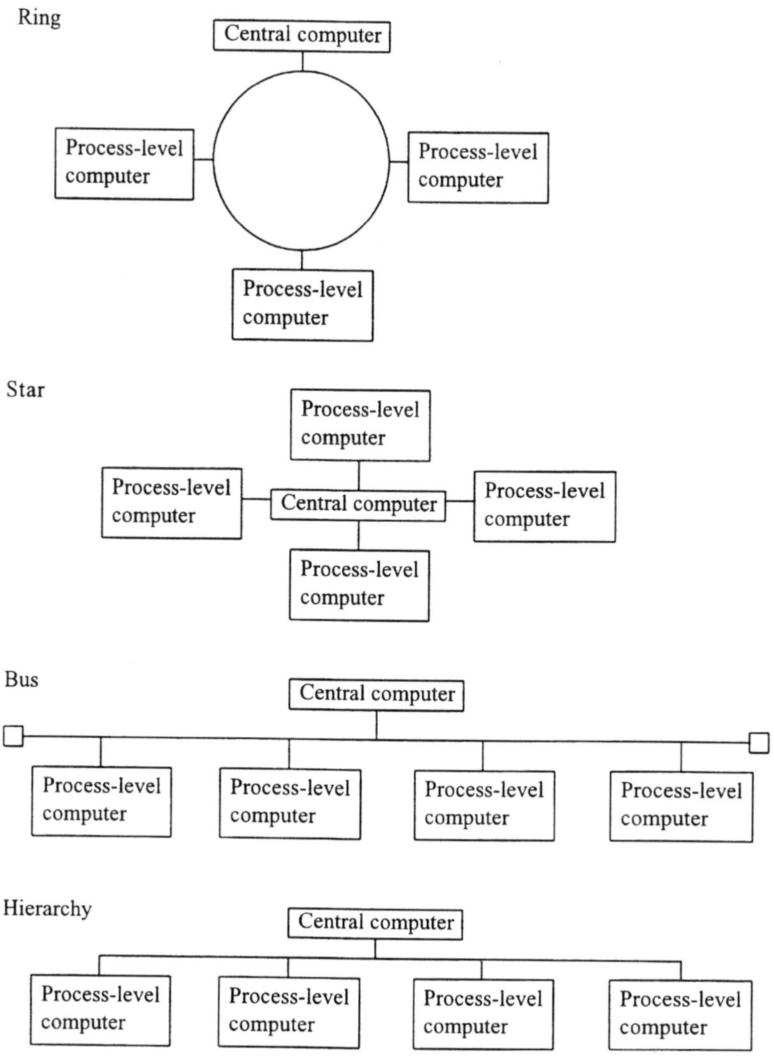

Fig. 11 Distributed computer system.

5. HUMAN ROLES IN PROCESS CONTROL

Some of activities performed by the human (i.e., engineer, technician, and operator) in designing and maintaining the process control are presented below.

1) Define the physical configuration of the process control, including measurements to be taken, actuations to be set up, controller modes to be chosen, and connections between inputs, controllers, and outputs.
2) Determine and calibrate control parameters such as the control gain and sampling interval.

3) Specify control requirements and program the control sequence.
4) Monitor the process control and adjust the set points.
5) Respond to abnormality in the process through fault diagnosis and operation recovery.

Among the above human responsibilities, diagnosis of process faults such as failures or malfunctions presents considerable difficulty to the human (Ye, 1994a; Ye, 1994b; Ye, Zhao, and Salvendy, 1993; Zeng and Wang, 1991). Some common causes of process faults are hardware failures, design flaws, environmental interferences, variations in processed objects, and human mistakes. Difficulty of diagnosing faults may be attributed to the complexity of the process itself and its automation, the unpredictable nature of many faults, and limitations of human cognitive capacity.

When a fault occurs in a complex process, many intermediate states of the process are undetectable and thus unknown. If one wants to reason from fault symptoms backward to trace down root causes, one needs to consider an overwhelming number of possible cause-effect paths. Since unknown intermediate states of the system do not help in eliminating some possible combinations of cause-effect paths, one faces the problem of combinatorial explosion. That is, the complexity of the process results in information overload over the limited cognitive channels of the human. Furthermore, fault symptoms often do not reflect the condition of the process at the time of fault occurrence but many steps after that. There is usually a time delay in the presentation of fault symptoms. This aggravates the problem of establishing the connection between fault symptoms and fault causes.

The human is often not able to perform reliably under stress induced by a fault event, sometimes an emergency situation. Under stress, the human concentrates on just one hypothesis of fault causes and corresponding remedial actions. Even if remedial actions do not eliminate fault symptoms, the human simply tries harder rather than giving up the hypothesis and searching for another hypothesis. This phenomenon is often referred to as "cognitive lockup."

Although process automation relieves the human from worrying about the normal operation of the system, process automation reduces requirements for the human to understand the process. Without adequate knowledge of the process and awareness of the process operation context, the human may be in a bigger trouble than before when the process malfunctions and needs human intervention. Moreover, unpredictable fault events make many forms of human training ineffective.

Since process faults result in production downtime and cost, fault diagnosis will become increasingly critical to the overall control of production costs. Research and development of diagnostic functions will attract more and more attention. Attention should also be paid to the integration of diagnostic functions with control functions. Detailed discussions of human roles in process control and diagnosis can be found in the references (Rasmussen and Rouse, 1981; Rasmussen, 1986).

References

J.G. Balchen and K.I. Mumme, *Process Control*, Van Nostrand Reinhold Company, New York (1988).

J.G. Bollinger and N.A. Duffie, *Computer Control of Machines and Processes*, Addison-Wesley, Reading, Massachusetts (1988).

R.P. Hunter, *Automated Process Control Systems: Concepts and Hardware*, Prentice Hall, Englewood Cliffs, New Jersey (1987).

J.S. Medith, *Stochastic Optimal Linear Estimation and Control*, McGraw-Hill, New York (1969).

E.A. Parr, *Industrial Control Handbook*, BSP Professional Books, Oxford, London (1986).

J. Rasmussen, *Information Processing and Human-Machine Interaction*, North-Holland, New York (1986).

J. Rasmussen and W.B. Rouse, *Human Detection and Diagnosis of System Failures*, Plenum Press, New York (1981).

W.B. Rouse, *Systems Engineering Models of Human-Machine Interaction*, North Holland, New York (1980).

S. Soloman, *Sensors and Control Systems in Manufacturing*, MacGraw-Hill, New York (1994).

N. Ye, Computer simulation of manufacturing systems for exploration of human information needs and strategies in system fault diagnosis, in *Proceedings of 1994 IEEE International Conference on Systems, Man and Cybernetics*, IEEE Press, Piscataway, New Jersey, pp. 2190–2194 (1994a).

N. Ye, Fault diagnosis of advanced manufacturing systems: Operator strategies and performance support with neural networks, in *Proceedings of The Fourth International Conference on Human Aspects of Advanced Manufacturing & Hybrid Automation*, pp. 445–448 (1994b).

N. Ye, B. Zhao and G. Salvendy, Neural-networks-aided fault diagnosis in supervisory control of advanced manufacturing systems. *International Journal of Advanced Manufacturing Technology*, **8**, 200–209 (1993).

L. Zeng and H.-P. Wang, Machine-fault classification: A fuzzy-set approach. *International Journal of Advanced Manufacturing Technology*, **6**, 83–94 (1991).

CHAPTER 11

Rapid Manufacturing Systems Modeling with Integrated Design Models

DAVID KOONCE[a], ROBERT P. JUDD[b] and CHARLES M. PARKS[c]

[a,c]Industrial and Manufacturing Systems Engineering Department, Russ College of Engineering and Technology, Ohio University, Athens, Ohio 45701, USA; [b]School of Electrical Engineering and Computer Science, Russ College of Engineering and Technology, Ohio University, Athens, Ohio 45701, USA

Manufacturing system design is a complex process involving the integration of multiple systems designed by several designers each optimizing their own subsystems. The life cycle of manufacturing systems design is also problematic because it is: incomprehensible by a single person, interdisciplinary in approach, asynchronous in operation, and constantly evolving. Manufacturing systems also have long development times with requirements that are not well understood in the beginning and that change during the design/build cycle. These issues are not appropriately addressed by the traditional serial design process. In view of these issues, we suggest that the manufacturing systems design process must become less serial and more concurrent. But, coordinating the design efforts of the multi-disciplinary teams involved in the design of manufacturing systems is a difficult task. Various designers have their own tools for executing their methodologies. Solving this problem requires a new technology that enables prototyping of complex multimodel manufacturing systems. We propose an integration technology that enables the systematic integration of new design tools and thus is readily extendible to include new design domains, methodologies, and technology. In this chapter, we characterize the manufacturing systems design problem and propose an architecture that supports manufacturing engineering in this complex, data-rich environment.

This chapter starts with a discussion of the current manufacturing systems design environment that exists today. The first section identifies the need to create an integrated environment for manufacturing design. Section 2 describes some techniques to analyze the information contained in design tools which cover different, but overlapping domains. These techniques are used in the next section which describes an architecture for an integrated design environment. The architecture provides a system in which tools covering many of the domains concerned with manufacturing design can be easily plugged together to create a highly effective design environment. The chapter concludes with an example of how a: process definition tool, a capacity planning tool, a layout tool and a simulation tool are integrated using this approach. To keep the example concise, the data representation of the tools have been greatly simplified.

1. BACKGROUND

This section describes many of the problems of the current manufacturing design environment. To solve these problems an integrated environment which allows virtual prototyping is proposed. The section continues to present some of the issues the environment addresses. Finally, related work in developing integrated environments both in the manufacturing domain and others, like VLSI and CAD, is described.

1.1. The Need for an Integrated Environment for Manufacturing Systems Design

Modern manufacturing system design shares many of the problems associated with the design of any large, complex system. It is an intricate, data rich process that produces multiple designs. These designs are expressed as models such as: system requirements, capacity plans, layout designs, simulation models, and information processing models (Air Force, MANTECH, 1994). A common theme in these models is that they all seek to represent the same manufacturing system. However they differ in their domain perspective, level of abstraction, and stage of the life cycle. Because of these differences, the elements and relationships in each model impact the elements and relationships in other models. First, the data in one model may be an abstraction of the data in another model. For example, a production line in a capacity plan is an abstraction of the machine tools specified in a layout design. This abstraction creates a parent child relationship that must be known and understood in order to propagate changes to models representing the same entities. Second, different data attributes in separate models may represent the same real world entity. For example, data about a machine tool must be represented in a capacity planning model, a layout model, and a simulation model. Appropriate changes in one model's attribute set must be propagated to instances of the affected entity in other models. Third, models can place constraints on other models. For example, the capacity plan for the production line specifies a production rate that must be met by other models and verified by a simulation model.

To integrate the various design models into a unified model of the system, the architecture allows designers to use the tools and methods best-suited to their models. Such an environment functions in the following way. An analyst enters the requirements of the proposed manufacturing system. Next, the requirements model is used to initialize the high level design models, such as the capacity planning and layout models. These models are then further refined. The unifying environment allows the constraints imposed by the requirements on these design models to be automatically checked and validated. Further, the requirements model can be automatically refined, based on the finished design models.

1.2. Integration Concepts

Three fundamental concepts are important in manufacturing systems engineering and design for large, complex manufacturing systems: iterative life cycle issues, integrated design methods and the types of integration among design methods. These concepts provide the foundation for a totally integrated design environment.

1.2.1. Iterative Life Cycles

Iterative life cycles address several of the problems in implementing complex systems. Iterative life cycles have risk identification and risk reduction as central concerns. The Spiral (Boehm, 1988) and Dynamic Systems Engineering (DSE) (Wiskerchen, 1989) life cycles start with these activities, viewing the traditional life cycle steps as either a degenerate case of risk reduction (Spiral) or as a final step, once all the significant risks have been identified and contained. Both these life cycle views recognize prototyping as the primary way to reduce technical risk. Prototyping allows the user to experiment with a relatively low-cost example of the developer's understanding of the requirements, and modify those requirements before much of the development budget has been expended. Unfortunately, physical prototyping and mockups are often prohibitively expensive and time consuming. With modern computing technology however, virtual prototyping and simulation can replace physical prototypes. In fact, with virtual prototypes, more experimentation and optimization is feasible.

A key requirement for iterative life cycles is the ability to move quickly between requirements definition, model design, virtual prototypes, and developing system specifications. Lessons learned while prototyping often imply changes in designs and even requirements. The effects of late design and/or technical changes must flow easily and quickly back to the requirements, designs and virtual prototypes. Even in the late stages of the system's life cycle (operation and maintenance), maintaining valid design and prototype models will help further testing and optimization of proposed changes to the system. Therefore, an environment which enables the integration of design methods across the stages of the life cycle will have a significant impact on the development and operation of manufacturing systems.

1.2.2. Integrated Design Methods

The realization of a manufacturing system involves the cooperation of designers and engineers from several disciplines. Information about requirements, costs and constraints must flow freely among these engineers. To design their systems effectively, each of these cooperating disciplines have developed domain specific methods and modeling techniques to help them analyze the system from their unique perspectives. Instead of requiring a consistent modeling approach for every discipline, we use an integration methodology which melds the existing methods and models. However, such a methodology requires a technology of its own which is the foundation for this chapter. To begin, we categorize design models along the following axes: domain view, level of abstraction and life cycle phase. These categories and the requirements they impose on an integrated environment are detailed below.

1. *Domain views.* Models of a manufacturing system are generally be done from the perspective of some engineering or design domain. To assure the system's design is consistent the architecture requires integration among these different discipline views of the system (e.g., software, data, mechanical, electrical, industrial, and systems).
2. *Levels of abstraction.* Models of a manufacturing system will conform to some level of detail and scope. Integration among the different levels of detail (from a description of the entire system down to the detailed models of individual components of the system) verifies that the constraints imposed by higher models are satisfied by lower level models.
3. *Life cycle phases.* Models and prototypes are also characterized by the developmental stage in the system's life cycle. Integration among the phases of the design life cycle (requirements models, design models, prototypes, and specifications) allows designers to incorporate feedback from the later design phases into the earlier models. Thus, the prototyping process often refines and adds new system requirements.

1.2.3. Types of Integration

We define three types of integration techniques which exist among the models with various domain views, levels of abstraction and life cycle phases. These techniques are static translation, dynamic integration, and model transformation.

1. Two similar (with respect to domain view, level of abstraction, and life cycle phase) design tools may exchange static information through an agreed upon file format. This is *static translation*. Examples in this category include CAD tool file sharing. This type of integration is well understood and enables mapping directly the entities in one model to the appropriate entities in another. It usually involves setting file formats standards (IGES, PDES, DXF, IDL, CDIF, etc.) for the tools to read to or write from. A number of commercially available tools perform static translation.
2. Suppose two tools contain different portions of a dynamic (executable) model. Then, state information must be shared across the tools. The ability to exchange state information at critical times during the execution of the model is *dynamic integration*. An example of this type of integration would be a distributed simulation environment in which state changes in each model are propagated to the other models for verification. Lamport (1978), Misra (1986) and Judd *et al.*, (1993) provide a theoretical basis for dynamic integration.
3. When two tools have very different domain views, levels of abstraction, or life cycle phases, then the information that they have about the model only partially overlaps. These are called heterogeneous databases. *Model transformation* transfers only the shared portions of the models. Great care must be taken to make sure consistency remains among the data in the two tools, as data is moved between the tools. For example, consider a model of a factory created by a simulationist. Many control issues, like queue discipline, routings, resource contention, etc., are decided by the simulationist while the model is being built. These decisions are buried in the simulation code and are not usually well documented. Later, the control and software engineers develop the controlling software for the manufacturing system. Again, system control issues are decided, but with a new group of engineers. Typically, they will not use the simulation model to guide them in the selection of the control strategies. As a result, the expected and actual performance of the system may be different. Model transformations occur among models in different disciplines, at different levels of abstraction and at different stages of the life cycle. Model transformation is a little-researched field, though some narrow variants have received considerable attention. One activity in this area has focused on the transformation of design models into implemented systems using software engineering (such as fourth generation languages or tools to convert structure charts into code shells).

1.3. Related Work

Many computerized tools exist which aid in the design and prototyping of manufacturing systems. These tools operate at different levels of abstractions, along different disciplines and at different stages of the life cycle. This section discusses some of the integration technology existing now, which facilitates the sharing and exchange of information between all of the different design methods and tools.

1.3.1. VLSI Success

An important success story in integrating design methods is the rapid prototyping of VLSI chips. There are now tools that exchange information about VLSI chip design from the algorithmic level to the logic and gate level and finally, to the device and layout design level. The models these tools generate exist at different levels of abstraction and use completely different notations; however, they describe the same device – the chip. The models at the high layers of abstraction impose constraints on the lower-level models. Engineers at the lower level (highly detailed design) know when these constraints are violated and can take corrective measures. The impact of changes at any level can be translated to the other levels. Finally, the standard solutions of lower-level models can be archived and maintained in libraries, creating the opportunity for rapid development of the gate and device designs once the algorithmic designs are complete. A successful integrated manufacturing design system must have similar capabilities.

1.3.2. Product Design Tools

New, powerful tools exist that facilitate the product design process. In particular, there are many tools that allow much of the design to be tested before the hardware and/or software is built. The kinematics, control, programming operation, and geometrical aspects of the system can be computer simulated. For instance, Williams and Pentland (1992) have developed a "prototype design tool" that encompasses an agile CAD system with tightly coupled physics. Arai and Iwata (1991) describe the development of a kinematic simulation system used in assembly planning that simulates physical phenomena including the effects of gravity. With the use of the product model presented by 3-D solid geometry, possible movements of each part in the product are calculated, and the removal possibilities from the product can be searched. These design tools focus on the integration of different domains in the design of products, however integrated tools for manufacturing system design have similar characteristics.

1.3.3. Tool Integration for Manufacturing System Design

Tool and model integration for manufacturing system design is very complex. Two problems occur as information is transferred among the tools. The first problem is simply that of syntax; two languages can say different things which requires translating one format into another format (static translation). The second problem is that of semantics; for example, "socket" has very different meanings in mechanical, electrical and software engineering domain models. Thus, semantic unification is very important (model transformation). Tools which allow these translations and semantic unifications are beginning to emerge from research following different approaches for model integration (Fox, 1992 and Woelk *et al.*, 1992). However, their application to discrete manufacturing systems design is not proven and constitutes a very important research issue.

Another dimension of this problem is represented by the degree of concurrence. The traditional product realization model is a series of specialized tasks: designing, prototyping, capacity planning, manufacturing, assembling, finishing, testing, and marketing (Gadh *et al.*, 1991). The mere fact that design had to be performed before prototyping, and prototyping before manufacturing suggests a serial process. Design divisions rarely had expertise in prototyping or manufacturing, so non-manufacturable designs were released to the downstream stages. Recently, concurrent engineering systems have tried to perform a significant amount of downstream analysis in a concurrent or simultaneous fashion during product design (Desa *et al.*, 1991). Such systems are based on a variety of inputs such as geometric data, product feature data, process information, cost knowledge, etc. Again, an integrated tool set is needed to enable this concurrency (Mc Ginnis, 1991).

Several national and international efforts are promoting frameworks to integrate design tools. These efforts focus primarily on the development of a common object-oriented data repository that forms the basis for sharing information across tools. Prominent initiatives are: Engineering Information System (EIS), CAD Framework Initiative (CFI), CASE Integration Services (CIS), Portable Common Tool Environment (PCTE+), CIM Open Systems Architecture (CIM-OSA) and the DARPA initiative on Concurrent Engineering (DICE).

The integration efforts in these approaches, however focus on product realization and not the integration of design methods and tools for manufacturing systems. In addition, these efforts only address the exchange of static (structural) information. The integration of dynamic (state) information among tools which have executable models is not addressed. This is crucial to the virtual prototyping of manufacturing systems. Another drawback of these approaches is a reliance on a common data repository approach to store the product design information. This common model constraints the tools to a core set of representable objects and relationships.

1.3.4. Previous Work in Integrating Design Tools

An early effort at the integration of diverse design tools is the Integrated Information Systems Evolution Environment (IISEE) (Mayer, 1989) which focused on the development of technologies for enabling the planning, definition, development and maintenance of integrated information systems. This effort uses meta-models of information methodologies to analyze their similarities and the possibility of exchanging information among them.

Also, there have been two recent efforts which show promise in enabling the integration of executable tools. XSpec (Judd *et al.*, 1991) is an object-oriented method used to specify the behavior of a manufacturing system. It is particularly useful in accurately specifying the interaction of simulation models of different portions of a manufacturing system that may be written in different languages and on different tools. XFaST (Judd *et al.*, 1993) provides a framework in which executable models can be developed independently, but executed concurrently. It allows various executable models written in different languages to be built and executed together to determine system behavior. A synchronization module is used to exchange state information among the elements of the model and to synchronize the simulation time among the tools.

ENVISION (Heim, 1994) is an integrating environment which seeks to coordinate multiple executable models of a manufacturing facility, similar to XSpec. These models are independently developed models such as simulation models of the materials handling system. A model network is then built from these individual models by creating an event signal map which indicates state changes that affect other models. Although this approach can identify inconsistencies in the dynamic behavior of system models, a more comprehensive architecture is needed to coordinate all models in a manufacturing design environment.

The Integrated Tool Kit and Methods program (Sauter and Judd, 1993), (Sauter *et al.*, 1993) integrated the IDEFO functional analysis tool with XSpec. This approach uses a central data repository to store the model data. The common data meta-model was developed by integrating data models of the individual tools. ITKM has the following limitations: it integrates only two tools, it uses a central repository, to add tools to the system requires a complete redesign of the data repository, and individual interfaces must be written for each tool in the system.

IDEM (Wang *et al.*, 1993) proposes a new integrated design methodology for manufacturing systems. Similar to the IDEF initiative of the U.S. Air Force, IDEM follows the three model classifications of function, information and dynamic. Functional models are developed using an extended IDEFO methodology. An object-oriented modeling approach is utilized for the information entities in a system model and dynamic behavior is modeled using a rule base. While this approach provides the links necessary to identify many of the design conflicts, it requires an entirely new approach to systems design and constrains designers to use its design methodologies instead of familiar methods.

2. TECHNIQUES

This section describes several techniques to describe and manipulate databases. The techniques are very useful to describe and understand how the integrating architecture works. Throughout the rest of this chapter we will use a relational description of the data within a tool. The first part of this section describes IDEF1X which is one of several methods used to model entity relationships. Using IDEF1X the common data that must be shared among very different tools can be identified. The second portion of the section describes methods that move data among databases whose information only partially overlap.

2.1. Information Modeling

Dr. Peter Chen's seminal paper, "The Entity-Relationship Model – Toward a Unified View of Data" (Chen, 1976) presents the principles and methods for entity relationship models. This section summarizes these ideas. In developing an information model, we seek to identify the types of information that will be represented in the information system. This information model is based on the world as the application will see it. The boxes and lines in the model represent things in the real world that the information system needs to know. The information in the model is classified as one of two types: entities or the relationships between these entities.

Entities are the things that we wish to record data about. There are six primary classes of entities: people, things (objects), places, organizations, events and concepts (Fertuck, 1995). It is these things that will be recorded in the database. Entities describe types of data to be stored in a database. When the database is populated there will be many records for every entity. Each record is called an *instance* of the entity. Entities however, are not enough to express all the information. Relations are the links that tie entities together.

2.1.1. IDEF1X

This chapter uses the IDEF1X syntax to express entity-relationship models. However, the choice of a specific ER notation has no bearing on the model or the techniques used to link entities among models. IDEF1X was developed by the U.S. Air Force's Integrated Computer Aided Manufacturing program. IDEF1X proposes to support integration in CIM environments through development of a single semantic definition of the data in a system (FIPS 184, 1993). The IDEF1X approach uses a formal development process that eliminates N:M cardinalities and n-ary relationships ($n > 2$). This produces a data model that can be transformed into a normalized set of relations (a desirable state) that an integrated information system can utilize.

We will use the following example to review syntax of this modeling method. Figure 1 shows an entity-relationship diagram of a simple capacity planning tool. At the top, an entity SYSTEM represents the manufacturing system under consideration. SYSTEM has a unique ID (S-ID) for each of its instances, this ID is called the *primary key*. The primary key is used to identify each instance of the entity. Every entity must have a primary key. SYSTEM's name (S-Name) is also unique for every instance and is called an *alternate key* (AK1). A manufacturing system has to manufacture something and this is represented as the entity PRODUCT. PRODUCT instances have a product ID (P-ID) that when combined with the system ID produce a unique identifier for each product. The link between SYSTEM and PRODUCT forces the SYSTEM's primary key (S-ID) to migrate to the PRODUCT entity. This link is called a *relationship*. For this relation, the SYSTEM entity is called the *parent* entity and PRODUCT entity is termed the *child*. Notice that like entities, relationships are titled. LABOR and RESOURCES are necessary to produce products, and therefore, are also included as entities in the information system.

Each entity has a set of attributes that describe the properties of the entity. For example, the PRODUCTS entity has an individual production requirement to be fulfilled (ProductionReq) a cost (P-Cost), and a description (P-Description).

Each relation has an implied cardinality. That is, how many child instances are there for each parent instance. The solid dot on RESOURCE signifies that a system may have zero, one or many resources. The implication is that a system has many

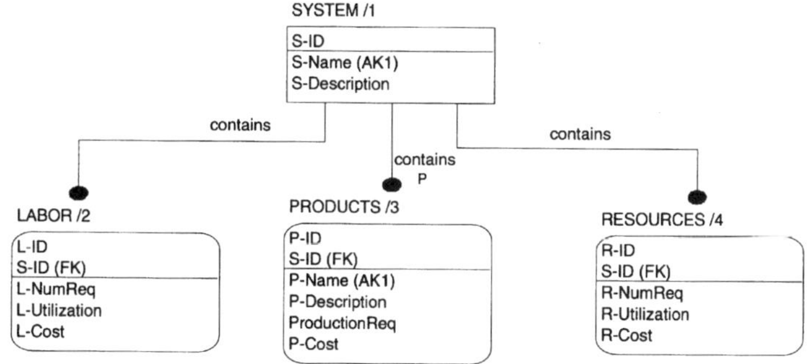

Fig. 1 Simple ER diagram (IDEF1.X syntax).

resources but, a single resource can only be placed in one system. The P next to the dot relating products to system forces a system to have at least one product.

2.1.2. Other Information Modeling Languages

IDEF1X however is not the only method of modeling information in CIM environments. Wang *et al.* (1993) use an extension of IDEF0 function modeling to model the relational information flows in a CIM environment. The limitations of the ER modeling scheme in modeling data for object-oriented data systems have led to the extension of ER techniques for object-oriented data modeling (Rumbaugh, 1992 and 1994). With this approach, entities are objects that belong to classes. They inherit both attributes and operations. Another approach, IDEF4 is an object oriented design (OOD) methodology that allows for objects, their structures (relations) and their behaviors (Mayer *et al.*, 1992). IDEF4 however, is a modeling scheme for implementing full object oriented information systems.

Information in a CIM environmennt exists in different formats and may represent system views that are not compatible in a single ER diagram. As noted above IDEF1X constrains the information model to ER modeling constructs. Other efforts however, have focused on the diversity of information in CIM environment. CIM-OSA, CIM Open Systems Architecture (Jorysz and Vernadat, 1990a and 1990b) and (Klittich 1990), is an approach to CIM modeling and integration from four views: functional, information, resource and organization.

Information modeling views in CIM-OSA use an Object-Entity Relatioship Attribute (OERA) approach (Jorysz and Vernadat, 1990b). In OREA, an object-oriented high level model, focusing on semantic unification of the model, drives the decomposition to a more traditional ER model. The ER model is extended to handle static and dynamic properties of the data, integrity constraints and database transactions (logical units of work) for consistency.

2.2. Consistency Among Databases

In the field of heterogeneous database systems, methodologies have been proposed for integrating heterogeneous relational databases (Reddy *et al.*, 1994). The literature defines eight semantic inconsistencies that can arise between data in multiple databases: name conflicts (Casanova and Vidal, 1983) data type/representation conflicts (Batini and Lenzerini, 1984), primary/alternate key conflicts (Batini *et al.*, 1986), referential integrity behavior conflicts (Batini *et al.*, 1986), missing data and null values (Breitbart *et al.*, 1986), level of abstraction (Mannino and Effelsberg, 1986), identification of related concepts (Reddy *et al.*, 1994), and scaling conflicts (Dayal and Hwang, 1984). Reddy *et al.* (1994) present an integration based on the semantic unification of entities by establishing the real world objects they represent in their local databases. Local models of the respective databases are unified based on the real world objects in the system modeled. If two entities model the same real world object, they can be matched with appropriate linear translation between scalar attribute values. The approach works if the databases model the system from different perspectives, but from similar reality. That is, two discrete event simulation models view the system from similar perspectives and could be unified to form a consistent view; while a simulation model and a capacity planning tool would require extending these methodologies. A good source for unification methodologies in heterogeneous databases is in Gupta (1989).

The techniques presented in this section extend the heterogeneous database integration approaches in literature. Further, these techniques can be effectively used even when the heterogeneous databases are from tools with very different domain views, stages of the life cycle, and/or level of detail. It is assumed that the information in these databases can be represented by a formal entity relation model. An entity relation model or schema of a database can be expressed as the triple $D = (E, R, A)$, where E is the set of entity classes, R the set of relations, and A the set of attributes. This section explores methods in the following two areas:

1. Two data bases are *consistent* if all the overlapping data describes the same physical reality.
2. A heterogeneous database is *synchronized* with another database when data is moved between the two databases is such a way that consistency is maintained.

2.2.1. Consistency

This section formally defines the notion of consistency among heterogeneous databases. Let $D_i\, i = 1, 2, \ldots, N$ be N different relational databases. Attributes $a_i \in A_i$ and $a_j \in A_j$ are *matching* attributes if they represent the same data in both models. For example, the names of products in two manufacturing databases are matching attributes, while the name of a machine does not match the name of an employee. Entities $e_i \in E_i$ and $e_j \in E_j$ are said to be *peer* entities if their primary keys (or a combination of appropriate alternate keys) consist only of matching attributes. Two peer entities in different databases are said to be *coupled.* The determination of peer entities can be established from just the IDEF1X data schema for different databases. However, due to the semantic differences among different design domains, this may be a difficult process (Reddy *et al.*, 1994).

The actual set of information modeled by an entity that populates a database is called an *instantiation* of the entity. A single entry of an instantiation of an entity is called an *instance* of the entity. Let e_i and e_j be peer entities, then an instance of entity e_i is *equivalent* to an instance of entity e_j if the value of their primary keys (or a combination of appropriate alternate keys) is equal. The following properties are defined for instantiations of peer entities residing in two databases D_i and D_j.

1. An instantiation of entity e_i *intersects* an instantiation of entity e_j, if there exists at least one instance of each of the entities which are equivalent.
2. Two instantiations are *disjoint* if they do not intersect.
3. An instantiation of entity e_i *contains* the instantiation of entity e_j, $e_i \supseteq e_j$, if for every instance of entity e_j there is an equivalent instance of entity e_i.
4. Two instantiations of peer entities are *equal,* $e_i = e_j$, if and only if, $e_i \supseteq e_j$ and $e_j \supseteq e_i$.

Let $e_i \in E_i\; e_j \in E_j$ and $e_k \in E_k$. From the above definitions it is easy to show that

if $e_i = e_j$ **and** $e_j = e_k$ **then** $e_i = e_k$.

Likewise,

if $e_i \subseteq e_j$ **and** $e_j \subseteq e_k$ **then** $e_i \subseteq e_k$.

The equals and contains properties are called *relational characteristics*.

The specification of the coupling between two peer entities can include a constraint on the type of relational characteristic which must be enforced between their instantiations. That is, if a coupling between peer entities e_i and e_j is constrained to be equal (contained), then the instantiations of these entities should be equal (contained). The instantiations of two peer entities are said to be *consistent* if their relational charcteristic conforms with the coupling constraint.

Consistency among the non-key attributes of two equivalent entity instances is now defined. Non-key matching attributes of peer entities are classified into one of the following types:

1. The attribute is a *constraint* if it is a desired limit on a value associated with the actual system value. A constraint attribute can either be an upper or lower value. For example, system requirements may set an upper limit constraint on the number of resources of the manufacturing system. A constraint type attribute is always consistent with all values of a matching attribute on any equivalent entity instance.
2. An attribute can be an *estimate* of the actual system value. For example, a capacity planning tool may calculate estimates for the number of resources of the manufacturing system. To be consistent, the value of an estimate type attribute must satisfy a constraint type matching attribute on any equivalent entity instance. However, two consistent estimate type attributes may have different values.
3. If an attribute corresponds exactly to some parameter of the actual system, then it is an *exact* type attribute. For example, a discrete simulation model is used to calculate the exact number of resources needed by the manufacturing system. To be consistent, the value of an exact type attribute must satisfy all the constraint type matching attributes and must be equal to all other exact type matching attributes. However, exact type attributes may have different values from estimate type matching attributes.

Finally, given the schemas for two databases, D_i and D_j, the coupling among the peer entities in the databases, and the types of all matching non-key attributes, then two heterogeneous databases are consistent if the instantiations of all the peer entities are consistent and if all the values of the non-key matching attributes are consistent. It is straight forward to develop an algorithm to check consistency between two databases.

2.2.2. *Synchronization*

Another area to explore is the translation of data between heterogeneous databases. This translation is termed the *synchronization* of the two databases. Synchronization can be specified by a series of rules for each entity class in the data schema. This section will present the rules for several examples of common synchronization situations. In the examples, rules to translate the attirbutes are developed in both directions. However, in specific applications this may not be desirable. For example, a discrete event simulation calculates the utilization of resources. Usually, there is no reason to copy these values from another tool's database into the database which holds the simulation model.

Figure 2 gives the notation is used in the description of the translation rules.

Notation	Description
Entity(DB)	A pointer to an instance of the entity class *Entity* in the database *DB*.
Entity(DB).Attribute	Reference to the attribute *Attribute* for the entity instance pointed to by *Entity(DB)*.
FOR EVERY *Entity(DB)* *Body* END FOR	Executes the *Body* for every instance of entity *Entity*. Each time through the loop *Entity(DB)* is set to be a pointer to the current entity instance.
FIND *Entity(DB)* WITH PK I AKx (*value*)	Sets *Entity(DB)* to the instance of entity *Entity* whose primary (or alternate) key equals *value*. *Entity(DB)* is set to NULL if the instance does not exist.
CREATE *Entity(DB)*	Creates a new instance of *Entity* and sets *Entity(DB)* to point to it.
DELETE *Entity(DB)*	Deletes the instance of *Entity(DB)*.

IF *Expression* THEN
Body
END IF
	Executes the *Body* if *Expression* is true.

UNIQUE	Creates a unique value for the attribute.

Fig. 2 Translation notation for chapter examples figure.

2.2.3. *Example 1*

This example describes the translation between entities in two databases which hold nearly identical information; a very common type of translation. The translation rules account for the following situations: creating and deleting entity instances, matching the primary key of one entity with the alternate key of its peer entity, properly translating the relations, and ignoring non-key attributes that are not matching.

The data schemas of two simple databases describing products made in various manufacturing systems are shown in Figure 3. Assume that both the SYSTEM and PRODUCT entities have an equals coupling constraint. Figure 4 presents the rules that translate between the peer entities. The first loop in the rules creates any new SYSTEM entities that are required and updates the appropriate attributes. The second loop deletes any extra SYSTEM entities and associated children PRODUCT entities. The third loop creates the required PRODUCT entities. Finally, the fourth loop deletes the extra PRODUCT entities.

2.2.4. *Example 2*

This example examines the case when one database contains entities that are not in the other databse. Translating from the database which contains the extra entities is simple, the information supplied by the extra entitties is simply not copied. However, there are two cases to consider when translating in the other direction:

1. an extra entity is the child of one of the peer entities, and
2. an extra entity is a parent of a peer entity (assuming NULLS for the parent are not allowed).

In the first case, if a peer instance does not exist, a new instance is created, but no child entities are created. If a peer instance does exist, care should be taken so that all instances of the child entity remain intact. Finally, if a peer instance is to be deleted, then all its children also must be deleted.

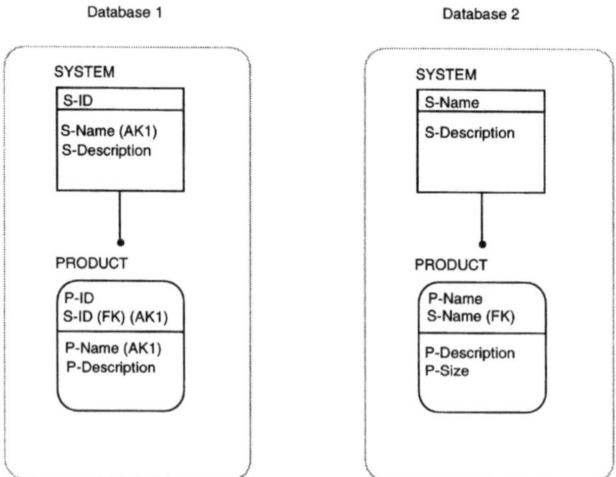

Fig. 3 Data schema for example 1.

In the second case, if a translated peer entity instance does not exist, a new instance of both the peer and parent entities are created, and the attributes for the parent entity are defaulted. Or, if it can be determined that the parent already exists, then just a link is created. If the translated peer entity instance does exist, care should be taken so that the parent entity remains intact. Finally, if the peer instance is to be deleted, the parent entity is deleted only if its entity class is related to the peer class and the parent instance is not related to any other peer instances. Figure 5 contains the data schemas for two databases which illustrate this situation. The databases contain peer entities SYSTEM and PRODUCT, but Database1 also contains information on the processing steps required to produce the product and the manufacturing cells that process the products. Assume that both the SYSTEM and PRODUCT entities have an equals coupling constraint, Fig. 6 presents the rules required to translate between the PRODUCT peer entities. The first loop creates the required new entity instances and the second loop deletes obsolete instances. If the values for the primary keys of equivalent entity instances do not have identical values then logic can be added to the rules as was done in Example 1.

2.2.5. *Example 3*

In the two previous examples, the peer entities in the two databases were assumed to be equally constrained. This last example explores translation between peer entities whose coupling in contained constrained. Let e_1 and e_2 be peer entities, such that $e_1 \supseteq e_2$. Without any other information, the synchronization of the databases can either delete entity instances of e_2 which do not have an equivalent instance in D_1, or add instances of e_1 which do not have an equivalent instance in D_2. If necessary, the choice of which method to employ is specified in the definition of the coupling. However, restricting the range of values of the non-key attributes of e_1 often creates a subset of instances of e_1 which do have a one-to-one relation to instances of e_2. In this case meaningful translation can still occur.

Database1 --> Database2

```
FOR EVERY SYSTEM(1)
   FIND SYSTEM(2) WITH PK (SYSTEM(1).S-Name)
   IF (SYSTEM(2) == NULL) THEN
      CREATE SYSTEM(2)
      SYSTEM(2).S-Name = SYSTEM(1).S-Name
   END IF
   SYSTEM(2).S-Description = SYSTEM(1).S-Description
END FOR

FOR EVERY SYSTEM(2)
   FIND SYSTEM(1) WITH AK1 (SYSTEM(2).S-Name)
   IF (SYSTEM(1) == NULL) THEN
      FOR EVERY PRODUCT(2)
         IF (PRODUCT(2).S-Name == SYSTEM(2).S-Name)
THEN
            DELETE PRODUCT(2)
         END IF
      END FOR
      DELETE SYSTEM(2)
   END IF
END FOR

FOR EVERY PRODUCT(1)
   FIND SYSTEM(1) WITH PK (PRODUCT(1).S-ID)
   FIND PRODUCT(2) WITH PK (SYSTEM(1).S-Name,
      PRODUCT(1).P-Name)
   IF (PRODUCT(2) == NULL) THEN
      CREATE PRODUCT(2)
      PRODUCT(2).P-Name = PRODUCT(1).P-Name
      PRODUCT(2).S-Name = SYSTEM(1).S-Name
   END IF
   PRODUCT(2).P-Description = PRODUCT(1).P-Description
   PRODUCT(2).P-Size = Default Value
END FOR

FOR EVERY PRODUCT(2)
   FIND SYSTEM(1) WITH AK1 (PRODUCT(2).S-Name)
   FIND PRODUCT(1) WITH AK1 (SYSTEM(1).S-ID,
      PRODUCT(2).P-Name)
   IF (PRODUCT(1) == NULL) THEN
      DELETE PRODUCT(2)
   END IF
END FOR
```

Database2 --> Database1

```
FOR EVERY SYSTEM(2)
   FIND SYSTEM(1) WITH AK1 (SYSTEM(2).S-Name)
   IF (SYSTEM(1) == NULL) THEN
      CREATE SYSTEM(1)
      SYSTEM(1).S-ID = UNIQUE
   END IF
   SYSTEM(1).S-Name = SYSTEM(2).S-Name
   SYSTEM(1).S-Description = SYSTEM(2).S-Description
END FOR

FOR EVERY SYSTEM(1)
   FIND SYSTEM(2) WITH PK (SYSTEM(1).S-Name)
   IF (SYSTEM(2) == NULL) THEN
      FOR EVERY PRODUCT(1)
         IF (PRODUCT(1).S-ID == SYSTEM(1).S-ID) THEN
            DELETE PRODUCT(1)
         END IF
      END FOR
      DELETE SYSTEM(1)
   END IF
END FOR

FOR EVERY PRODUCT(2)
   FIND SYSTEM(1) WITH AK1 (PRODUCT(2).S-Name)
   FIND PRODUCT(1) WITH AK1 (SYSTEM(1).S-ID,
      PRODUCT(2).P-Name)
   IF (PRODUCT(1) == NULL) THEN
      CREATE PRODUCT(1)
      PRODUCT(1).P-ID = UNIQUE
      PRODUCT(1).S-ID = SYSTEM(1).S-ID
   END IF
   PRODUCT(1).P-Name = PRODUCT(2).P-Name
   PRODUCT(1).P-Description = PRODUCT(2).P-Description
END FOR

FOR EVERY PRODUCT(1)
   FIND SYSTEM(1) WITH PK (PRODUCT(1).S-ID)
   FIND PRODUCT(2) WITH PK (SYSTEM(1).S-Name,
      PRODUCT(1).P-Name)
   IF (PRODUCT(2) == NULL) THEN
      DELETE PRODUCT(1)
   END IF
END FOR
```

Fig. 4 Translation rules for example 1.

Figure 7 contains the data schemas for two databases which illustrate this situation. Let Database2 be a subset of the information model for a capacity planning tool. In these tools resources are considered to be the machine tools that process the part. Let Database1 represent the information in a simulation tool. In simulation models, resources would include not only the machines, but labor, transportation devices, sources, sinks, etc. Clearly, in this case $e_1 \supseteq e_2$. However, if we restrict e_1 to be only those resources that transform (machine) parts, then meaningful translation can occur. Figure 8 presents the rules required to translate between the RESOURCE/TRANSFORM entities. Again, the first loop creates the required new entity instances and the second loop deletes obsolete instances. It will be assumed that the values for the primary keys of equivalent entity instances have identical values.

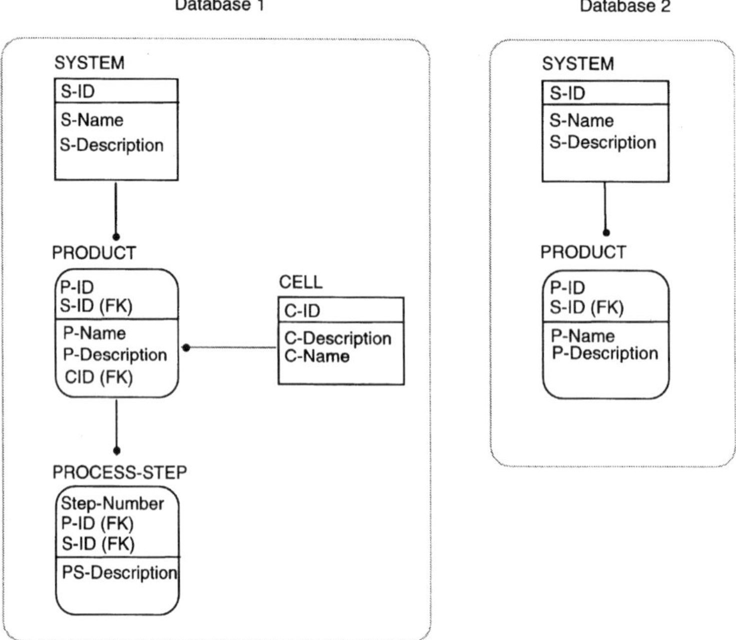

Fig. 5 Data schema for example 2.

PRODUCT(1) --> PRODUCT(2)	PRODUCT(2) --> PRODUCT(1)

FOR EVERY PRODUCT(1)
FIND PRODUCT(2) WITH PK (PRODUCT(1).S-ID,
PRODUCT(1).P-ID)
IF (PRODUCT(2) == NULL) THEN
CREATE PRODUCT(2)
PRODUCT(2).S-ID = PRODUCT(1).S-ID
PRODUCT(2).P-ID = PRODUCT(1).P-ID
END IF
PRODUCT(2).P-Name = PRODUCT(1).P-Name
PRODUCT(2).P-Description = PRODUCT(1).P-Description
END FOR
 	
FOR EVERY PRODUCT(2)
FIND PRODUCT(1) WITH PK (PRODUCT(2).S-ID,
PRODUCT(2).P-ID)
IF (PRODUCT(1) == NULL) THEN
DELETE PRODUCT(2)
END IF
END FOR
	FOR EVERY PRODUCT(2)
	FIND PRODUCT(1) WITH PK (PRODUCT(2).S-ID,
	PRODUCT(2).P-ID)
	IF (PRODUCT(1) == NULL) THEN
	CREATE PRODUCT(1)
	PRODUCT(1).P-ID = PRODUCT(2).P-ID
	PRODUCT(1).S-ID = PRODUCT(2).S-ID
	Create CELL(1)
	CELL(1).CID = UNIQUE
	CELL(1).C-Name = CONCAT(PRODUCT(2).P-Name, "CELL")
	CELL(1).Description = NULL
	END IF
	PRODUCT(1).P-Name = PRODUCT(2).P-Name
	PRODUCT(1).P-Description = PRODUCT(2).P-Description
	PRODUCT(1).CID = CELL(1).CID
	END FOR
	
	FOR EVERY PRODUCT(1)
	FIND PRODUCT(2) WITH PK (PRODUCT(1).S-ID,
	PRODUCT(1).P-ID)
	IF (PRODUCT(2) == NULL) THEN
	FOR EVERY PROCESS-STEP(1)
	IF (PROCESS-STEP(1).S-ID == PRODUCT(2).S-ID &&
	PROCESS-STEP(1).P-ID == PRODUCT(2).P-ID) THEN
	DELETE PROCESS-STEP(2)
	END IF
	END FOR
	DELETE PRODUCT(2)
	END IF
	END FOR

Fig. 6 Translation rules the PRODUCT entity in example 2.

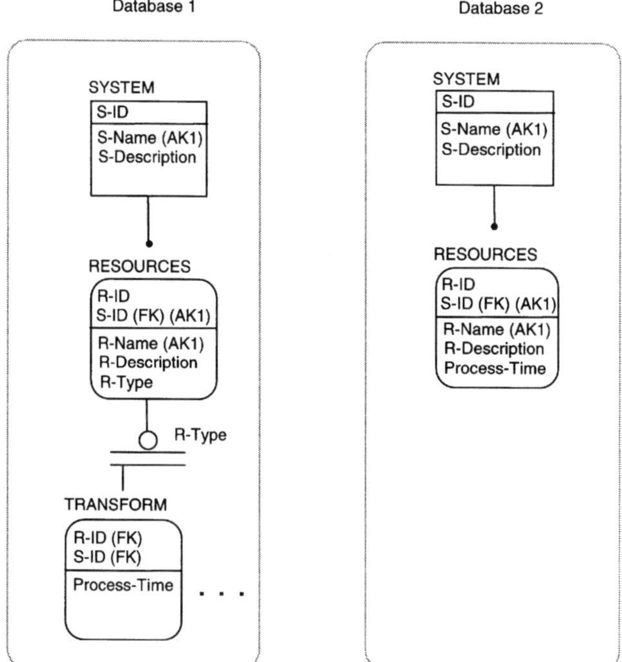

Note: The symbol of a circle with two lines in an IDEF1X diagram signifies that the child entity is one of a set of recognized categories of the parent. In Figure 4, the TRANSFORM entity is one of a set of resources in the model. A TRANSFORM entity is therefore one type of RESOURCE which has its own attributes describing the TRANSFORM RESOURCE.

Fig. 7 Data schema for example 3.

3. THE IMDE ARCHITECTURE

To address the needs and special requirements placed upon the design process for manufacturing systems we present an architecture, the Integrated Manufacturing Design Environment (IMDE), which will maintain and verify constraints and relationships among the multiple subsystem models. This architecture is an extension of the one developed by Parks *et al.* (1994). The characteristics of the IMDE are as follows:

- The IMDE architecture coordinates design and modeling in an environment that allows designers to operate with familiar tools,
- The IMDE architecture allows data to be stored by the local tools in the format/structure specified by the tool vendor,
- New modeling tools can be added to and removed from the IMDE architecture through an automated registration mechanism,
- A standardized interface design will minimize the effort that is required to attach tools, and
- A message system is incorporated to notify all concerned parties about conflicts, data updates, constraint violations, etc.

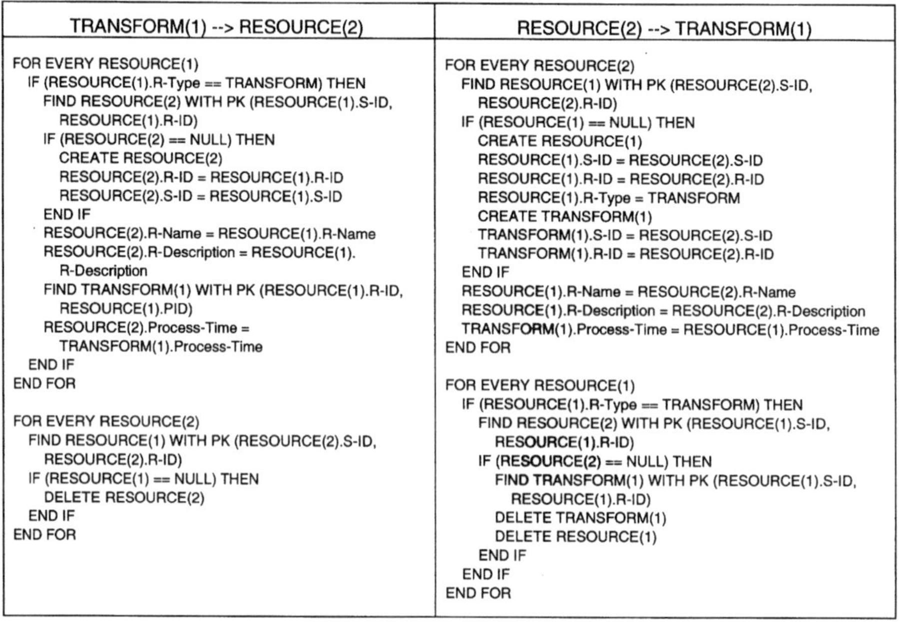

TRANSFORM(1) --> RESOURCE(2)

```
FOR EVERY RESOURCE(1)
  IF (RESOURCE(1).R-Type == TRANSFORM) THEN
    FIND RESOURCE(2) WITH PK (RESOURCE(1).S-ID,
      RESOURCE(1).R-ID)
    IF (RESOURCE(2) == NULL) THEN
      CREATE RESOURCE(2)
      RESOURCE(2).R-ID = RESOURCE(1).R-ID
      RESOURCE(2).S-ID = RESOURCE(1).S-ID
    END IF
    RESOURCE(2).R-Name = RESOURCE(1).R-Name
    RESOURCE(2).R-Description = RESOURCE(1).
      R-Description
    FIND TRANSFORM(1) WITH PK (RESOURCE(1).R-ID,
      RESOURCE(1).PID)
    RESOURCE(2).Process-Time =
      TRANSFORM(1).Process-Time
  END IF
END FOR

FOR EVERY RESOURCE(2)
  FIND RESOURCE(1) WITH PK (RESOURCE(2).S-ID,
    RESOURCE(2).R-ID)
  IF (RESOURCE(1) == NULL) THEN
    DELETE RESOURCE(2)
  END IF
END FOR
```

RESOURCE(2) --> TRANSFORM(1)

```
FOR EVERY RESOURCE(2)
  FIND RESOURCE(1) WITH PK (RESOURCE(2).S-ID,
    RESOURCE(2).R-ID)
  IF (RESOURCE(1) == NULL) THEN
    CREATE RESOURCE(1)
    RESOURCE(1).S-ID = RESOURCE(2).S-ID
    RESOURCE(1).R-ID = RESOURCE(2).R-ID
    RESOURCE(1).R-Type = TRANSFORM
    CREATE TRANSFORM(1)
    TRANSFORM(1).S-ID = RESOURCE(2).S-ID
    TRANSFORM(1).R-ID = RESOURCE(2).R-ID
  END IF
  RESOURCE(1).R-Name = RESOURCE(2).R-Name
  RESOURCE(1).R-Description = RESOURCE(2).R-Description
  TRANSFORM(1).Process-Time = RESOURCE(1).Process-Time
END FOR

FOR EVERY RESOURCE(1)
  IF (RESOURCE(1).R-Type == TRANSFORM) THEN
    FIND RESOURCE(2) WITH PK (RESOURCE(1).S-ID,
      RESOURCE(1).R-ID)
    IF (RESOURCE(2) == NULL) THEN
      FIND TRANSFORM(1) WITH PK (RESOURCE(1).S-ID,
        RESOURCE(1).R-ID)
      DELETE TRANSFORM(1)
      DELETE RESOURCE(1)
    END IF
  END IF
END FOR
```

Fig. 8 Translation rules for the resource entity in example 3.

The IMDE environment has the following components: a unified data meta-model, intelligent interfaces, a link manager, a message board, and a simulation manager. These components provide the means to accomplish the static translation, model transformation and dynamic integration among the tools and methods. Figure 9 shows the relationships between the major components in the architecture.

The IMDE architecture uses an integrating approach similar to a distributed database. Data resides locally with the creating tools and a management system is responsible for linking entities to maintain consistency among the local databases. Current approaches to distributed database systems address issues of concurrency, local autonomy, transaction processing, locks, roll backs and commits in support of sharing relational data distributed in horizontally-fragmented, vertically-fragmented, and replicated relations. While the IMDE architecture addresses many of these same issues, it integrates data in an architecture that has no initial links defined between entities in the multiple databases. While replication must be strictly controlled in a distributed database, the IMDE architecture identifies replication and allows the users to resolve inconsistencies among the replicated entities by posting messages. The following sections describe the functioning of the components and how they provide for model integration and coordination.

3.1. Unified Data Meta-Model

The unified data-meta-model (UDMM) defines a data schema for all information that is to be shared across the tools. It is important to note that a database using the

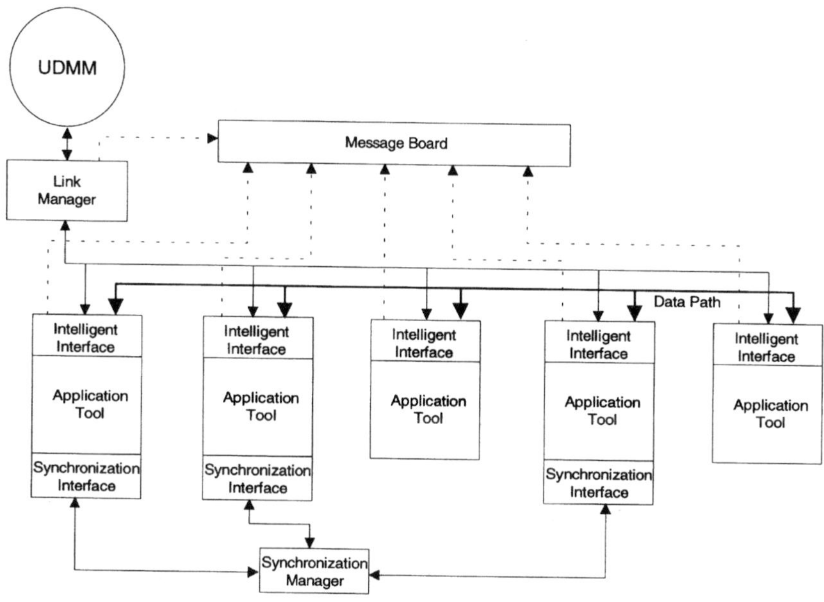

Fig. 9 IMDE system architecture.

UDMM schema is not implemented in the IMDE. Rather, only the schema itself is stored in the link manager system. This is done so that proper links among the various local databases are established by the link manager when a tool registers. The UDMM has a representation for every entity, attribute and relation that is under the verification and propagation control of the IMDE.

The UDMM contains the entities, the relationships among the entities, and the constraints that are commonly shared among the methods used by manufacturing design systems. The UDMM specifies the format for the entities, the attributes and the relations it contains. Further, the UDMM defines a standardized data format for translation and analysis of information across the tools. Development of a suitable UDMM will be a large research project involving universities, manufacturers and tool vendors, not unlike the current development of the PDES standards.

It is not feasible, nor necessary, for the UDMM to specify all the data and relations for all methods and tools for manufacturing design. The UDMM specifies only those entities and relationships common to more than one method or tool. Individual design tools have data that is not specified in the UDMM, as necessary, to perform their particular functions. Figure 10 shows the relation of the UDMM to the data models from the various design tools. Each tool in the diagram represents some aspect of the overall system design. Data in one tool may be duplicated in some format in another tool. The UDMM must be robust enough to capture all information that overlaps between independent tools.

3.2. The Link Manager

The establishment of links is the function of the link manager. However, the actual links are stored in each intelligent interface. When a tool is brought on-line, its

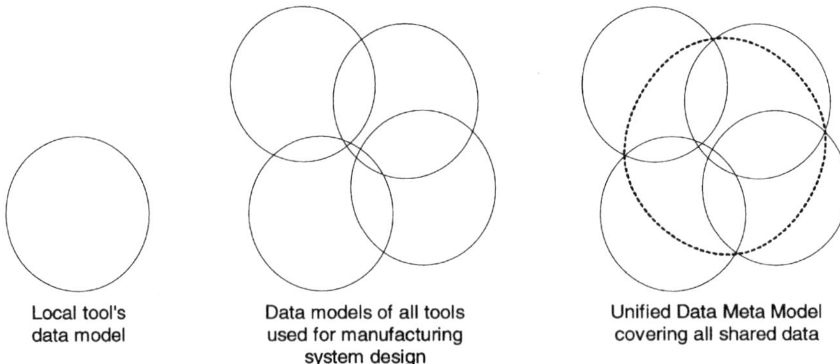

Fig. 10 UDMM's relation to the local data models.

intelligent interface sends the link manager the portion of the UDMM schema that is represented in its local model. Using the UDMM, the link manager determines all other tools in the IMDE whose local data share the same entities as the new tool. The link manager creates links to all shared entities and sends the links to the intelligent interface.

3.3. The Message Board

Any inconsistencies, unresolvable constraints, or errors are posted to the message board. Serving as a human interface, the message board provides the designer with the ultimate control over the domain and model. Messages are generated automatically by the respective intelligent interfaces and directed to persons with authority or knowledge to resolve the conflict.

3.4. The Synchronization Interface and Manager

System constraints may also be based on the performance characteristics of executable models of the system. These constraints may themselves be dynamic, imposed by system states of the other executable models of the system. The synchronization interface and manager dynamically integrates the executable tools and coordinates the execution of simulation models of the manufacturing system. This coordination can use any of the current approaches to distributed simulation (Misra, 1986) or the XFaST technology developed by Judd *et al.* (1993).

3.5. The Intelligent Interface

A key component to the functioning of the IMDE architecture is the intelligent interface. Funtioning as an intelligent agent for the local database, each intelligent interface must understand its tool's local data model and the tool's relationship to the UDMM. Additionally, the intelligent interface shares data from the local model with other tools. The role of the intelligent interface is similar to Dilts and Wu's (1991) alpha knowledge base. Their intelligent interfaces translate directly between relational databases and an integrated interface. They propose an individual interface that has knowledge of the domain of the database, conceptual data model and logical structure of the database, methods of data access, and methods of changing the database to suit changes in the system. To these requirements, we add knowledge of the relationship between local objects, the UDMM, and the methods to the translate them into a format specified by the UDMM.

A block diagram of the interface is illustrated in Figure 11. It consists of the following components:

1. *Local UDMM Schema.* The portion of the UDMM schema that overlaps the schema of the local data is the local UDMM schema.
2. *Local Portion of the UDMM Database (LPU).* The LPU is a database that stores, using the Local UDMM schema, information about the local model that other tools can access. Information from the local database is moved into this database only when the user requests his model to be posted to IMDE.
3. *Data Links.* This database stores information about the coupling of peers to the entities in the local UDMM schema. The information includes the location of peer entities, the constraints on the coupling between the peers, and existence and types of all matching attributes. The information in this database is generated by the link manager when the tool registers to the system.

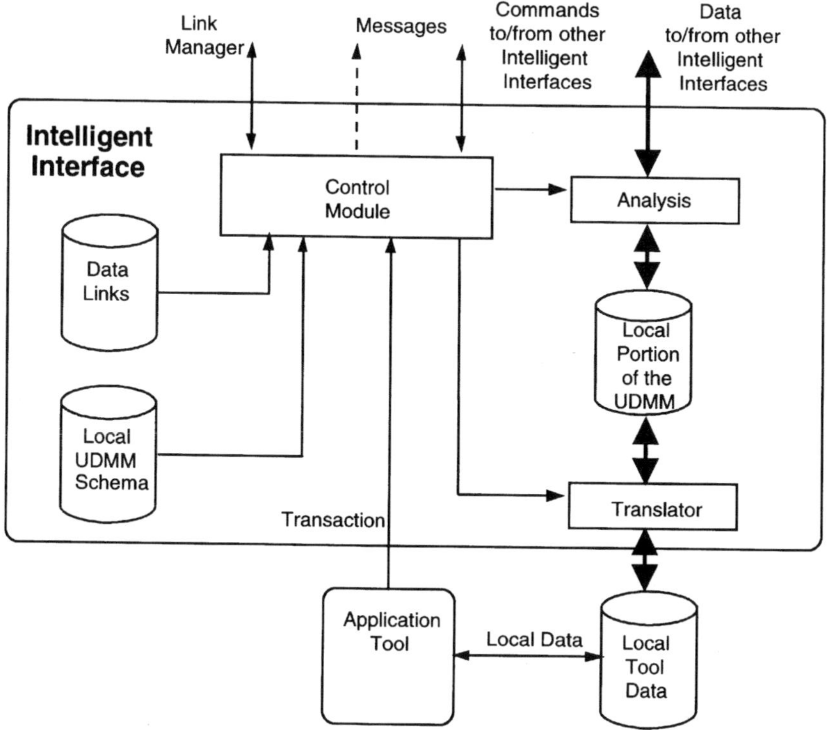

Fig. 11 Intelligent interface.

4. *Translator Module*. This module translates information from the local database to the Local Portion of the UDMM. This translation is defined using rules such as those presented in Section 2.2.
5. *Analysis Module*. This module checks for consistency among models, using the analysis techniques presented in Section 2.2.1, stored on different intelligent interfaces.
6. *Control Module*. This module coordinates the operation of the components within the interface.

When a design tool is brought into the system, the intelligent interface broadcasts to the link manager the Local UDMM schema used by its tool. The link manager uses this information to create the logical links among peer entity classes and matching attributes which are stored among the various tools. The link manager sends this information back to the interface, where it is stored in the Data Link database. Once the links to the remote data elements have been established, whenever the intelligent interface requires information on another tool, it sends a request directly to the appropriate tool for this information. The information will be received in a standard format and then is translated and used locally.

The interface between the tool and the intelligent interface occurs through transaction requests. The following transactions are supported by the intelligent interface:

1. **Create**, This transaction creates a new design model and populates it with related entity instances from the other tools. During a **Create**, the interface interrogates the other tools for existing data about the system. If related entities exist, they are translated into the LPU database and the data is translated and stored in the Local Tool database.
2. **Post**, When the designer is ready, s/he invokes **Post**. **Post** then reads the local model, translates it to the LPU database and invokes **Check**. The entities in this model are then compared to all their peers, constraint violations are posted to the message board, missing data is determined and requests for corrective action are posted, and inconsistencies in shared data on the different tools are determined and requests to resolve the inconsistencies are posted.
3. **Update**, This transaction updates a selected portion of the local model with data from the other tools. Used during the design process, **Update** works in a similar fashion to the **Create** transaction.
4. **Check**, This transaction allows user to check for violations of cons-traints imposed by other tools anytime during a modeling session. **Check** does not change the local model like **Create** or **Update**.
5. **Query**, This transaction collects data on another tool and displays it. **Query** does not change the local model.

Note that in this architecture, the local application and the designer have ultimate control over the composition and elements in the local design tool model.

In a typical design session, an engineer first **Create**s his model. **Create** brings in all the information that is known about this model from other tools. The engineer then refines and adds detail to the model. These changes are done locally and the IMDE is not made aware of these intermediate phases of the modeling process. The reason for this off-line design is to allow the designer freedom to create designs that are inconsistent with other designs or inconsistent internally without the overhead of the IMDE verifying the model. When the engineer is satisfied with the model, it is **Post**ed to the system. Data integrity and constraints with other models are now checked and requests for corrective action are posted to the message board. As the other models evolve, the engineer can import new features from the models on other tools by **Updat**ing the model. At any time the engineer may **Check** constraints and **Query** other models. This set of transactions provides a powerful and controlled integrated environment for manufacturing systems design.

4. EXAMPLE IMDE IMPLEMENTATION

This section presents an IMDE for a set of typical manufacturing systems design tools. The data models presented in this section are greatly simplified and are intended for illustrative purposes only. However, they are sufficient to explain how the IMDE architecture operates. The first section describes the UDMM. The subsequent sections develop simple information models for the following: Process Definition Tool, Capacity Planning Tool, Layout Tool, and a Discrete Event Simulation Tool. Each of these sections ends with a description of the portion of the UDMM that each tool will use. The section concludes with a description of a design session using the IMDE.

4.1. Simplified UDMM

The Unified Data Meta-Model must represent all data that can be represented by more than one design tool. Thus, the UDMM is a conceptual schema of the data at a higher level than the tools. While the individual tool's data models abstract reality to provide a single view of the system for their specific algorithms, the UDMM provides an abstract view of the entire system to support the reality of all tools.

Figure 12 shows a UDMM design which supports four domain specific design tools. At the highest level, we specify a parent entity of SYSTEM. The system entity represents the manufacturing system under design. It is the parent that owns most of the other entities in the model. The attribute S-Name uniquely identifies the system and is the primary key. There are two main classes of entities that the UDMM recognizes as being in a system: PRODUCTs and RESOURCEs.

Product names are unique to the system and have a description, production schedule and may have some calculated unit costs. Resources also have unique names and descriptions. Depending on the state of the system design, they may have an assigned location and an estimated or projected utilization.

There are two types of resources identified in the UDMM: resources that TRANSFORM material into PRODUCTs and LABOR to operate the system. These must be matched to the two parent classes of RESOURCES-AVAILABLE and LABOR-CLASS. RESOURCES-AVAILABLE must also come from a set of recognized resource classes (RESOURCE-CLASS).

PRODUCTs have some positive number of required processes (PROCESS-STEP) for realization. These PROCESS-STEPS are linked to the RESOURCEs that

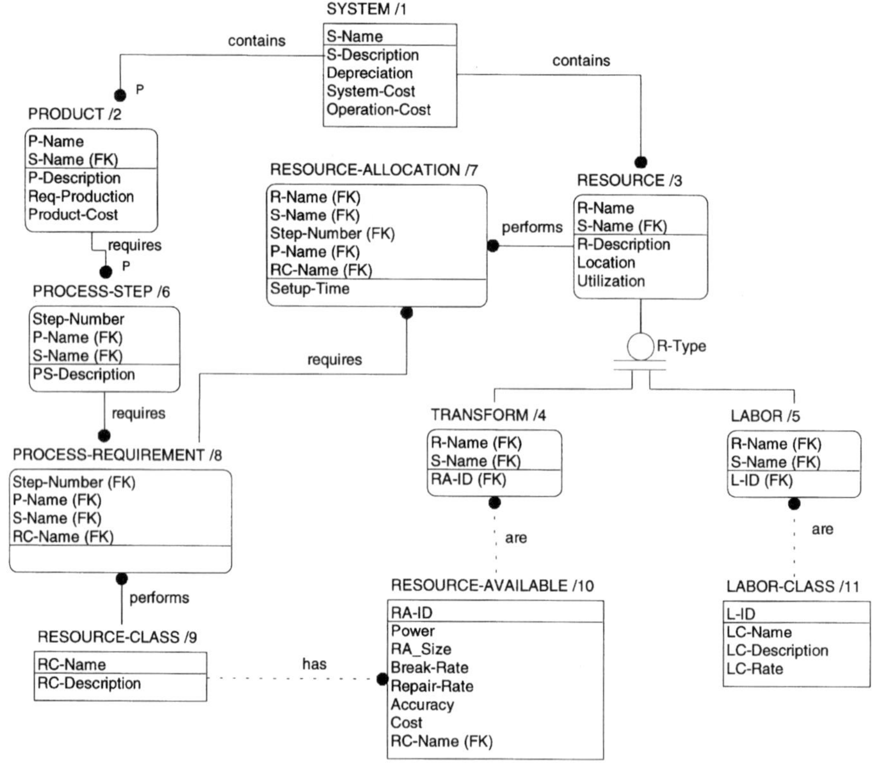

Fig. 12 Simplified UDMM.

perform them in the PROCESS-REQUIREMENT entity. To satisfy this process requirement, a RESOURCE-ALLOCATION is used to link the required resource class to a resource of the same class.

It is important to recognize that the UDMM is not a database but, is a conceptual schema for the shared data in several disparate databases. When developing an Intelligent Interface, the UDMM is examined and the portion of the UDMM which overlaps the tool data is determined. This is the stored in the Local UDMM Schema, and the LPU database is formatted according to this schema. Thus, data conforming to the UDMM format is stored only in the LPUs of each Intelligent Interface. To finish the construction of the interface a translator is built, using rules such as those developed in Section 2.2, to synchronize data between the local tool data and the LPU. At this point, a fully functional interface is created. It is easy to see that tools can be easily added (removed) to the system at any time. Also, when new tools are developed, an Intelligent Interface can be configured for them from just the knowledge of the UDMM; and, the new tools will be able to interact completely with the tools already on the system.

The next several sections will demonstrate how the Local UDMM schema can be determined for several diverse manufacturing tools.

4.2. Process Definition Tool

To design any manufacturing system, you must know the products and their required manufacturing processes. This specification is usually done on a process modeling tool. The data model for such a tool is illustrated in Figure 13. For a given manufacturing SYSTEM, there are a set of associated PRODUCTs to be manufactured. Each product will have a set of PROCESS-STEPs necessary to transform raw materials in to the final product. Each step will have a set of PROCESSING-REQUIREMENTs that tie specific RESOURCE-CLASSes to the process steps. However, the specific machines and tools are not specified. The resource classes are generic (e.g., lathe or grinder). This is an adequate classification at this stage, since we are simply assessing what tools are necessary for the system under design.

Figure 14 shows that all of the entities of the process modeling tool are mapped into the UDMM. Therfore, the Local UDMM schema in the intelligent interface will consist of: SYSTEM, PRODUCT, PROCESS-STEP, RESOURCE-CLASS, and PROCESS-REQUIREMENT. The IMDE ensures that if any other tool reference any of these entities (attributes), it will be consistent with the data in this process modeling tool or errors are generated.

4.3. Capacity Planning Tool

Capacity planning identifies the specific machines and the number of resources required to produce a set of products at a specified production rate. Figure 15 is an

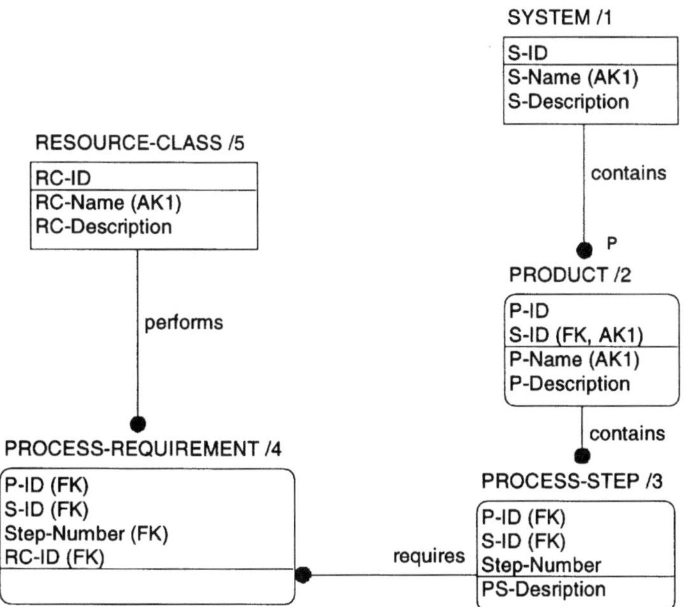

Fig. 13 Process modelling data model.

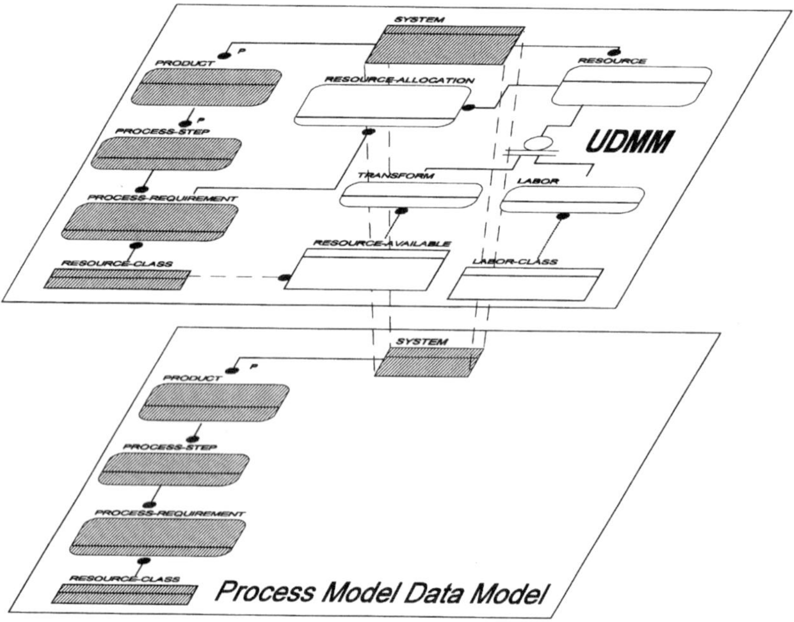

Fig. 14 Coupling between the process model's and UDMM's entities.

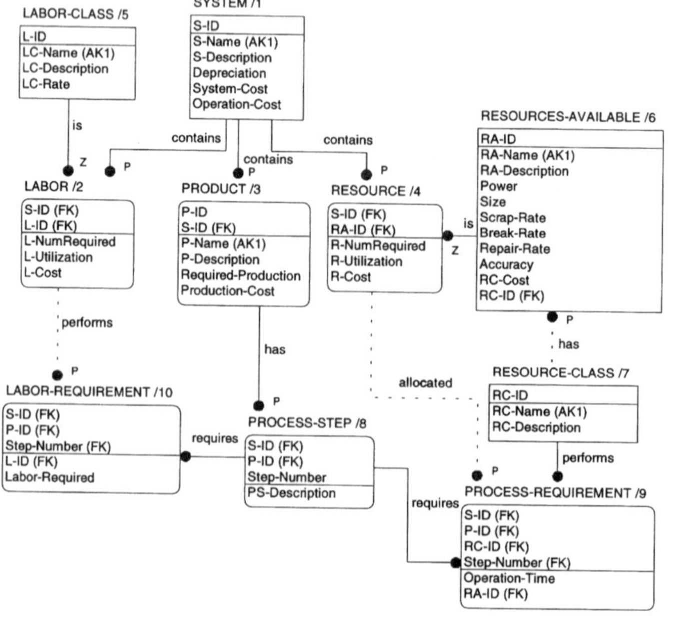

Fig. 15 Capacity planning data model.

example data model of a capacity planning tool. Under the parent SYSTEM entity, there are three entities in the manufacturing system under consideration: LABOR, PRODUCTs and RESOURCEs. Each product has an associated set of PROCESS-STEPs that identify the LABOR-REQUIREMENTs and the PROCESS-REQUIREMENTs. LABOR-REQUIREMENTs are used to identify the quantity of each LABOR-CLASS utilized. PROCESS-REQUIREMENT is linked to RESOURCE-CLASS to identify the basis resource type necessary for the operation. Each RESOURCE-CLASS has an instantiated set of AVAILABLE-RESOURCEs that are selected to perform the processes.

Capacity planning links to the UDMM by mapping the following capacity plan entities: SYSTEM, LABOR-CLASS, LABOR, PRODUCT, RESOURCE, RESOURCE-CLASS, RESOURCES-AVAILABLE, PROCESS-REQUIREMENT and PROCESS-STEP into the UDMM

4.4. Layout Tool

In the facilities layout data model (Figure 16), the physical location of each resource is identified. Under the parent SYSTEM entity, the layout tool has three entities that are identified: RESOURCE, STATION and PRODUCT. RESOURCEs are the machines to be placed in the facility. STATIONs are physical locations where one or more RESOURCEs may be located. It is at the STATIONs where a PRODUCT's PROCESS-STEPs are performed.

Facilities layout is linked to the UDMM by using the layout entities: SYSTEM, RESOURCE, PRODUCT and PROCESS-STEP.

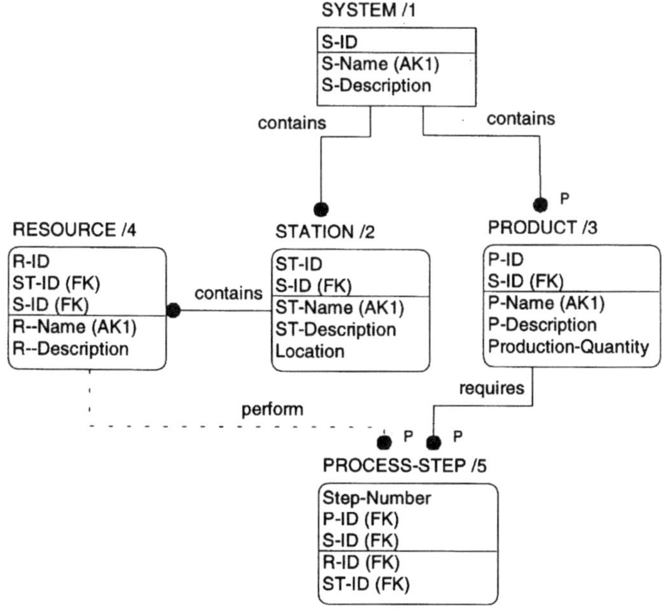

Fig. 16 Facilities layout data model.

4.5. Simulation Tool

In the simulation tool's data model (Figure 17), the entire system is represented from the perspective of how material flows in the system. Under the parent SYSTEM entity, the simulation tool has three entities that are identified: RESOURCE, RESOURCE-PROCESS and WIDGET. The entity PATH expresses an allowable flow by linking all possible machine-to-machine product flow combinations. RESOURCES are used perform RESOURCE-PROCESSES by allocating them with a PROCESS-ALLOCATION. A RESOURCE-PROCESS in most cases has an input widget IN-WIDGET and an output widget OUT-WIDGET. RESOURCEs can be one of six recognised types. STORE resources are buffers where widgets can remain between processing. TRANSFORM resources are the machines which change a widget's properties or geometry. LABOR is the manpower necessary for some resources to perform processes. TRANSPORT moves the widgets. SOURCE and SINK are the origin and termination of widgets in the system.

The simulation model is mapped into the UDMM by: SYSTEM and RESOURCE. LABOR resources are represented by the UDMM's LABOR entity and TRANSFORM is represented by the UDMM's RESOURCE entity.

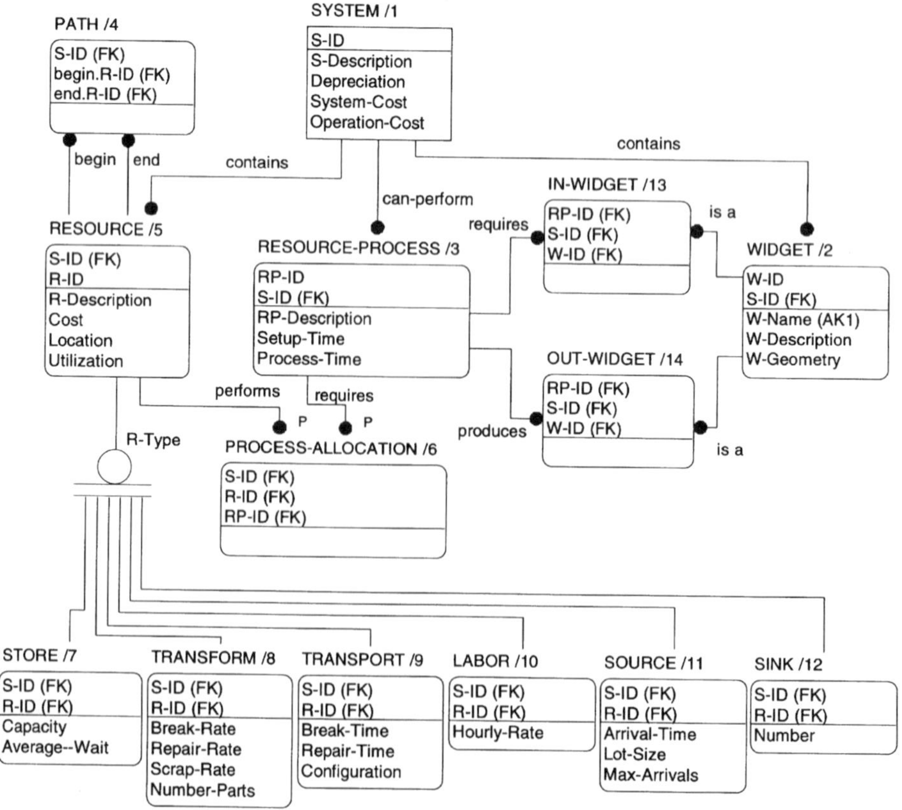

Fig. 17 Simulation data model.

4.6. Linking Entities Through the UDMM

With the individual data models developed, we use the UDMM to develop the local portion of the UDMM for each tool. In Figure 18, the layout model is added to Figure 18. Each tool's portion of the UDMM is shaded. Where the two local portions overlap, links to peer entities can be formed by the intelligent interfaces.

In the PROCESS-STEP(LDM) to PROCESS-STEP(PMDM) coupling, PROCESS-STEP(LDM).P-ID must be equal to PROCESS-STEP(PMDM).P-ID, as must S-ID and STEP-NUMBER. (See Figure 19.) Once instantiations have been equated, the relationship between PROCESS-STEP(PMDM).PS-DESCRIPTION and PROCESS-STEP(LDM).R-ID and ST-ID must be established. The rules governing this relationship are established using the procedures given in example 2 from section 2.2.2. For example, both PROCESS-STEP(LDM) and PROCESS-STEP(PMDM) are children of PRODUCT entities. Therefore if a new PROCESS-

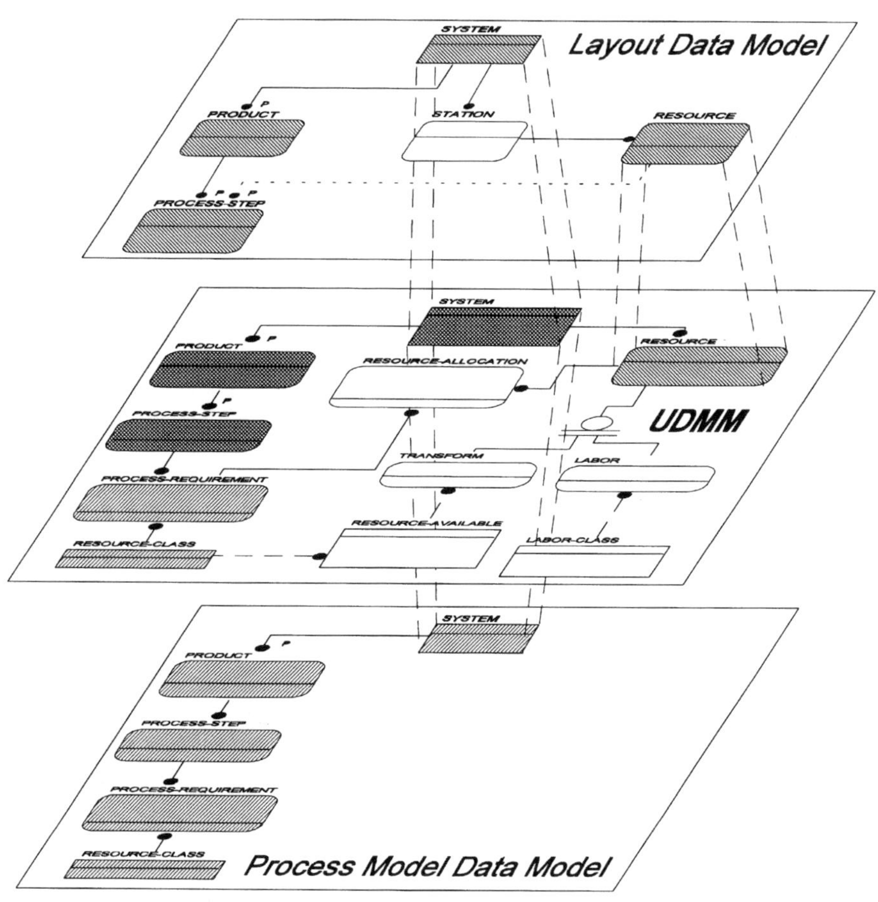

Fig. 18 Linking data models with the UDMM.

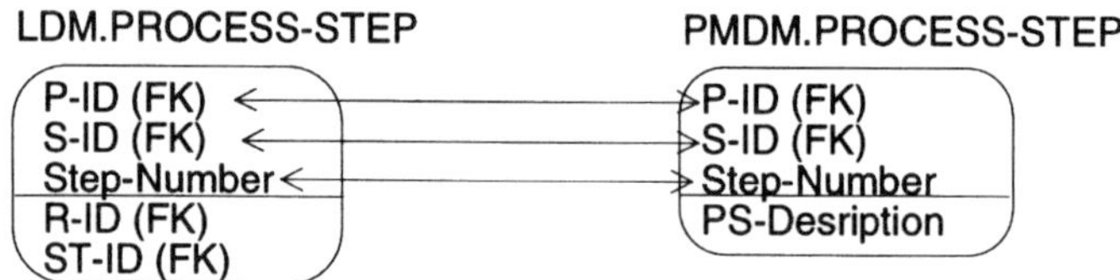

Fig. 19 Peer enitites.

STEP(LDM) instance is added, the parent PRODUCT in the layout model must exist in, or be added to the process modeling database.

4.7. System Usage

In the design of a typical manufacturing system, the processing required for each of the products of the system is specified first. This is usually done on a process definition tool. Next, requirements on the production rates of these products are determined. This information can be entered as constraint attributes within the IMDE. The process definition and production constraints information is copied across the IMDE to initialize the data for the capacity planning tool. Engineers use this tool to help select and determine the number of resources required to process the products. Labor requirements and estimated costs are also calculated. However, all these calculations are done using estimates for machine and labor utilization. Next a layout is determined, and finally, a simulation is performed. Again data initializing these models is created from the other models using the IMDE.

At this point several alternative scenarios can occur. First, assume that the simulation indicates that the planned proudction can be accomplished within the resource constraints imposed by the capacity planning tool. However, the simulation tool will have much better predictions of the utilization of the resources. This new information can be read by the production planning tool and now refined cost estimates of the system can be generated.

In another scenario, suppose the simulation cannot get the required production rates with the given resources, then conflicts are posted to the message board. Appropriate design engineers and managers can discuss the situation and corrective action decided upon. Perhaps it is determined to add more resources. A simulation is performed and the production constraints are now satisfied. When the new simulation model is posted, however, it invalidates the information in the capacity planning and layout tools. These inconsistencies are broadcast to the message board and the appropriate system design engineers update the models. The majority of the updates can be done automatically through the IMDE **Update** command.

In a third scenario, suppose all the models are consistent, but the processing requirements of a particular product are changed. Assume this is done on the process definition tool. When the new model is **Posted**, inconsistencies will occur in all the other models. But again, the appropriate design engineers are notified though the message board and corrective action can be taken. Information from the new process definition model is then used to **Update** these other models. It is important to note that using the database synchronization method described in Section 2.2., none of the specialization performed on the other tools for the unchanged process definitions is affected. Therefore, when these models are **Updated**, the development effort expended for the unchanged process definitons will not have to be redone. Only the appropriate specialization for the new process definitions needs to be added.

This bi-directional flow of information is what make the IMDE so powerful. Model development and changes can occur on any tool in the system, regardless of the level of detail, stage in the life cycle, or engineering discipline. The effects of the changes on the proposed manufacturing system are immediately determined. All engineers involved in the design of the system are immediately informed when another engineer makes a change that affects their portion of the design. The IMDE provides an environment that will significantly speed up the design process, provide better communication among the different designers assigned to the system, and thus reduces the system implementation errors.

5. SUMMARY AND CONCLUSIONS

Manufacturing systems share many of the well-documented design issues of other large, complex systems. These issues center on changing requirements during the long systems design life cycle and thus, the communications required among the multiple engineering disciplines who design the various sub-systems. The needed communication between and among these design disciplines is currently hampered by the islands of methods and computer-aided design tools which designers of each discipline use. Thus, we argue that these methods and tools are not integrated to the extent needed to facilitate the manufacturing systems design environment. We defined three types of integration; static translation, dynamic integration, model transformation and showed they were all needed at different levels of abstraction and at different stages of the design life cycle.

We suggest the need for a virtual, seamless environment for manufacturing systems design and define it as the integrating environment for the many existing manufacturing systems design tools. This environment ensures that all the sub-systems designs are consistent and valid. It further provides a test-bed for analyzing proposed changes. Other design disciplines, notably VLSI, have successfully integrated design methods and tools. Although these integration efforts focus on product design, their successes demonstrate that design systems integration can work.

The architecture for the Integrated Manufacturing Systems Design Environment addresses the needs and special requirements of the design process for manufacturing systems. It coordinates the design and modeling on familiar tools, allows data storage in tool vendor formats, provides for the addition or removal of tools, minimizes tool addition effort, and its, messaging capability provides for an appropriate level of communications. The heart of this architecture is the Synchronization Manager and the Unified Data Meta-Model coupled with the Intelligent Interfaces. These two features provide the integrating mechanism that enables efficient, effective communications among the different design tools. Implementing this architecture will require considerable research effort focused primarily on developing a more fundamental understanding of the three types of integration. However, global competitive pressures are creating a pressing need for faster, more effective methods of manufacturing systems design and re-design. Thus, the Integrated Manufacturing Design Environment is based on the practical, workable philosophy of using existing software-based design support tools (and integrating new tools when necessary).

References

C. Batini, M. Lenzerini and S. Navathe, A Comparative Analysis of Methodologies for Database Schema Integration. *ACM Computer Surveys*, **18**(4), 323–364 (1986).

C. Batini and M. Lenzerini, A Methodology for Database Schema Integration in the Entity Relationship Model. *IEEE Transactions on Software Engineering*, **SE-10** (6), (1984).

B.W. Boehm, Spiral model of software development and enhancement. *Computer* **5**(12), 61–72 (1988).

Y.J. Breitbart, P.L. Olson and G.R. Thompson, Database Integration in a Distributed Heterogeneous System. *Proceedings of the IEEE International Conference on Data Engineering* (1986).

M. Casanova and M. Vidal, Towards a Sound View Integration Methodology. *Proceedings of the Second ACM SIGACT/SIGMOD Conference on Principles of Database Systems* (1983).

P.P. Chen, The Entity-Relationship Model – Toward a Unified View of Data. *ACM Transactions on Database Systems*, **1**(1), 9–36 (1976).

U. Dayal and H. Hwang, View Definition and Generalization for Database Integration in MULTI-BASE: A System for Heterogeneous Distributed Databases. *IEEE Transactions on Soft-ware Engineering*, **SE-10**(6), (1984).

D.M. Dilts and W. Wu, Using Knowledge-Based Technology to Integrate CIM Databases, *IEEE Transactions on Knowledge and Data Engineering*, **3**(2), 237–245 (1991).

R. Gadh and F. Prinz, Abstraction of manufacturing features from design. *Proceedings of the Winter Annual Meeting of the American Society of Mechanical Engineers*, pp. 1–6 (1991).

A. Gupta, *Integration of Information Systems: Bridging Heterogeneous Database Systems*, IEEE, Piscataway NJ (1989).

J. Heim, Integrating distributed models: the architecture of ENVISION. *International Journal of Computer Integrated Manufacturing* **7**(1), 47–60 (1994).

H.R. Jorysz and F.B. Vernadat, CIM-OSA Part 1: total enterprise modeling and function view. *Int. J. Computer Integrated Manufacturing*, **3** (3 and 4), 144–156 (1990).

H.R. Jorysz and F.B. Vernadat, CIM-OSA Part 2: information view *Int. J. Computer Integrated Manufacturing* **3** (3 and 4), 157–167 (1990).

R.P. Judd, J.A. Sauter and R.S. VanderBok, XSpec: a methodology for the integrated design of manufacturing systems. *Proceedings of the 1991 International Conference on Design Productivity*, pp. 393–398 (1991).

R.P. Judd, J. White and P. Hickman *et al.*, System for combining originally software incompatible control, kinematic, and discrete event systems into a single integrated simulation system, U.S. Patent 5,247,650 (1993).

L.K. Keys and C.M. Parks, Mechatronics, systems, elements, and technology: a perspective. *IEEE Transactions on Components, Hybrids, and Manufacturing Technology*, **14**, 457–461 (1991).

M. Klittich, CIM-OSA Part 3: CIM-OSA integrating infrastructure – the operational basis for integrated manufacturing systems, *Int. J. Computer Integrated Manufacturing*, **3** (3 and 4), 168–180 (1990).

L. Lamport, Time, clocks and the ordering of events in a distributed system. *Communications of the ACM*, **21**(7), 558–565 (1978).

M.V. Mannino and W. Effelsberg, A Methodology for Global Schema Design, *Technical Report TR-84-1* Computer and Information Science Department, University of Florida (1984).

R. Mayer, Analysis of Methods, Internal report #KBSL-89-1001, Department of Industrial Engineering, Texas A&M University, College Station, TX 77843 (1989).

J. Mills, B. Huff and T. Criswell, *et al.*, Virtual manufacturing workstation. *Proceedings of the 4th International Conference on Design Theory and Methodology*, Scottsdale, AZ, 35–39 (1992).

J. Misra, Distributed discrete-event simulations. *Computing Surveys* **18**(1), 39–51 (1986).

C. Parks, R. Judd and D. Koonce *et al.*, Model-based Manufacturing Integration: A Paradigm for Virtual Manufacturing Systems Engineering. *Computers and Industrial Engineering* **27**(1–4), 357–360 (1994).

M.P. Reddy, B.E. Prasad and P.G. Reedy *et al.*, A Methodology for Integration of Heterogeneous Databases. *IEEE Transactions of Knowledge and Data Engineering*, **6**(6) 920–932 (1994).

W. Wang, K. Popplewell and R. Bell, An integrated multi-view system description approach to approximate factory modeling. *International Journal of Computer Integrated Manufacturing*, **6**(3), 165–174 (1993).

M.J. Wiskerchen and R.B. Pittman, Dynamic systems-engineering process: the application of concurrent engineering. *The Journal of Engineering Management*, **1**(2), 27–34 (1989).

N. Yannoulakis, S. Joshi and R. Wysk, Manufacturability evaluation and improvement system. *Proceedings of the 3rd International Conference on Design Theory and Methodology*, Miami, FL, 217–226 (1991).

CHAPTER 12

A Framework for Robust System and Process Design Through Simulation

SUSAN M. SANCHEZ[a], PAUL J. SANCHEZ[b] and JOHN S. RAMBERG[c]

[a]School of Business Administration, University of Missouri-St. Louis, 8001 Natural Bridge Road, St. Louis, Missouri 63121-4499, USA; [b]Parks College of St. Louis University, St. Louis, Missouri 63156-0707, USA; [c]Systems and Industrial Engineering, The University of Arizona, Tucson, Arizona 85721, USA

1. INTRODUCTION AND SCOPE

The design and analysis of complex systems, such as manufacturing facilities, communications networks, transportation systems, and service organizations, are not straightforward tasks. One technique which designers and analysts often employ is simulation. This is a powerful tool with many potential benefits. It is typically faster, easier, and less expensive to build a simulation model and investigate its behavior than to perform similar experimentation on a real system. Simulation can also be used to study systems in the planning stages, before they are implemented. However, despite the rapid gains in computing power and speed over the last two decades, simulation has its limitations: individual runs can be time-consuming for complex models, and trial-and-error approaches are not efficient ways of improving or seeking to understand the system's behavior. Statistical design and analysis techniques provide effective tools: one way that the system or process performance can be characterized is by building statistical models of the simulation output. Since the simulation model is itself a model of a real (or prospective) system, such models of the simulation's response are called *metamodels* of the system's behavior. In order to be useful, a metamodel should be an accurate approximation of the true behavior over an appropriate range of the conditions (i.e., those which might be faced in the real world setting). It is then often faster and easier to evaluate system behavior by using the metamodel than by rerunning the simulation. Experimental design is necessary for the efficient construction of response surface metamodels.

Experimental design has recently played a greater role in the areas of quality planning and robust product design, where Genichi Taguchi has made innovative contributions through the integrated use of loss functions and orthogonal arrays (Taguchi, 1986, 1987; Taguchi and Wu, 1980). He found that it was frequently cheaper to make a process insensitive to variations in the inputs than to attempt to control causes of manufacturing variation (see Ramberg, 1980, for manufacturing examples). Taguchi's three-stage approach for quality improvement activities consists of *system design, parameter design* and *tolerance design*. Taguchi's system design is the application of scientific and engineering knowledge to produce a functional prototype model, rather than the design of a system (such as a manufacturing line or customer processing network) or enterprise. This prototype model defines the product/process design characteristics (parameters) and their initial settings. The goal of parameter design is the identification of settings that minimize variation in the performance characteristic and adjust its mean to an ideal or nominal value. Tolerance design is a method for scientifically assigning tolerances in order to minimize total product manufacturing and lifetime costs, and assessing the system's sensitivity to improvements or degradations in the sources of noise.

In the simulation context, system design might correspond to building and validating a functional model of an existing real-world system or a prospective new facility, process, or product. Parameter design is appropriate for attempting to "optimize" or "improve" the performance of the simulation model by judiciously selecting settings for some of the decision factors in the model. Tolerance design is useful for characterizing the simulation response and determining its sensitivity to sources of noise in the system. It can also be used to evaluate the model's sensitivity to the modeling assumptions and approximations, such as the values of distributional characteristics used to generate random inputs. Throughout the rest of this chapter, when we refer to "designing a system" we mean determining settings for the decision factors in the simulation model.

A robust metamodeling approach to simulation analysis has potential benefits in a variety of contexts. First and foremost, it may improve decision-making over alternative methodologies. It systematizes the data collection effort, since experimental design is preferable to trial-and-error. The robustness of the resulting design also means that implemented systems are more likely to conform to expectations of their performance. Additionally, it provides a framework in which the Taguchi strategy can be applied to a simulation model for system or process design. While this may be a novel concept for many manufacturing engineers, product designers have used computer models (CAD/CAM tools) for experimentation in place of physical prototypes, particularly in the semiconductor industry (Sacks *et al.*, 1989a, 1989b). These applications have typically been Monte Carlo simulations, but it is not surprising that the robust design approaches can also be applied in discrete event contexts. However, many simulation analysts, as well as managers and management scientists unfamiliar with recent advances in quality engineering, may be unaware of the power of this approach. Robust design methodology has typically not been employed in the investigation of complex systems via discrete event simulation.

As we will demonstrate, the synergy of response-surface modeling and simulation as the medium of experimentation yields benefits which go beyond those obtained by a straightforward application of Taguchi's methodology to the design of a process or system. Analysts have an added degree of control when building the simulation model and collecting the data. This provides opportunities for further exploiting the model structure, with potential gains in efficiency for metamodel estimation. The complexity of the response surface and the inherent importance of randomness to system behavior are two additional areas where robust design for discrete event simulation is notably different than for manufactured products. The experimental setting in Taguchi's approach gains efficiency in sampling by removing virtually all randomness from the experiment. While this is possible for Monte Carlo simulation experiments as well, it is unreasonable in the context of discrete event simulation of inherently stochastic systems or processes. For simulation analysts attempting to describe or design such systems or processes, the response surfaces which characterize process behavior are usually nonlinear. This means that experimental designs which only allow estimation of main effects (as Taguchi's orthogonal designs) will not accurately estimate the response surfaces, which may negate the benefits of robust design. The tolerance design stage must also be handled differently for process design in a simulation framework to account for the randomness inherent in the response at each design point.

While the approach in this chapter is technically straightforward, it represents a distinct change in viewpoint relative to traditional simulation analyses. This permeates the data collection, analysis, and interpretation processes. It is our hope that this approach will facilitate the design of systems and/or processes by capturing crucial aspects of the system/process behavior which might otherwise go undetected, as well as by easing communication between the analysts and management. Those already familiar with the use of robust design methods for product design and improvement will recognize these benefits (Pignatiello and Ramberg, 1991). In addition, by expanding robust design techniques to address the areas of process and system design, a firm can create a unified, integrated approach to design and improvement efforts.

In the next Section, we discuss the concepts common to all three stages of the improvement process: performance measures and their evaluation, the classification of controllable and uncontrollable variation, and experimental designs and analysis. In Section 3, we present an example of parameter design for a dynamic stochastic simulation model of a job shop. This demonstrates that, for some systems, robustness considerations may lead to the selection of a different configuration than that suggested by analyzing only the mean response. In Section 4, we apply tolerance design to the job shop example. This offers insights into two types of information which the analyst can obtain: (1) an assessment of the simulation's sensitivity to modeling characteristics, which may indicate the level of detail at which various random aspects of the system need to be modeled, and (2) insights into economical ways of improving or optimizing system performance. Alternative data collection plans, which may be useful or even necessary for evaluating certain systems accurately and efficiently, are discussed in Section 5. In Section 6, we return to the product design problem and point out where the simulation metamodeling framework might provide new opportunities for improved performance or new insights into system behavior. Our conclusions are provided in Section 7.

2. THE ROBUST DESIGN FRAMEWORK

Taguchi's philosophies and teachings have sparked a great deal of both criticism and support in quality-engineering and related circles. Kackar (1985) (with discussions), Nair *et al.* (1992), and Pignatiello and Ramberg (1991) are just a few of the many articles in the literature that reflect this mixed reaction. Much of this controversy focuses on Taguchi's tactics, rather than the underlying strategy. As such, we do not advocate strict compliance with Taguchi's tactical recommendations. Our presentation is based on enhancements which make use of data-analytic and graphical techniques (Nair *et al.*, 1992; Ramberg *et al.*, 1992; Sanchez *et al.*, 1993; Vining and Myers, 1990).

2.1. Performance Characteristics

The first step in designing an experiment is to identify the *performance characteristic(s)* of interest. In queueing system simulations, for example, the steady-state mean waiting time of customers in the queue is one widely-used measure. Percentiles of the steady-state distribution of waiting time are another type of performance characteristic which might be studied.

In order to determine the degree of satisfaction with the performance characteristic, an ideal state must be specified for comparison purposes. Such an ideal state is called the *target value* for the performance characteristic. Three types of targets, and associated goals, are considered. First, one could strive to make the performance characteristic as small as possible. For example, in a manufacturing simulation, one might seek to identify the system which minimizes the time between order arrival and delivery of the product. Second, one could strive to make the performance characteristic as large as possible. For example, the goal of exploring a model of a nuclear power plant which runs until failure (a terminating simulation) might be to identify systems where the time until failure is quite large. Finally, one could strive to make the performance characteristic as close to some ideal state as possible. An economic simulation model, designed to assist the Federal Reserve Board in manipulating money supplies and interest rates to exercise control over the level of inflation, would fall into this category.

In systems where some stochastic variation is present, the performance characteristic will exhibit random fluctuation or variation around its target value. The cost of variation in the performance characteristic is thus something which must be measured. Taguchi advocates using *loss to society*, and has been commended for looking beyond manufacturing costs and considering the costs incurred by end-users of the product. The *expected loss* is the expected value of the monetary losses that an arbitrary user of the product is likely to suffer during the product's life span due to performance variation. Unfortunately, it is usually difficult or impossible to specify an exact form for the underlying loss function. Consequently, a quadratic loss is often utilized. Letting Y denote the performance characteristic (a random variable), τ denote the target value, and $\ell(\cdot)$ denote the loss function, we have

$$\ell(Y) = K(Y - \tau)^2 \tag{1}$$

for some constant K. (Loss is assumed to be zero when Y achieves the ideal state.) It follows that the expected loss, where expectation is taken across the noise factor space, is

$$L = E[\ell(Y)] = KE[(Y - \tau)^2] = K[Var[Y] + (E[Y] - \tau)^2] \equiv K[MSE] \tag{2}$$

If K converts deviations from the target value into monetary units, then the expected loss can be used directly in the cost/benefit analyses of various alternatives. As we show in Section 3, this may lead to different recommended solutions than conversions based solely on mean performance. If a monetary conversion factor is not available, then the analysis can proceed by working with the scaled losses L/K. This makes it difficult to perform cost/benefit analyses, but does allow the analyst to rank the performances associated with different system configurations to determine those which are 'good' relative to the others.

Quadratic loss functions have been used in a variety of other settings. For example, minimizing mean squared error loss is the basis for least squares linear regression. It is also intuitively appealing: small deviations from the target value τ have little impact on the loss, while large deviations from τ will result in extremely large losses. Thus, in the context of system evaluation, the expected loss will tend to be low if most noise in the system inputs is not transmitted to the response, and tend to be high for systems where the response variability is large.

Despite its apparent simplicity, the concept of expected loss radically alters the idea of optimality for simulation experiments. Attaining the ideal state in expectation is neither sufficient nor necessary for optimality: joint consideration of the mean *and variability* of the system response is necessary. As we shall show, the use of the loss function in conjuction with separate metamodels of the performance mean and variability provides insights into system behavior which cannot be obtained by analyzing either the mean performance or the expected loss alone.

Before proceeding, we emphasize the distinction between two types of variance: system variance and estimator variance. A well established field in simulation is that of variance reduction, where researchers have established methods for reducing the variance of the *estimators* of mean responses. This increases the power for hypothesis testing purposes, and allows the analysts to obtain tighter confidence intervals if the primary goal is estimation of the mean response. Unequal system response variance at different system configurations is recognized as pervasive. It often influences the experimental design and analysis, as by leading to different run lengths for different system alternatives, but has rarely been incorporated into the system evaluation. When we speak of identifying system configurations with low variance, we are referring to low *system* variance rather than variance reduction techniques.

2.2. Parameter Design Versus Noise Factors

In order to achieve systems/products with little variation in performance characteristics, several steps are necessary. First, one must identify factors in the system which potentially affect the system response. Factors are classified as *parameter design factors* (hereafter called *parameters*) or *noise factors*, where the noise can result from sources either internal or external to the system. In a real-world setting, the parameters are all those factors over which it is possible to exercise control, while the noise factors are not easily controllable or controllable only at great expense. Although all factors are truly controllable when the system under investigation is a computer simulation, the classification into parameter and noise factors can and should still be made on the basis of their controllability in the real-world system for which the simulation is a model. This distinction has been noted by simulation analysts (Law and Kelton, 1991) but common practice does not make use of it when designing a simulation experiment.

2.3. Experimental Design and Analysis

Potential model configurations result from changing the settings of some or all parameters in the system (simulation model). Sometimes there may be managerial reasons for limiting investigation to a few alternatives. For example, the simulation might have been written to assess which station within a multi-server queueing network would yield the greatest improvement in processing time via the addition of another server. If, due to budgetary constraints, these are the only alternatives which can be considered (at least in the short term), then simulations of all alternatives could be conducted and evaluated to determine which is preferable. This has been the basic thrust of several techniques proposed for comparing and contrasting a limited set of system designs. These techniques include pairwise or multiple comparison methods, and selection and ranking. (For a survey article, see Goldsman, Nelson and Schmeiser, 1991).

Alternatively, experimental designs can be utilized to observe how the expected value of the performance measure varies as a function of the system factors. If the parameters are quantitative, then experimental design and analysis can lead to metamodels — mathematical models of the simulation model of the system response. Metamodels provide much more information about the underlying system than haphazard investigation of a few specific configurations. Thus, if the goal of the analyst is to optimize or improve the model's performance, and flexibility exists in the settings of the parameter levels (as in prospective studies), then building a metamodel is appropriate. The actual number of configurations studied, and the form (linear, quadratic, etc.) of the resulting metamodel are dependent on the experimental design chosen. Clearly, there are tradeoffs between the metamodel accuracy and the experimental effort: higher-order metamodels can approximate nonlinear response surfaces more accurately, but require sampling at a larger number of configurations.

Traditionally, the parameter space has been the focus of experimental design, and analysts have assumed that all randomness between replications at a single design configuration is due to the influence of random, uncontrollable noise factors. Experimental designs which make use of variance reduction techniques are often used to improve the precision of estimating the expected responses. Taguchi, on the other hand, incorporates the identifiable noise factors into the experimental design by crossing a parameter matrix with a noise factor matrix. Output from each configuration in the parameter space (commonly a simulation run) is then calculated across the noise factor space, in order to determine the parameter levels which result in the smallest sensitivity of the response to the noise factors. For example, suppose we have a single noise factor W and two possible parameter configurations, C_1 and C_2 of Figure 1. In this figure, the relationships between W and the responses can be described as

$$Y = \begin{cases} 6W^2 - 2 & \text{for configuration } C_1 \\ 3W^2 + 5 & \text{for configuration } C_2. \end{cases}$$

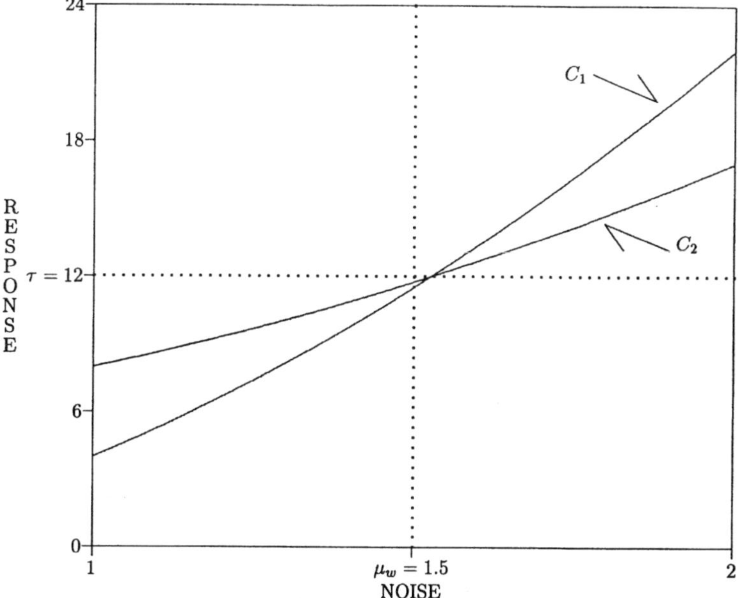

Fig. 1 Response sensitivity to noise.

If W is uniformly distributed over the range $[1, 2]$, then integrating the two response curves confirms that both result in the same mean response of 12. However, much less of the noise variability is transmitted to the response for C_2 than for C_1. (The standard deviations of the response are 5.22 and 2.61 for C_1 and C_2, respectively.) This is readily identified graphically by noting that the response for C_2 is flatter than that for C_1.

Taguchi advocates the use of orthogonal arrays for the parameter and noise designs, and the use of an appropriately selected signal-to-noise ratio to systematically evaluate the tradeoffs between a performance characteristic's variability and its deviation from the target value. These tactical issues are the basis of most of the controversy concerning Taguchi's methods (Pignatiello and Ramberg, 1991; Box, 1988). We advocate joint analysis of the response mean and variance, and implement the Taguchi strategy using conventional factorial designs for the parameters since these are among the most common designs used in industrial experimentation. Typically, either two or three levels are chosen for each parameter: designs with two levels are simpler and require fewer experimental runs, but designs with three levels allow one to model quadratic as well as linear effects. Complete factorials (CFs) allow estimation of all interaction effects (or all but one if a single replicate is used). Fractional factorials (FFs) allow one to reduce the number of experimental runs if one is willing to assume that high order interactions are less important. (Elementary concepts of factorial designs are discussed in *Askin, Ramberg and Sodui*, 1997. See Box, Hunter, and Hunter, 1978 or Montgomery, 1991 for detailed presentations of fractional factorial designs.)

An advantage of the parameter/noise factor distinction is that it facilitates the evaluation of a greater number of alternatives by reallocating data collection effort from the noise factors to the parameters. There is much more flexibility in the design for the noise factors, since we will not be explicitly fitting these terms in the metamodels for the response mean and variance. This is further discussed in Section 5.

3. PARAMETER DESIGN FOR SYSTEMS AND PROCESSES

The system we use to illustrate parameter design for processes is loosely based on a dynamic, stochastic simulation model described in Law and Kelton (1991) and analyzed in Sanchez *et al.*, (1997). We present the problem in Section 3.1, and discuss alternative model configurations and design considerations in Sections 3.2 and 3.3. In Section 3.4 we discuss issues related to the simulation metamodel and preparation of the simulation output for analysis. Results are summarized in Section 3.5.

3.1. Initial Job Shop Simulation Model

A manufacturing shop consists of five machine stations, and at present stations 1 through 5 consist of 3, 3, 4, 3, and 2 identical machines, respectively. The second machine in station 5 was purchased recently to alleviate a bottleneck at this station, and sits idle much of the time. At the present time, two types of jobs are processed, and they are batched in sets of five according to job type before proceeding to the shop floor. However, the shop soon plans to begin processing a third type of job.

After the changeover occurs, jobs are anticipated to arrive at the shop according to independent, identically distributed exponential interarrival times with a mean of .30 hours. It is expected that arriving jobs will be of type 1, 2, or 3 with respective probabilities .25, .50, and .25. The shop has a ten hour workday.

Batch setup times are fixed but allowed to vary by job type and station, and are presented in Table 1. Processing times are assumed to be independently distributed gamma random variables (the sum of two exponential random variables, also known as 2-Erlangs). The routings and mean processing times at a station differ by job type, and are also provided in Table 1. For example, jobs of type 1 are first routed to station 3, where they require an average of .25 hours each in processing. They then proceed to station 1, and so forth. If a job arrives at a particular station and finds all machines in that station already busy, then the job joins a single FIFO queue at that station. The target time in the system is ten working hours (one ten hour day).

3.2. Alternative Model Configurations

The given description represents only one possible configuration of the stochastic job shop model, such as might be constructed to determine whether an existing job shop configuration would be capable of responding to the pattern of demand described. A metamodel of this job shop simulation would provide additional insights into the system performance, by allowing the analyst to compare several different potential configurations (e.g., by changing the number of machines in one or more stations). Alternatively, experimental design techniques could be used to construct metamodels of the expected value of the system (simulation model) response (Kelton, 1988; Law and Kelton, 1991).

As a first step, the performance characteristic of interest must be specified. We choose to examine the time elapsed from a job arrival to the completion of that job (hereafter referred to as the *time in system*) for all three job types. The average, steady-state value of this random variable will be our performance measure, and the associated target value is 10 hours.

The job shop model has several additional factors which could be investigated in more detail. We list several of these in Table 2, and classify each as either a parameter (e.g., controllable variable in the real system) or a source of noise (internal/external). Although it is not the focus of this chapter, it is important to recognize that when improvement efforts are an ongoing process, factors which are uncontrollable in the short term may be considered parameters in the long term. For our example, purchasing new equipment may affect the mean and/or variability of

TABLE 1
Setup and processing requirements by job type.

Job Type	1	2	3
Machine Groups in Routing	3, 1, 2, 5	4, 1, 3	2, 5, 1, 4, 3
Mean Times (hrs), Successive Tasks	.25, .30, .425, .25	.55, .40, .375	.60, .125, .35, .35, .50
mean Setup (hrs), Successive Tasks	.90, 1.08, 1.20, .36	1.20, 1.20, .90	1.20, .36, 1.08, 1.20, .90

TABLE 2
Potential factors for the job shop model.

Factor	Description	Classification
M_i	Number of machines in station i	Parameter
B_j	Batch size for jobs of type j	Parameter
T_{ij}	Time to process job of type j at station i	Noise (Internal)
S_{ij}	Batch setup time for job type j at station i	Noise (Internal)
ID_i	Time between breakdowns of machines in station i	Noise (Internal)
D_i	Down-time for a machine in station i	Noise (Internal)
IA	Inter-arrival time for jobs	Noise (External)
PJ_j	Proportion of arriving jobs of type j	Noise (External)

the required processing time, while maintenance scheduled off shift may reduce or eliminate the problem of machine breakdowns. Even external noise factors, such as the product mix, may be influenced in the long term by pricing and advertising strategies.

The size of the experiment depends on the numbers of parameters and noise factors of interest. If we used all of the factors described in Table 2, then we would have 8 parameters and 37 noise factors. If we allowed batch size to vary by station as well as machine type, or if stations were composed of machines with slightly different characteristics, the numbers of factors would increase still further. However, since the goal of this section is to illustrate the robust design method rather than to study the sytem *per se*, we shall limit the experiment to five parameters and two noise factors. The parameters are the number of machines in the first four stations (M_i, $i = 1, \ldots, 4$) and the batch size B_3 for job type 3. We fix $M_5 = 2$ since this station currently has excess capacity. B_1 and B_2 are fixed at 5, corresponding to batch sizes currently used on the floor, by a managerial decision to leave production of existing jobs unchanged as much as possible. The noise factors PJ_1 and PJ_2 describe the job mix (note that PJ_3 is fixed by the values of the other two). The remaining factors are fixed at the levels or distributions according to the original problem statement.

3.3. Experimental Designs

As previously discussed, we will use conventional factorial designs. For our example with five parameters and two noise factors, a single replicate of a two-level CF (complete factorial) for the parameters has 32 configurations, while a three-level CF design has 243 configurations. Clearly, a three-level design imposes a much greater data gathering burden than does a two-level design. We select instead a 2^{5-1} resolution V design (16 configurations) for the parameter space, which insures that main effects and two-factor interactions are not aliased with one another, and augment it with one or more center points. This efficient design allows us fit the model above with many fewer runs than would be required by a 3^k CF, and still assess potential lack-of-fit. The augmented FF for the parameter space is crossed with a 2^2 CF for the noise space. (We use a CF for the noise space only because the number of noise factors is small. Interactions between noise factors and parameters are a critical component of unequal response variability across parameter configurations, but neither such interactions nor interactions between noise factors will be explicitly modeled. Thus, one can use saturated or nearly saturated designs for the noise factors for reasons of efficiency if the number of noise factors is large.)

The levels must then be set for each of the factors. Two-level designs often suffice for noise factors, since we will not explicitly model their effects. For continuous noise factors (or discrete noise factors which can be closely approximated by continuous distributions), the two levels are set to one standard deviation above and one standard deviation below the mean so that each of these discrete distributions has a mean and standard deviation equal to those of the underlying distribution. Levels for other discrete noise factors can similarly be chosen to reflect the underlying mean and variance — even for unusually shaped distributions, a method as simple (though potentially inefficient) as sampling each level proportionally to its probability of occurrence can be used. However, Kelton (1988) observes that the choice of parameter level settings is not nearly so clear, and requires some knowledge of how the system behaves. In general, settings ought to be selected such that the response surface contained with the parameter boundaries can be adequately described by a linear or quadratic equation. The settings established for our example are shown in Table 3. Since the numerical center of the low and high levels for M_2 is fractional (3.5 machines), two "center points" augment the design: one with $M_2 = 3$ and the other with $M_2 = 4$.

3.4. Results

A program for the system described above was implemented in C+ + and verified by a structured walk-through. The performance characteristic is the average steady-state time in the system. Data are collected for each configuration for 10,000 simulated hours (approximately 5 simulated years of plant operations) past a truncation point determined by correlogram analysis for the high traffic intensity configuration.

The mean and variance of the response for parameter configuration i and noise configuration j, denoted by $\bar{Y}_{ij}$ and s_{ij}^2, are estimated by the sample average and sample variances of the output from the corresponding run. Note that the first portion of the output stream should be discarded prior to calculation of these values in order to remove any initialization bias (Law and Kelton, 1991). Although the data are correlated, s_{ij}^2 is an asymptotically unbiased estimate of the variability of the

TABLE 3
Factor level settings for the job shop experiment.

		Low	Center	High
Parameters	M_1	3	4	5
	M_2	3	n.a.	4
	M_3	4	5	6
	M_4	3	4	5
	B_3	1	3	5
Noise	PJ_1	0.23	0.25	0.27
	PJ_2	0.47	0.50	0.53

performance characteristic for a particular combination of parameter and noise factor values (Ceylan and Schmeiser, 1993).

Crossing the parameter matrix and the noise matrix means that for each configuration of the parameters, a run is made for each of the n configurations in the noise factor plan ($n = 4$ in our experiment). Then, for each parameter configuration i, we compute the following summary measures of performance mean and variation:

$$\bar{Y}_{i\cdot} = \frac{1}{n}\sum_{j=1}^{n} \bar{Y}_{ij}, \quad \bar{V}_{i\cdot} = \frac{1}{n-1}\sum_{j=1}^{n}(\bar{Y}_{ij} - \bar{Y}_{i\cdot})^2 + \frac{1}{n}\sum_{j=1}^{n} s_{ij}^2 \tag{3}$$

This variance measure has two components. The first is the measure of variance computed across the noise points, as is often used in Taguchi analyses. This is augmented by an average of asymptotically unbiased estimates of within-run variability to comprise the total variability observed at design point i, similar to the Total Sum of Squares used in ANOVA.

We evaluate the configurations using the scaled loss (L/K from equation (2) based on an underlying quadratic loss function, which eliminates the need to ascertain a value of the conversion constant K. The mean, variance, and scaled loss for each of the 18 initial parameter configurations are presented in the 'First Experiment' portion of Table 4, along with the total number of machines ($\Sigma_{i=1}^{5} M_i$). Note that the

TABLE 4
Simulation results for the job shop experiments.

Configuration Run	Parameter Settings $(B_3, M_1, M_2, M_3, M_4)$	Total Machines	Mean $\overline{Y}_i$	Variance $\overline{V}_i$	Scaled Loss
First Experiment					
1	(1, 3, 3, 4, 5)	17	11.74	9.66	12.68
2	(1, 3, 3, 6, 3)	17	12.30	10.58	15.89
3	(1, 3, 4, 4, 3)	16	12.28	10.84	16.04
4	(1, 3, 4, 6, 5)	20	11.65	9.84	12.56
5	(1, 5, 3, 4, 3)	17	11.39	8.31	10.24
6	(1, 5, 3, 6, 5)	21	10.49	7.36	7.61
7	(1, 5, 4, 4, 5)	20	10.42	7.50	7.68
8	(1, 5, 4, 6, 3)	20	11.26	8.49	10.08
9	(5, 3, 3, 4, 3)	15	12.91	10.68	19.14
10	(5, 3, 3, 6, 5)	19	12.94	10.74	19.40
11	(5, 3, 4, 4, 5)	18	12.95	10.73	19.44
12	(5, 3, 4, 6, 3)	18	12.89	10.68	19.05
13	(5, 5, 3, 4, 5)	19	12.81	10.70	18.62
14	(5, 5, 3, 6, 3)	19	12.75	10.66	18.22
15	(5, 5, 4, 4, 3)	18	12.76	10.64	18.25
16	(5, 5, 4, 6, 5)	22	12.80	10.70	18.53
17	(3, 4, 3, 5, 4)	18	11.55	4.36	6.77
18	(3, 4, 4, 5, 4)	19	11.54	4.35	6.74
Second Experiment					
19	(1, 4, 3, 4, 4)	17	10.67	7.31	7.76
20	(3, 3, 3, 4, 4)	16	11.74	4.42	7.45
21	(3, 4, 3, 4, 3)	16	11.65	4.35	7.08
22	(3, 4, 3, 4, 5)	18	11.56	4.36	6.80
23	(3, 5, 3, 4, 4)	18	11.55	4.36	6.76
24	(5, 4, 3, 4, 4)	17	12.78	10.65	18.35

center points (configurations 17 and 18) yield the best results among the observations collected; as we demonstrate later in this section, this does not mean it is the best point in the parameter space. Initial metamodels with linear terms, a single quadratic term (in standardized units), and all two-way interactions were then constructed for $\bar{Y}_{i\cdot}$ and $ln(\bar{V}_{i\cdot})$ using linear regression. (The logarithmic transformation is used for stability purposes.) The results are presented in Table 5.

The fits are excellent ($R^2 > .99$) for both full models. However, we only have a single degree of freedom for error, so reliance on t-statistics for identifying important effects is questionable. Low degrees of freedom for error estimation is a common problem in designed experiments where data are costly and the analyst is interested in extracting a large amount of information from a small sample. One way of dealing with this problem, valid if the number of fitted model terms is greater than or equal to 15, is the use of a normal probability plot. If there are h (standardized) regression coefficients, $\hat{\beta}_1, ..., \hat{\beta}_h$, sort them to obtain $\hat{\beta}_{[1]} \leq ... \leq \hat{\beta}_{[h]}$. Then construct a normal probability plot by plotting $\Phi^{-1}((2i-1)/2h)$ versus $\hat{\beta}_{[i]}$ for $i = 1, \ldots, h$, where Φ is the cdf of the standard normal distribution. Under the null hypothesis of no term effects, the standardized coefficients should be normally distributed with mean zero, hence the plotted points will fall on or near a straight line. Points which appear to be outliers from a straight line are judged to correspond to significant terms in the model. A reasonable estimate of error can then be obtained by pooling the effects for the remaining "inactive" factors in order to assess statistical significance.

The normal probability plot for the coefficients in the metamodel of the mean response is given in Figure 2. The magnitude of the standardized B_3 coefficient is by far the largest, but five other terms are potentially important (M_1, M_4, and the M_1B_3 and M_4B_3 interactions, as well as the quadratic effect). After removing these points from the plot (Figure 3), the coefficients conform more closely to a straight line. Pooling the effects of inactive terms allows us to increase the error degrees of

TABLE 5
Initial metamodels of the mean and variability of total time in system.

	μ		$ln(\sigma^2)$	
Term	Coefficient	t-value	Coefficient	t-value
Constant	12.1463	819.4822	2.2825	1112.1509
$B_3 - 3$	.7050	141.0000	.0870	42.3731
$M_1 - 4$	−.3113	−62.2500	−.0651	−31.7179
$M_2 - 3.5$	−.0183	−3.8891	.0043	2.2367
$M_3 - 5$	−.0113	−2.2500	−.0000	−.0119
$M_4 - 5$	−.1713	−.34.2500	−.0258	−12.5929
$(M_1 - 4)(B_3 - 3)$	.2400	48.0000	.636	30.9773
$(M_2 - 3.5)(B_3 - 3)$	.0188	3.75	−.0054	−2.6138
$(M_3 - 5)(B_3 - 3)$	.0050	1.0000	.0004	.1836
$(M_4 - 4)(B_3 - 3)$	.1950	39.000	.0283	13.7892
$(M_1 - 4)(M_2 - 3.5)$	−.0050	−1.0000	−.0002	.1038
$(M_1 - 4)(M_3 - 5)$	.0013	.2500	.0006	.2841
$(M_1 - 4)(M_4 - 4)$	−.0038	−6.7500	−.0037	−1.7831
$(M_2 - 3.5)(M_3 - 5)$	−.0350	7.0000	.0035	1.7270
$(M_2 - 3.5)(M_4 - 4)$	.0000	1.2500	.0001	.2269
$(M_3$ - 5)(−4)	−.6013	−40.0833	−.8111	−0466
quadratic	−.6013	−40.0833	−8.111	−131.7443
Regression F	1905.8		1342.8	
Regression p	.018		0.21	
R^2	.999967		.999953	

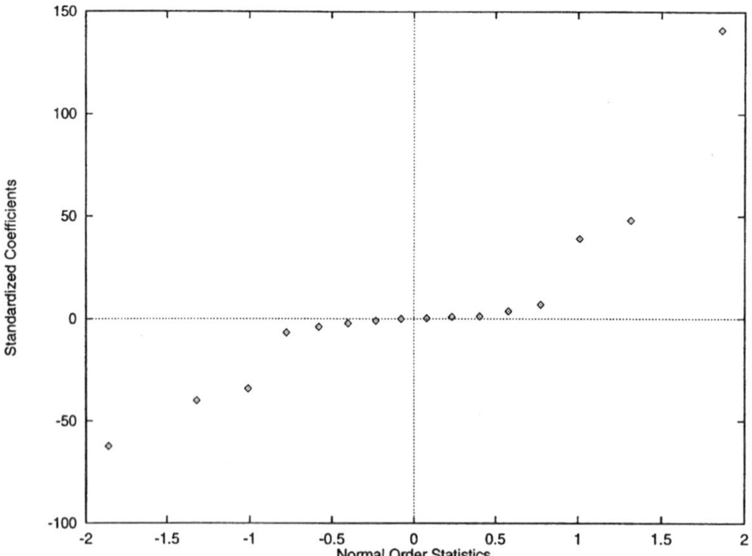

Fig. 2 Normal probability plot for μ metamodel.

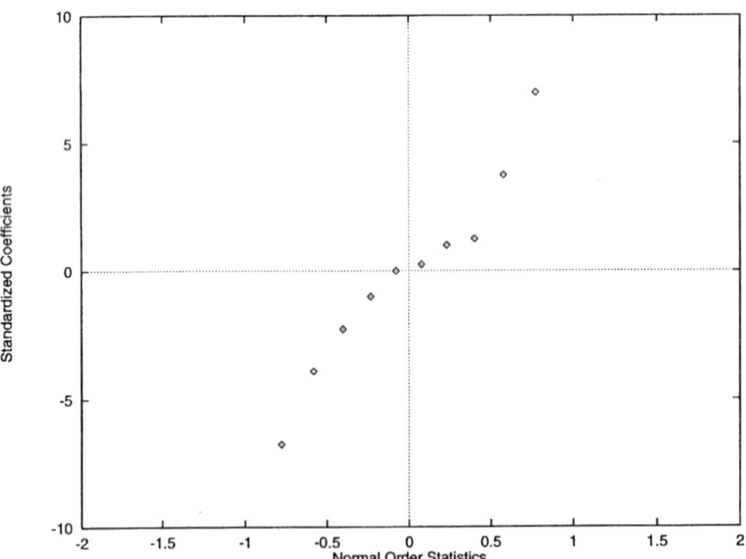

Fig. 3 Normal probability plot for μ metamodel after removing important effects.

freedom. Since R^2 is so high, we elected to use a cutoff of $p < .01$ to assess significance. The resultant model (with 6 independent variables) still accounts for over 99.5% of the variation in the mean performance. For the *ln*(variance) metamodel, the normal probability plot (not shown) and numerical results yield similar findings. M_2 and M_3 are not statistically significant at $p = .01$, and the quadratic term again enters the model.

Taken together, the results for the two metamodels indicate a need for further experimentation in order to identify the factors associated with the quadratic term. Since M_2 and M_3 affect neither the mean nor variance of the response over the ranges specified, we set them at their lowest (most economical) levels for all remaining runs. (Although they could contribute to the quadratic effect, empirical evidence suggests this is unlikely.) Viewing the parameter space of the remaining factors in three dimensions, the inital 18 configurations provided information about the behavior at the corners of the cube, as well as at the center. We augment this design with six additional parameter configurations: one at the center of each face of the cube (as in a central composite design with the distance from the center to the axial points set to one). Once again, each parameter configuration is run under four noise conditions to obtain overall estimates of the mean and variance. When these additional points are added to the initial 18, after suitable refinement the regression models (in natural units) are:

$$\begin{aligned}\mu \approx\ & 11.5081 + .3719(B_3 - 3) - .2872(M_1 - 4) - .1572(M_4 - 4) \\ & + .1200(M_1 - 4)(B_3 - 3) + .0975(M_4 - 4)(B_3 - 3) \\ & + .0877(B_3 - 3)^2 + .2707(M_1 - 4)^2\end{aligned} \tag{4}$$

$$\begin{aligned}ln(\sigma^2) \approx\ & 1.4740 + .0491(B_3 - 3) - .0586(M_1 - 4) \\ & + .0318(M_1 - 4)(B_3 - 3) + .1992(B_3 - 3)^2.\end{aligned}$$

The standardized coefficients for both models are provided in Table 6. For the μ metamodel we have $R^2 > .968$ all variables significant with $p < .015$, and no others significant at level $p < .10$. For the $ln(\sigma^2)$ metamodel we have $R^2 = .979$, all variables are significant with $p < .001$, and no others are significant at level $p < .05$. (The adjusted R^2 values are .955 and .974, respectively.)

Predictions based on these two equations can be calculated over an integer grid. Restricting factor levels to the ranges specified in Table 3, we need search only over a

TABLE 6
Final metamodels of the mean and variability of total time in system.

	μ		$ln(\sigma^2)$	
Term	Coefficient	t-value	Coefficient	t-value
constant	11.5081	147.6576	1.4740	60.0251
$B_3 - 3$	.3719	18.2172	0.0491	6.9269
$M_1 - 4$	−.2872	−7.0338	−.0586	−4.1348
$M_4 - 4$	−.1572	−3.8502		
$(M_1 - 4)(B_3 - 3)$	.1200	5.5413	.0318	4.2277
$(M_4 - 4)(B_3 - 3)$	.0975	4.5023		
$(B_3 - 3)^2$	.0877	3.5707	.1992	28.1009
$(M_1 - 4)^2$	.2707	2.7562		
Regression F	70.34		281.15	
Regression p	.000		.000	
R^2	.968528		.97869	

grid of size $5 \times 3 \times 3$ since only three factors appear in the metamodels. The results suggest that the best alternative has not yet been tested: the job shop with $B_3 = 2, M_1 = 5$ and $M_2 = 5$ has a lower predicted scaled loss (5.33) than any of the configurations already investigated. A validation experiment indicates that the metamodels are still providing a close fit over this portion of the response surface. The mean and variance from the validation experiment are 10.95 and 4.76, respectively. These fall within the 90% confidence intervals computed from the respective regression metamodels for response mean and variance (shown in Table 7, Low Loss configuration), hence well within 90% prediction intervals. Validation is advisable whenever untested configurations are the predicted optima; it is most critical when untested configurations extrapolate outside the range of initial factor levels.

Note that the recommended configuration makes use of three fewer machines than the maximum allowed according to factor level settings, and only four more than the initial production setting. It is interesting to note that $B_3 = 2$, which is neither the common (hence convenient) batch size of the other two jobs, nor that associated with a Just-In-Time approach ($B_3 = 1$).

3.5. Discussion

The general process of planned data collection, metamodel construction and refinement, and configuration selection and confirmation described previously was meant to serve as a guide for the use of metamodels and robust design philosophy in the analysis of discrete event simulation experiments. The analysis differs from ordinary response-surface methodology because the results are averaged *over the noise space* before metamodel building begins. However, the specific data collection plans employed should not be viewed as a prescription for all applications. If it is felt that a linear metamodel will suffice, then the parameter plan can be much simpler; if it is felt that quadratic and/or three-way interactions are likely to be important, or if the parameter levels are very wide, then a larger parameter plan (such as a 3^k CF or central composite design) may be appropriate. In all cases, the experimenter must trade off the cost of additional information against the potential benefits of that information. This assessment is highly contextual.

Further discussion of Tables 4 and 7 is in order. These results underscore the importance of including variance in the decision-making process and the benefits of using metamodels which can suggest alternatives not yet tested. In queueing systems one generally expects that the mean and variance are positively correlated, but the introduction of batch size as a decision parameter affects that relationship. For example, configurations 1 and 20 (Table 4) have the same mean response to two decimal places, yet the associated variances differ by over a factor of two. Note also that if the decision were being made on the basis of mean performance alone then configurations 6 and 7 (with scaled losses of 7.36 and 7.50, respectively) are clearly the best among the first 16 runs (Table 4). Without the metamodeling approach, the analyst might stop and select one or both of these configurations as best; although with the metamodeling approach and a small number of additional runs, a configuration with a much lower predicted loss (5.66) is found. Note also that configurations 6 and 7 use more machines than does the low loss configuration (Table 7).

TABLE 7
Metamodel predictions and performance by selection criteria for the target $\tau = 10$.

Selection Criteria	Parameter Configurations		Mean	Varience	Scaled Loss
Low Loss:	$B_3 = 2, M_1 = 5, M_4 = 5$	Prediction	10.83	4.64	5.33
	$(M_2 = 3, M_3 = 4)$	(range‡)	(10.656 – 11.010)	(4.434 – 4.846)	(4.434 – 5.865)
		Validation	10.95	4.76	5.66
Low Mean:	$B_3 = 1, M_1 = 5, M_4 = 5$	Prediction	10.51	7.77	8.03
	$(M_2 = 3 - 4,\ M_3 = 4 - 6)$,	(range‡)	(10.327 – 10.685)	(7.393 – 8.168)	(7.500 – 8.638)
		Validation (6 configurations)	10.35 – 10.55	7.37 – 7.52	7.64 – 7.67
Low Variance:	$B_3 = 3,\ M_1 = 5$	Prediction	11.65	4.12	6.84
	$(M_2 = 3 - 4, M_3 = 4 - 6, M_4 = 3)$	(range‡)	11.453 – 11.845	3.921 – 4.325)	(6.031 – 7.729)
		Validation (6 configurations)	11.61 – 11.68	4.35 – 4.41	6.99 – 7.16
	$B_3 = 3, M_1 = 5$	Prediction	11.49	4.12	6.34
	$(M_2 = 3 - 4, M_3 = 4 - 6, M_4 = 4)$	(range‡)	(11.309 – 11.674)	(3.921 – 4.325)	(5.634 – 7.127)
		Validation (6 configurations)	(11.52 – 11.58)	(4.36 – 4.28)	6.76 – 6.93
	$B_3 = 3, M_1 = 5$	Prediction	11.33	4.12	5.90
	$(M_2 = 3 - 4, M_3 = 4 - 6, M_4 = 5)$	(range‡)	(11.138 – 11.531)	(3.921 – 4.325)	(5.217 – 5.621)
		Validation (6 configurations)	11.51 – 11.59	4.36 – 4.43	6.65 – 6.95

†Factors in parentheses do not appear in the corresponding metamodels. Economical settings are used for the low loss configuration, otherwise settings vary between the low and high values in Table 3.
‡See Section 3.5 for a discussion of how the intervals are computed.

The recommended configuration (low scaled loss) can also be compared to those which would be obtained using other selection criteria: low mean and low variability. By construction, the losses predicted by the metamodels will change whenever parameter settings corresponding to terms included in one or more metamodels (i.e., statistically significant decision variables) are altered. However, in keeping with the robust design philosophy we might not want to distinguish among two parameter configurations if the resulting losses were not significantly different. In order to compare alternative configurations, it is necessary to construct intervals for the loss and its two components. The interval ranges shown in Table 7 are generated as follows. Let $[g_1, g_2]$ and $[h_1, h_2]$ denote 90% confidence intervals for μ and $ln(\sigma^2)$, respectively, based on the corresponding metamodels. The first interval can be converted to an interval on $(\mu - \tau)^2$ by taking

$$[r_1, r_2] = \begin{cases} [\min\{(g_1-\tau)^2, (g_2-\tau)^2\}, \max\{(g_1-\tau)^2, (g_2-\tau)^2\}] & \text{if } g_1, g_2 \text{ are either both above or both below } \tau; \\ [0, \max\{(g_1-\tau)^2, (g_2-\tau)^2\}] & \text{if } g_1, g_2 \text{ are on opposite sides of } \tau. \end{cases}$$

A 90% confidence interval for σ^2 is found by exponentiation, and an approximate 90% interval for loss is then $[r_1 + \exp(h_1), r_2 + \exp(h_2)]$. From Table 7, we see that the low loss configuration is associated with neither the lowest mean nor the lowest variance. Its loss differs significantly from the metamodel predictions for the loss of *all* of the low mean configurations. While the low loss is not significantly different from the losses corresponding to some of the low variance configurations, it is significantly different from that which would be chosen if the analyst set all non-important variables (i.e., M_2, M_3 and M_4) at their most economic levels. In fact, the optimal configuration according to mean performance (the traditional simulation selection criterion) has a loss 35% higher than the recommended configuration, and over 8.5% higher than the configuration with lowest variance. One cannot even argue that the low mean configuration is better in terms of capital investment: the low mean configurations use no fewer machines at any individual machine station than the low loss configuration uses.

Another benefit of the robust approach is that the loss function lends itself to joint consideration of multiple performance criteria. The scaled losses reported in Tables 4 and 7 could be converted to dollar figures (e.g., by considering the cost of a one-day backlog and solving for K in equation (1)). The prorated costs of adding machines could then be included in the loss calculations to directly assess the value of adding equipment. In general, this would result in a partial re-ranking of the alternatives. We remark that the conversion of physical behavior into monetary cost is one of Taguchi's major triumphs in improving the communication between manufacturing/design and management (Pignatiello and Ramberg, 1991).

For our job shop, recall that M_m is the number of machines in station m. Let C_m denote the per-production-unit cost of a machine in station m, derived by prorating its cost over the expected number of production units over its anticipated lifetime. Let **x** explicitly denote the parameter configuration. Then the loss associated with configuration **x** is

$$L_x = K[Var[Y] + (\mu_Y - \tau)^2] + \sum_{m=1}^{5} C_m M_m \qquad (5)$$

where $Var(Y), \mu_y$ and $M_m, m = 1, \ldots, 5$ are dependent on the parameter configuration x. Then configuration x_1 is preferred to configuration x_2 if $L_{x_1} < L_{x_2}$. Consider the following two scenarios for the job shop example: configuration $x_1 = (2, 4, 3, 4, 5)$ with a scaled loss of 6.01 and $x_2 = (2, 5, 3, 4, 5)$ with a scaled loss of 5.33. If $C_2/K < .68$ then x_2 is the better alternative; if $C_2/K > .68$ then the reductions achieved in the mean and variance of time in system are not sufficiently large to justify the addition of another machine to station 1. Note that the calculation of C_m could be more complex if circumstances warranted. For example, the cost of new machines could be converted to a per-production-unit representing the amortized value of capital, then converted to a per-production-unit figure by dividing by the expected number of jobs exiting the system per year. This is a much more detailed objective than typically considered in comparisons of alternative job shop designs, where a single factor like mean throughput or mean waiting time is used to make comparisons.

Finally, we advocate building separate metamodels of the performance mean and performance variability, and then combining this information using the loss function (see also Vining and Myers, 1990). While Taguchi's signal-to-noise ratio loss function destroys simple assessment procedures based on linear models (Barton, 1992), the use of a quadratic loss function facilitates analysis in conjunction with metamodel building. Although one could use loss as the performance measure and generate a metamodel of expected loss directly, this approach would provide less information than joint analysis of the performance mean and variability. If a metamodel shows that expected loss decreases as factor X increases, the root cause remains unknown. Perhaps the response mean is closer to the target. Perhaps the response variance is smaller. It could be that both the mean and variance improve, or that an improvement in one aspect is partially offset by a degradation in the other. This uncertainty about the cause of improvement limits the ability of the analyst to identify new designs which may be even better than those considered in the experimental framework.

4. TOLERANCE DESIGN FOR SYSTEMS AND PROCESSES

In manufacturing, tolerances for manufactured products are ideally set to reflect the anticipated variation in the performance characteristic resulting from uncontrollable variation in the manufacturing process and component parts. Random fluctuation in the response is then attributable solely to noise factors. In simulation, we can view tolerance design as a means of understanding system performance for a single configuration of parameters. This is a form of sensitivity analysis with respect to the noise parameters.

Taguchi breaks down tolerance design into three phases (D'Errico and Zaino, 1988). In the first phase, *system evaluation*, a metamodel of the system response is constructed and the overall response mean and variance are determined. In the second phase, *noise factor assessment*, the sensitivity of the response to variation in each of the noise factors is determined. The final phase of tolerance design is *system optimization*. If some parameters are no longer considered fixed (i.e., alternative configurations are possible), then noise factor assessments can be combined with cost information to determine whether or not adjustments to the system will decrease expected loss. We will illustrate all three phases for the 'low loss' job shop configuration chosen in the previous section ($B_3 = 2, M_1 = M_4 = 5, M_3$ and M_4 fixed at their low levels, and $M_5 = 2$). For simplicity, we shall limit the number of noise components while still illustrating the types of analyses possible.

4.1. Detailed Job Shop Configuration Information

Suppose the following statements describe additional information regarding the job shop system (or simulation model thereof) in the low loss scenario.

1. Although setup time was modeled as fixed for given job/machine station combinations, it is actually a random variable.
2. The mean interarrival time is an estimate.
3. The mean of the processing time distributions of machines in station 1 (Table 1) is an estimate. This might result from data obtained from observing the real system, or from expert judgment or supplier's literature if the simulation is being performed for a prospective system.
4. The head of marketing belives the volatility of demand for Product 1 may increase. How will results be affected if the true standard deviation of demand is wider than the specified 2%?
5. How would performance suffer if we removed one machine from station 3?
6. The machines in station 1 can be refurbished for a cost of $8,000 each, which would reduce the mean of the associated 2-Erlang variables for processing times by 10%; they can be replaced at a cost of $23,000 each, which would reduce the mean processing times by 30%; or new fixtures which reduce the mean setup times can be installed throughout the shop, which will decrease the mean setup times by 20%. Which alternative is best?

The first two items are related to model complexity and validation. Consider the setup times: one could make the simulation model more complex by modeling this component as a random variable. Similarly, another noise factor (mean interarrival time) could have been introduced into the model during the parameter design stage. However, there are drawbacks to including all potential sources of variability in the simulation model. First, additional data collection and/or expert judgment would be needed to model the randomness appropriately. If the contribution of the additional model complexity to the system performance is negligible, this has been largely a wasted effort: at some point, simplifying assumptions must be made in order to get a tractable model. A second drawback is simulation run times. As more complexity is added to a given model, the amount of computer time expended often increases dramatically, at times to the point where current technology cannot deal with the model (Schruben and Yücesan, 1993). A third drawback relates to data collection effort. If during parameter design of our job shop we included four noise factors rather than two, then instead of four runs per parameter configuration we would need eight (if a partially saturated noise plan was used) or 16 (if a CF noise plan was used). Rather than deal with *all* potential noise factors during the parameter design stage, we can use system evaluation and noise factor assessment to determine the extent to which uncertainty in the underlying components translates to variation in the response.

The third item, uncertainty in the distributional characteristics used to generate the random processing times in the model, is also a model validity issue. If the system is highly sensitive to those characteristics then the simulation may not mimic the true system behavior adequately. Once again, the analyst can use tolerance design to obtain feedback regarding the modeling process. If a factor is relatively unimportant, then modeling it closely is not required; we believe that, given modeling budget constraints, model refinement efforts should be expended in accordance to factor sensitivity.

The final three items deal primarily with the system, rather than the validity of the simulation model. A change in noise factor values is described in case (4), which might occur over the life-cycle of the product. In case (5), the actual value of a parameter is changed from the 'optimal' setting chosen in the parameter design stage. In the short term, the production supervisor may wish to understand how this will affect the performance. In the long term, this might be useful information for deciding how closely to control (i.e., how much effort/time/money to spend) to assure that all machines are available during working hours. An alternative system configuration is described in case (6): M_3 was not found to affect the system performance when varied over the range 4–6, but we are now interested in determining whether or not it could be decreased further without a degradation in performance.

One way of addressing questions regarding the simulation model validity (cases (1)–(3)) is to build in the uncertainties, and perform sensitivity analysis. Tolerance design is a special case of sensitivity analysis in which we fix sources of variation at specific values rather than sample randomly over the possibilities. While care must be taken to avoid common pitfalls, such as modeling the effects of machine down-times as fixed increments in processing times in manufacturing simulations (Law and McComas, 1989), the sensitivity of the response (average time in system) to additional model detail can be determined more efficiently using experimental design than using random sampling. As we shall show, the results may indicate which aspects of the system must be modeled more closely, allowing the analyst to evolve an accurate simulation model as frugally as possible. With respect to simulation validation, the analyst will not include additional sources of variability into the model unless they are clearly important.

Questions regarding the system itself (cases (4)–(6)) could be addressed by making additional runs of the simulation model every time a new scenario is presented. The tolerance design approach we recommend may require a few more runs initially, but then allows the analyst to assess a broad range of alternatives without re-running the simulation every time.

4.2. System Evaluation

We now demonstrate how system evaluation can be carried out using a designed experiment. Two-level designs are sufficient for fitting a linear metamodel. Our design matrix is a nearly saturated two-level factorial, which allows us to estimate all six main effects in only eight experimental runs. Recommendations for selecting the low and high levels of the factors depend on the type of uncertainty.

- If the standard deviation of the factor *in the real world setting* is known, then the low and high levels should be set so that the discrete distribution (i.e., sampling points) has a mean and standard deviation equal to those of the underlying distribution. In the case of two-level sampling of continuous factors (or discrete factors whose distributions can be closely approximated by continuous distributions), this corresponds to one standard deviation below and one standard deviation above the mean. In the case of equally likely Bernoulli outcomes, this corresponds to the two factor levels. For discrete distributions where $\mu \pm \sigma$ does not yield valid factor levels, the outcomes can be sampled (approximately) proportional to their probability of occurrence. (Clearly, the smaller the number of levels per factor, the fewer the number of runs required.)
- If the factor is a mean estimated from data, then the upper and lower bounds of a 95% confidence interval for the mean can be used (Wild and Pignatiello, 1991).
- If the factor itself is qualitative (e.g., the brand of a machine) but describable by quantitative characteristics (e.g., its distribution of processing times) then these quantitative characteristics can be controlled in the experiment. Analogous to choosing the confidence interval bounds, one can select levels which cover the range of possible values, conduct the experiment, and then analyze various alternatives directly from the metamodels (see Section 4.4).

Table 8 gives the design matrix in terms of these natural values, along with the response means and variances from simulation runs after suitable truncation. Three of these factors were used during parameter design: M_3 (a parameter, whose variation from its nominal value is now included as a source of noise), as well as PJ_1 and PJ_2 (noise factors). The new factors are the mean interarrival time (IA), and multipliers for setup time across the shop (α_S) and processing time at station 1 (α_{T_1}). Thus, setup times differ from the initial specification by ±5%, and mean processing times at station 1 differ from the initial specification by ±10%. We emphasize that another layer of randomness has been removed from the experiment: we control the levels of the six noise factors rather than allowing them to vary. Note also that for

TABLE 8
Tolerance design experiment and results.

Factor Levels						Response	
α_{T_1}	PJ_1	PJ_2	M_3	IA	α_S	Mean	Variance
.9	.23	.47	4	.33	1.05	11.063	5.672
.9	.23	.53	4	.27	.95	10.517	4.195
.9	.27	.47	3	.33	.95	10.799	5.160
.9	.27	.53	3	.27	1.05	11.195	3.782
1.1	.23	.47	3	.27	1.05	11.591	4.855
1.1	.23	.53	3	.33	.95	11.164	5.801
1.1	.27	.47	4	.27	.95	10.814	4.201
1.1	.27	.53	4	.33	1.05	11.454	5.010

tolerance design we do not combine results across the noise runs–if so, we would have but a single point to analyze.

When conducting tolerance analysis for manufactured products, it is assumed that all important sources of noise have been identified and controlled during the experiment. It is also assumed that the response can be adequately described by a linear function of the noise factors over their ranges. (This restricts the order of the metamodels from those considered in the parameter design stage, but may be reasonable since experimentation occurs over a much smaller region.) Let $\{W_1, \ldots, W_w\}$ denote the noise factors, and suppose a total of n runs are conducted. The runs, although not a random sample of possible noise factor combinations, are a representative sample. Since tolerance design is conducted for a single parameter configuration, we suppress notation regarding the parameter configuration for ease of presentation. Let n denote the total number of noise factor configurations, and let Y_j denote the response from the jth run $(j = 1, \ldots, n)$. Then the overall response mean and variance for the parameter configuration under investigation can be estimated very simply as

$$\mu \approx \bar{Y} = \frac{1}{n}\sum_{j=1}^{n} Y_j, \quad Var(Y) \approx \frac{1}{n-1}\sum_{j+1}^{n}(Y_j - \bar{Y})^2. \tag{6}$$

A more detailed model of the system behavior can be found by fitting a linear metamodel to the n responses:

$$Y \approx \hat{\beta}_0 + \hat{\beta}_1 W_1 + \ldots + \hat{\beta}_w W_w, \tag{7}$$

where the $\{\hat{\beta}_j\}$ are the least-squares regression coefficients corresponding to this model. By treating the $\{\hat{\beta}_j\}$'s as fixed (which is reasonable when their respective standard errors are small relative to the noise factor variances), the overall variance can then be approximated by

$$Var(Y) \approx \sum_{j=1}^{w} \hat{\beta}_j^2 \, Var(W_j). \tag{8}$$

Substituting the noise factor means and variances into equations (7) and (8), we obtain estimates of the mean and variance of the overall performance. This methodology is also appropriate for Monte Carlo simulations, where the system variability is captured by the mean response for a particular run. Note that if the parameters were set such that $E(Y) \approx \tau$ (the target value) during the parameter design stage, then the variance in equation (6) or (8) is equal to the scaled loss.

The tolerance design approach outlined above must be modified for stochastic systems, such as queueing systems or discrete event simulations. Unlike the situations described above, it is not practical (or even desirable) to attempt to remove all variation from the simulation experiment during tolerance design. Discrete-event simulations, which are inherently stochastic, exhibit qualitatively different behavior than deterministic systems. A classic example is a single-server queue with traffic intensity equal to one. If interarrival and service times are deterministic the maximum queue size is bounded by its initial value plus one, but the maximum queue size is infinite if interarrival and service times are exponentially distributed. Thus, for our job shop example (as well as other models of inherently stochastic systems) randomness will remain an integral part of each simulation run. In such a case, the within-run variance contributes critical information about the system performance.

Recall from equation (3) that we calculated the performance variability for a particular *parameter* configuration as the sum of two components: the 'between configuration variability' (resulting from the variation of response means across the noise points) and the 'within configuration variability' (the response variance averaged over the noise runs). The first component is the only one Taguchi considers for tolerance design; the second component represents the noise inherent in the system at a particular parameter configuration. Let n denote the total number of noise factor configurations for the tolerance design experiment, and let $\bar{Y}_j$ and s_j^2 denote the response mean and variance (after suitable truncation) for run j. Then the overall response mean and variance can be approximated from the simulation results as follows:

$$\mu \approx \bar{\bar{Y}} = \frac{1}{n}\sum_{j=1}^{n} \bar{Y}_j, \quad Var(Y) \approx \frac{1}{n-1}\sum_{j=1}^{n}(\bar{Y}_j - \bar{\bar{Y}})^2 + \frac{1}{n}\sum_{j=1}^{n} s_j^2 \tag{9}$$

For our job shop example, the data in Table 8 yield estimates of 11.070 for the mean and .072 + 4.826 = 4.898 for the variance. The mean is slightly larger than the low loss configuration value because half the samples have one fewer machine in station 3 than nominal. The variance is slightly larger than that of the low configuration because we are introducing additional sources of noise.

4.3. Noise Factor Assessment

The overall performance mean and variance may help the analyst determine normal operating conditions of the system. However, a more precise model of the relationship between the noise factor variances and the performance variance (or loss) will provide additional information. For Monte Carlo simulation experiments, the term $\hat{\beta}_j^2 Var(W_j)$ in equation (8) is called the *transmitted variance* of noise factor W_j; the magnitude of $\hat{\beta}_j$ determines how much variability in the performance measure can be attributed to the variability in the noise component (Ramberg *et al.*, 1991). By examining the transmitted variances of each noise factor relative to the total, one can determine which factors are important determinants of noise in the system. An implicit assumption underlying this comparative analysis is that mean performance remains unchanged as the variance decreases.

For inherently stochastic systems, capturing the change in the variance *of the mean responses* is not sufficient for assessing the relative strength of the noise factors. For ease of calculation we fit two linear models – one for $\bar{Y}$ (as before) and one for the performance standard deviation, s. If the least-squares regression equation for the latter model is

$$s \approx \hat{\gamma}_0 + \sum_{j=1}^{w} \hat{\gamma}_j W_j \tag{10}$$

then, assuming independence among the noise factors, we can combine the results for the two components of variance:

$$Var(Y) \approx \hat{\gamma}_0^2 + \sum_{j=1}^{w} (\hat{\beta}_j^2 + \hat{\gamma}_j^2)\, Var(W_j). \tag{11}$$

The term $(\hat{\beta}_j^2 + \hat{\gamma}_j^2)\, Var(W_j)$ in equation (11) is the appropriate transmitted variance for a stochastic system. The *inherent variance* is the smallest variance achievable if all noise factors investigated in the experiment have variances driven to zero; we denote this as $\hat{\gamma}_0^2$.

The results of fitting the regression models (in natural units) for the job shop data follow.

$$\begin{aligned} \mu &\approx 11.0746 + 1.8109(\alpha_{T_1} - 1) - .2254(M_3 - 3.5) + 1.5132(IA - .3)(\alpha_S - 1), \\ s &\approx 2.1933 - .0336(PJ_1 - 25) + 4.3906(IA - .3) \end{aligned} \tag{12}$$

These models had all terms significant at the .05 level, and R^2 values of .996 and .904, respectively. Combining the models as per equation (11) yields the following model of variance:

$$\begin{aligned} Var(Y) \approx\ & 4.8106 + 3.2794\, Var(\alpha_{T_1}) + .0011\, Var(PJ_1) + .0508\, Var(M_3) \\ & + 21.567\, Var(IA) + 25.193\, Var(\alpha_s). \end{aligned} \tag{13}$$

The transmitted variances and contributions (as percentages of total loss) are tabulated in Table 9. These results indicate that the great majority of the noise remaining in the system is not attributable to the six noise factors studied: even when all are considered together, their combined transmitted variances account for less than 3% of the total variance. With respect to the questions (1)–(3) at the beginning of this section, it appears that further refinement of the distribution characteristics examined (mean job time for machines in station 1, mean interarrival time) is unnecessary. Similarly, little additional information *in terms of expected loss* would be gained be modeling setup time as a random variable.

The loss contributions as percentages of non-inherent loss are also provided in Table 9. For tolerance design involving a Monte Carlo experiment, there is no

TABLE 9
Transmitted loss multipliers.

Factor	Transmitted Variance	Percent of Total	Percent of Non-Inherent
Inherent	4.8106	97.32	–
α_{T_1}	.0328	.66	24.79
PJ_1	.0044	.09	3.33
M_3	.0127	.26	9.60
IA	.0194	.39	14.66
α_S	.0630	1.27	47.62
Total	4.9429	99.99	100.00

inherent noise if all important noise factors have been identified and included in the experiment (Ramberg *et al.*, 1991). The magnitude of the inherent noise for our job shop example is determined to some extent by our selection of noise factors to control during the tolerance design experiment. If, for example, processing times were modeled as two-parameter distributions rather than one-parameter 2-Erlangs, then the analyst could specify the processing time mean and processing time standard deviation as noise factors, and control them for the tolerance design experiment, while retaining the stochastic nature of the simulation. In such a case, the transmitted loss associated with processing time means or variances could be quite substantial, and the inherent loss might not overwhelm all other sources of variation.

Question (4), the anticipated effect if the variability in the demand for Product 1 increases, can also be addressed using noise factor assessment. Suppose the standard deviation associated with this probability doubles, which quadruples the transmitted variance of PJ_1. The total in Table 9 will increase from 4.9429 to 4.9561 – less than .3% of the base amount. This indicates that the system is robust to demand variability for Product 1. This factor is not an important determinant of the mean performance.

4.4. System Optimization

We now illustrate how the noise factor assessment can be used to drive system improvement efforts. The quadratic loss function means that losses will be minimized when the response is equal to the target value: in our example, when there are five machines at station 2 and the batch size is low. The basis for system optimization is the quantification of expected loss in dollars, so the benefits of altering the initial system configuration can be evaluated. A 'typical' alteration is one which would affect the variance but leave the mean response unchanged. In product design, this might correspond to selecting the level of consistency around the nominal value chosen during parameter design. For example, one could replace a 10 ± 2 ohm resistor with a 10 ± 1 ohm resistor if the reduction in loss offset the increase in price. Alternatively, one could replace a 10 ± 2 ohm resistor with a cheaper 10 ± 3 ohm resistor if it did not degrade product quality. For our example, increased training might reduce variability in setup or processing times, and better maintenance might reduce variability in the number of operable machines or processing times.

In order to assess the benefits (costs) of various alternatives, we must be able to convert the scaled loss to monetary units. Suppose the cost of a one day (ten hour) deviation from the target is assessed at \$300 per unit (a cost associated with storage or shop floor congestion if the job is completed early; a cost associated with customer goodwill or penalty costs incurred if the job is completed late). Substituting this into the right-hand side of equation (1) and solving for the constant K yields $K = \$3$ per hour^2 per unit.

Although the variance in setup time had a small effect on the overall performance variability, we can determine the effect of reducing the setup time variability for each station-job combination from the current level by 50% (i.e., standard deviation of α_S decreases from 5% to 2.5%). This would reduce the transmitted variance of α_S by a factor of four. The total variance would drop by $\frac{3}{4}(.0630) = .0473$ for an average savings of 3(.0473) = 14.2 cents per unit. If the training costs \$2000, the pay-back period would be 2000/.142 = 14,085 units or 1.7 years (assuming 250 ten-hour workdays per year).

If we consider the mean performance, as well as the variance, we can address question (5). This might be of interest for two reasons. First, suppose that after running the parameter design stage, we find out that one of the machines at this station is very unreliable and is operable only half the time. Second, suppose that machines of this type are crucial to the performance of another production line, and they have requested a transfer of one machine. M_3 was not an important term in the parameter design metamodels, so we cannot determine from them whether or not decreasing M_3 to 3 machines (outside the original range of experimentation) will have a noticeable adverse impact. Including M_3 as a factor in our tolerance design experiment allows us to examine the effect of removing this machine. If we reduce M_3 to three, then from equations (12) and (13) the new mean job time would be 11.1873 and the new variance would be 4.9302, for an average loss of \$19.02 per unit. Comparing this to the current loss of \$18.29 per unit, we find that removing this machine permanently would cost an additional \$.73 per unit or \$29 per ten-hour day. This cost could be compared with the cost of replacing or refurbishing this machine (in the case of an unreliable machine) or the improvement in loss for the other production unit (in the case of a machine transfer).

Similarly, we can use equations (12) and (13) to evaluate three possible enhancements to the system. Option (A) is to refurbish the machines in station 1 for a cost of \$8000/machine (\$40,000 total), option (B) is to replace them at a net cost of \$23,000/machine (\$115,000 total), and option (C) is to install fixtures throughout the shop to reduce average setup time at a cost of \$3000 per machine (\$57,000 total). Suppose that (A) will decrease the mean processing time by 10%, (B) will decrease the mean processing time by 30%, and (C) will decrease the mean setup time by 20%. Which of these options is most cost-effective?

The expected loss under each of the alternatives can be used to answer this question. Both the mean and the variance should be estimated from equation (12), and then used to determine expected loss for the initial model and the three alternatives. The results are:

$$Loss_{Init} \approx 3((11.0746 - \tau)^2 + 4.9429) = \$18.29 \text{ per unit}$$
$$Loss_A \approx 3((10.8935 - \tau)^2 + 4.9367) = \$17.21 \text{ per unit}$$
$$Loss_B \approx 3((10.5313 - \tau)^2 + 4.9262) = \$15.63 \text{ per unit}$$
$$Loss_C \approx 3((10.0707 - \tau)^2 + 4.9026) = \$14.72 \text{ per unit}$$

If the costs over a five year time horizon (41,667 units) are of interest, then the savings associated with each system are \$45,000 for (A), \$110,834 for (B), and \$148,751 for (C). When the costs of the three alternatives are taken into consideration, (A) results in a net savings of \$5,000; (B) results in a net loss of \$4,166; and (C) results in a net savings of \$91,751. Option (C) is clearly superior to the others. (Note that although (B) resulted in three times the reduction in processing time as (A) for less than three times the cost, it is not the better of these two alternatives!)

4.5. Discussion

The tolerance design metamodel differs from that developed in the parameter design stage in several ways. First, it has only noise factors: these include the noise factors studied in parameter design (the product mix variables), noise in the 'parameter' settings chosen during parameter design (e.g., the nominal number of machines available at station 3), as well as additional factors for purposes of assessing model sensitivity (random set-up times, processing time means). Second, it is valid over a much smaller range. If one were interested primarily in evaluating (or improving) a particular system configuration, then one could proceed to tolerance design without going through the parameter design stage.

As previously mentioned, the typical manufacturing application in tolerance design is analysis of alternatives where variance in the noise factors contributes to variance in the performance measure *without* affecting its mean. For example, if the processing times were modeled using two-parameter distributions, such as normal $(\mu_{ij}, \sigma_{ij}^2)$ random variables, then we could determine the extent to which reducing processing time variability (e.g., by increased maintenance, refurbishment, or operator training) would improve the overall performance even if the mean processing time remained unchanged. (We caution that mean performance might still change if the system is stochastic in nature.) Tolerance analysis could also be used to assess model sensitivity to a qualitative factor not easily converted to numerical characteristics, such as distributional forms (e.g., normal versus uniform). One can vary such a factor using indicator or dummy variables to represent the alternatives. Then if tolerance analysis is conducted including all dummy variables as potential metamodel terms, one can assess whether or not the factor significantly affects performance. Separate analyses, conducted after grouping the runs according to the indicator variables (e.g., distributional forms) would allow the analyst to explicitly study the system performance under the specific alternatives.

Note also that the importance or unimportance of the factors in noise factor assessment or system optimization efforts is not the same as statistical significance or insignificance for the factors in the metamodels determined using regression. The latter is largely a function of sample size: for sufficiently large n one will find all factors to be significant (since the underlying system is nonlinear) while if n is small then it may be the case that no individual factors are statistically significant. Even if a factor is statistically significant, if it has no appreciable impact on either the variance or the mean of the performance characteristic, then the metamodels can be simplified by its removal.

5. ADVANCED DESIGNS

Whether the analyst is seeking robust process designs or robust product designs, one additional benefit of using simulation as the experimental medium is the flexibility that results in the choice of the experimental design for the noise factors during the parameter design stage. Because of the parameter/noise distinction, noise factor terms are not explicitly included in the metamodels for response mean and variance. In addition, any artificial factors (or simulation-specific factors, such as run length, choice of random number stream, etc.) can be exploited to obtain efficient designs.

In this chapter, we have illustrated the methodology using complete and fractional factorials, and augmented these by centerpoints and axial points to assess quadratic terms in the model. In summary, the factor levels should be chosen to cover the range of interest for a two-level factorial or fractional factorial experiment. For noise factor plans, the levels should be chosen so that the mean and variance of the two-point sampling distribution are equal to the mean and variance of the underlying distribution. In the case of two-level sampling of continuous factors (or discrete factors whose distributions can be closely approximated by continuous distributions), this corresponds to one standard deviation below and one standard deviation above the mean. In the case of equally likely Bernoulli outcomes, this corresponds to the two factor levels. For discrete distributions where $\mu \pm \sigma$ does not yield valid factor levels, the outcomes can be sampled (approximately) proportional to their probability of occurrence. If the factor is a mean estimated from data, then the upper and lower bounds of a 95% confidence interval can be used.

Factorials and fractional factorials are often used in practice: among these, two-level designs are popular choices because of their simplicity and efficiency. They permit the evaluation of the linear parameter effects, as well as interaction or synergistic effects. However, in some process or product design situations, further efficiency and/or discriminatory power might be necessary in order to achieve reasonable metamodels of the process or product performance. We now provide references to other plans often used in response surface metamodeling for simulation experiments. We also briefly discuss some alternative plans, developed specifically for simulation, which may offer benefits in particular circumstances (Sanchez, 1994a).

5.1. Response Surface Modeling Plans

Several classes of data collections plans (or experimental designs) have been advocated for response surface modeling. Two-level plans are not sufficient if quadratic effects are anticipated, as is typically the case for complex systems or processes. Some commonly used second-order designs include 3^k factorials, central composite designs, Box-Behnken designs, and small composite designs. Entire texts devoted to response surface modeling include Box and Draper (1987), Khuri and Cornell (1987), and Myers (1976). Experimental design texts, such as Box, Hunter and Hunter (1978) and Montgomery (1991), also provide detailed explanations of the construction and properties of many of these designs. For texts dealing with designs specifically for simulation experiments, we refer the reader to Kleijnen (1987), Law and Kelton (1991) or Banks, Carson and Nelson (1996). Central composite designs in simulation experiments are discussed in Tew (1992) and Hood and Welch (1993). Designs for minimizing the bias or mean-squared error of the regression coefficients are presented in Donohue, Houck and Myers (1992).

5.2. Artificial Factor Plans

A well established field in simulation is that of variance reduction, where researchers have established methods of reducing the variance of the *estimators* of mean responses in order to increase power for hypothesis testing purposes. In the robust design context, variance reduction schemes hold promise for further increasing the efficiency of experimentation. Rather than using mutually independent random number streams, one might use a common/antithetic sampling strategy (Schruben and Margolin, 1978; Tew and Wilson, 1991, 1994). This reallocates variance among the coefficient estimates. The implications for parameter design are that the artificial factor plan should be chosen in order to induce correlations which reallocate variance from the interesting terms (parameters) to the uninteresting terms (noise factors) (Schruben *et al.*, 1992). The artificial factor plan is typically embedded in the noise factor plan, as in the choice of random number streams used during a simulation run. This means the potential exists for increasing the precision of the metamodel coefficients without increasing the number of runs required.

5.3. Frequency Domain Plans

If the number of noise factors is large, even a saturated factorial plan for the noise factors may result in an unwieldy experimental design after crossing it with the parameter plan. One way to cut down the size of the experiment is to first screen the noise factors and then employ a highly fractionated factorial design. Another efficient way to collect the data is to oscillate each noise factor sinusoidally within a simulation run at a unique, carefully selected frequency. This allows examination of the system across a range of noise factor combinations without a prohibitively large number of runs (Moeeni *et al.*, 1994, 1997; Sanchez *et al.*, 1994). Such oscillation forms the basis of frequency domain experimentation in the simulation field (Schruben and Cogliano, 1987; Sanchez and Buss, 1987, Sanchez, Konana and Sanchez, 1998), although the analysis differs. Indexing by time, rather than by entity, is recommended (Mitra and Park, 1991).

In the tolerance design stage, the analyst is interested in determining what portions of the total system variability can be attributed to the noise factors, and a frequency domain approach could be used for factor screening purposes. During the parameter design stage, we are interested in what the performance variability is at a particular parameter configuration: the fact that noise factors are varying across the noise space is important, while estimates of their specific effects are not. In both cases, care should be taken to select driving frequencies which will not result in confounding and to choose frequencies resulting in cycles sufficiently long to affect the system response (Jacobson *et al.*, 1991). Discrete factors can be handled either by oscillating their probabilities of realizing particular levels, or by discretizing the sinusoidal function (Sanchez and Sanchez, 1991).

5.4. Correlated Factor Plans

If the noise factors are correlated in the real world system, it might be that a factorial or fractional factorial design could not be conducted over the entire range of interest. For example, a queueing system might be unstable if all noise factors were held at their high levels. If this situation was unlikely to occur in practice because of correlation among the variables, then a sampling scheme which made use of the underlying dependence structure would seem more appropriate. If the noise factors are normally distributed, the analyst can sample at axial points on the elliptical contours of the joint distributions (Sanchez, 1994b).

5.5. Combined Array Plans

Recent work suggests that in some circumstances, a crossed parameter×noise factor plan may not be the most efficient in terms of the total number of observations (runs) required. An alternative is a combined plan, where a single design matrix (such as a factorial) is used with columns divided among parameters and noise factors. As Myers *et al.* (1992) suggest (see also Welch *et al.*, 1990) this can be used if one can specify *a priori* which of the many possible interaction terms are potentially important. For example, consider our experiment with five parameters and two noise factors. Complete factorials for both the parameter and noise factor plans would require $2^5 \times 2^2 = 128$ runs. The base plan we used, with a half-fraction for the parameters, required 64 runs. However, if it could be determined *a priori* that the only interactions would involve B_3, at most four two-way interactions must be estimated and the base experiment could be conducted in 16 runs. The design matrix for a 2^4 experiment has 15 columns. Five of these are used to set the levels for the parameters. These in turn define four specific additional columns for the suspected interactions. Two of the remaining six columns set the noise factor levels. The 'unused' columns can accomodate additional noise factors or be use to guard against additional potential interactions. Combined plans such as these may mean that the experiment can be conducted using a smaller total number of simulation runs than a crossed plan would require. However, if the system or process is not well-understood, such plans may not lead to accurate metamodels.

6. APPLICATION TO PRODUCT DESIGN

While our focus in this chapter is the use of simulation for robust process design, simulation is used as a surrogate medium of experimentation for many problems, including that of product design. Rather than presenting Taguchi's original tactics for generating off-line quality using industrial experimentation, we summarize areas where the simulation approach and analyses described in Sections 3 and 4 might be simplified for addressing a product design problem.

The major distinction that needs to be made relates to the inherent variability in the system. Conceptually, one can think of the following: is the simulation modeled using a Monte Carlo approach (i.e., numerical approximation or integration), or is it a discrete event simulation? (We leave continuous simulations which cannot be fit into either category aside.) If Monte Carlo simulation is used, then all sources of variability can be controlled during the course of experimentation if the noise factors are completely specified. This affects the analyses in both the parameter design and tolerance design stages. In the former, the variance within each parameter configuration (simulation run) is equal to zero, and so the estimate of the associated variance in equation (3) (with w noise factors) simplifies to

$$\bar{V}_{i\cdot} = \frac{1}{w-1}\sum_{j=1}^{w}(Y_{ij} - \bar{Y}_{i\cdot})^2. \tag{14}$$

Here Y_{ij} is used instead of $\bar{Y}_{ij}$ because each run results in a single value, rather than a stream of output.

During the tolerance design stage, both the construction of the metamodel for variance and the calculation of transmitted variances are simplified if the simulation uses Monte Carlo methods. The variance metamodel in equation (11) reduces to

$$Var(Y) \approx \sum_{j=1}^{w} \hat{\beta}_j^2 \, Var(W_j), \tag{15}$$

and (provided all important noise factors have been controlled) the inherent variance is equal to zero.

For those involved in product design who were previously unfamiliar with Taguchi's strategy, the insights obtained by assigning variability in the response to variability in the noise factors are invaluable for focusing system improvement and optimization efforts. We emphasize that the approach can be applied when experimentation takes place using real products, as well as with simulations. An additional benefit of the robust design approach is the use of carefully chosen data collection plans: a designed experiment may require many fewer runs than does the Monte Carlo distribution sampling approach. For example, Ramberg *et al.* (1991) show that comparable metamodels are obtained for a simple circuit using (i) 1000 independent random trials in a Monte Carlo experiment, and (ii) 8 trials in a designed experiment. Such savings in time and effort are substantial, particularly if evaluating product performance at a particular parameter/noise factor configuration is time-consuming or complex.

For those familiar with Taguchi's strategy but not metamodeling approaches, there are definite benefits of combining the two approaches. One is the opportunity to gain additional insight into system behavior; another is the ability to predict product designs (not previously examined) which will have low loss. In addition, if the behavior of the product is very complex, then the main-effects data collection plans advocated by Taguchi may not allow reasonable approximation of the product's performance. By using simulation as the medium of experimentation, rather than real products, higher order metamodels can be examined while the alternative designs discussed in Section 6 can be exploited to reduce the data requirements.

7. CONCLUSIONS

In this chapter, we have presented a robust design framework for studying products, processes and systems via discrete event simulation. Cornerstones of robust design are an experimentation plan that subjects the proposed system to conditions similar to those under which it is likely to operate, and the adoption of a loss function based on both performance mean and performance variability. This requires a shift in viewpoint from traditional optimality criteria that focus on mean performance. It provides better guidance for designing, developing and improving the operation of stochastic systems to achieve customer expectations. Robustness cuts to the heart of the formulation of design problems: if the wrong question is asked, then the answer may be of little use. Another benefit of the robust design framework relates to the ease of its implementation. Although the performance of products, processes or systems after implementation often differs from that predicted *a priori,* large discrepancies are less likely if robust considerations have guided the design.

The integrated simulation framework we have proposed also has a place in academic research. Computational results contribute to the literature in many fields, and simulation is often used to evaluate analytic or heuristic solutions to problems. Hooker (1994) has advocated the development of an empirical science of algorithms, stating "One occasionally sees tests [in scholarly publications] conducted according to the principles of experimental design, or results analyzed using rigorous statistical methods. But only occasionally". Even if statistical methods are used, in many instances the evaluations are performed considering only mean performance measures. The methodology we have presented can provide more complete characterizations of the behavior of algorithms and heuristics, and stronger insights into factors affecting that behavior, when used in conjunction with discrete event simulation.

Simulation is an effective vehicle for carrying out robust design efforts. It provides the analyst or engineer added opportunity for control over the experiment, such as by choosing levels for artificial factors or employing variance reduction techniques. These can increase the efficiency of the data collection plans. Simulation also allows robust design methodologies to be applied prospectively – at the inception and conceptualization phases of a system or process design project – rather than waiting until after implementation. The integration of response surface metamodeling into the robust design framework allows the analyst greater opportunities for identifying configurations which minimize expected loss, as well as targeting selected noise factors for improving system performance and/or decreasing total cost. In this chapter, we showed how to augment the tolerance analysis so its benefits can be realized for process and system design. This also provides a new tool for simulation model validation.

Although we have presented this framework in terms of experimenting on simulation models, most of our discussion can also applied when experimentation occurs on real or prototypical processes or systems. This type of industrial experimentation has drawbacks: control is often more difficult and costly to achieve, particularly for noise factors. However, our framework is appropriate for designing inherently stochastic systems or processes, such as queueing networks, as well as for design problems when it is not practical to control all sources of noise during experimentation. We presented tactics for analysis in the presence of inherent process variability. These tactics are necessary when not all noise is controlled in the experimental setting.

The net result is a unified approach that uses the increased control which simulation experiments make possible relative to industrial experiments. It provides an efficient mechanism for understanding, evaluating, and improving inherently stochastic systems, thereby expanding the application of robust design beyond the product design problem to the design of robust processes, systems and enterprises. This framework can lead to increased efficiencies in the design process, better performance following design efforts, and realistic performance assessments prior to implementation. It facilitaties the integration of design and improvement efforts across the organization.

Acknowledgements

John S. Ramberg was supported in part by Grant DDM 9006710 from the National Science Foundation.

Susan M. Sanchez was supported in part by Grant DDM 9110573/DFM 9396135 from the National Science Foundation.

An earlier version of the introductory portion of this paper and motivation for parameter design using the job shop example appeared in Ramberg *et al.* (1991), and the advanced experimental design discussion is adapted from Sanchez (1994a).

References

R.R. Barton, Metamodels for Simulation Input-Output Relations. *Proceedings of the 1992 Winter Simulation Conference,* J.J. Swain, D. Goldsman, R.C. Crain and J.R. Wilson (Eds.), 289–299. Institute of Electrical and Electronics Engineers, Washington, D.C. (1992).

G.E.P. Box, Signal-to-Noise Ratios, Performance Criteria, and Transformations (with discussions). *Technometrics,* **30**(1), 1–40 (1988).

G.E.P. Box and N.R. Draper, *Empirical Model-Building and Response Surfaces*, John Wiley and Sons, Inc., New York (1987).

G.E.P. Box, W.G. Hunter and J.S. Hunter, *Statistics for Experimenters*, John Wiley and Sons, Inc., New York (1978).

D. Ceylan and B.W. Schmeiser, Interlaced Variance Estimators of the Marginal Variance. *Proceedings of the 1993 Winter Simulation Conference*, G.W. Evans, M. Mollaghasemi, E.C. Russell and W.E. Biles (Eds.), 1382–1383. Institute of Electrical and Electronic Engineers, Los Angeles, California (1993).

J.R. D'Errico and N.A. Zaino, Statistical Tolerancing Using a Modification of Taguchi's Method, *Technometrics,* **30**(1), 397–405 (1988).

J.M. Donohue, E.R. Houck and R.H. Meyers, Simulation Designs for Quadratic Response Surface Models in the Presence of Model Misspecification. *Management Science*, **38**(12), 1765–1791 (1992).

D. Goldsman, B.L. Nelson and B.W. Schmeiser, Methods for Selecting the Best System, *Proceedings of the 1991 Winter Simulation Conference*, B.L. Nelson, W.D. Kelton, G.M. Clark (Eds.), 177–186. Institute of Electrical and Electronics Engineers, Phoenix, AZ (1991).

S.J. Hood and P.D. Welch, Response Surface Methodology and its Application in Simulation. *Proceedings of the 1993 Winter Simulation Conference*, G.W. Evans, M. Mollaghasemi, E.C. Russell and W.E. Biles (Eds.), 115–122. Institute of Electrical and Electronic Engineers, Los Angeles, California (1993).

J.N. Hooker, Needed: An Empirical Science of Algorithms, *Operations Research*, **42**(2), 201–212 (1994).

S. Jacobson, A.H. Buss and L.W. Schruben, Driving Frequency Selection for Frequency Domain Simulation Experiments. *Operations Research*, **39**(6), 917–924 (1991).

R.N. Kackar, Off-Line Quality Control, Parameter Design, and the Taguchi Method, *Journal of Quality Technology*, **17**(4), 176–188 (1985).

W.D. Kelton, Designing Computer Simulation Experiments, *Proceedings of the 1988 Winter Simulation Conference,* M.A. Abrams, P.L. Haigh and J.C. Comfort (Eds.), 15–18. Institute of Electrical and Electronics Engineers, San Francisco, CA. (1988).

A.I. Khuri and J.A. Cornell. *Response Surfaces, Designs, and Analyses*, Marcel Dekker, New York (1987).

J.P.C. Kleijnen, *Statistical Tools for Simulation Practitioners*, Marcel Dekker, New York (1987).

A.M. Law and W.D. Kelton, *Simulation Modeling and Analysis, 2nd Edition*, McGraw-Hill, New York (1991).

A.M. Law and M.G. McComas, Pitfalls to Avoid in the Simulation of Manufacturing Systems. *Industrial Engineering*, **21**, 28–31 (1989).

M. Mitra and S.K. Park, Solution to the Indexing Problem of Frequency Domain Experiments, *Proceedings of the 1991 Winter Simulation Conference*, B.L. Nelson, W.D. Kelton and G.M. Clark (Eds.), 907–915. Institute of Electrical and Electronic Engineers, Phoenix, Arizona (1991).

F. Moeeni, S.M. Sanchez and A.J. Vakharia, Implementing JIT in uncertain manufacturing environments. *DSI Conference Proceedings*, November 1994, 1494–1496.

F. Moeeni., S.M. Sanchez and A.J. Vakharia, A Robust Design Methodology for Kanban System Design. *International Journal of Production Research*, **35**(10), 2821–2838 (1996).

D.C. Montgomery, *Design and Analysis of Experiments*, John Wiley and Sons, Inc., New York (1991).

R.H. Myers, *Response Surface Methodology*, Allyn and Bacon, Boston (1976).

R.H. Myers. A.I. Khuri and G.Vining, Response Surface Alternatives to the Taguchi Robust Parameter Design Approach, *The American Statistician*, **46**(2), 131–139 (1992).

V.N. Nair, *et al.*, Taguchi's Parameter Design: A Panel Discussion, *Technometrics*, **34**(2), 127–161 (1992).

J.J. Pignatiello, Jr., An Overview of the Strategy and Tactics of Taguchi, *IIE Transactions*, **20**(3), 247–254 (1988).

J.J. Pignatiello, Jr., and J.S. Ramberg, Top Ten Triumphs and Tragedies of Genichi Taguchi, *Quality Engineering*, **4**(2), 211–235 (1991).

J.S. Ramberg, Improvements on Taguchi Methods for Semiconductor Design/Manufacture. *SME Semiconductor Manufacturing Conference Papers*, MS89–798, 1–16 (1989).

J.S. Ramberg, J.J. Pignatiello, Jr. and S.M. Sanchez, A Critique and Enhancement of the Taguchi Method. *ASQC Quality Congress Transactions*, May, 491–498 (1992).

J.S. Ramberg, S.M. Sanchez, P.J. Sanchez and L.J. Hollick, Designing Simulation Experiments: Taguchi Methods and Response Surface Metamodels. *Proceedings of the 1991 Winter Simulation Conference*, B.L. Nelson, W.D. Kelton, G.M. Clark (Eds.), 167–176. Institute of Electrical and Electronics Engineers, Phoenix, AZ (1991).

J. Sacks, S.B. Schiller and W.J. Welch, Designs for Computer Experiments, *Technometrics*, **31**, 41–47 (1989).

J. Sacks, W.J. Welch, T.J. Mitchell and H.P. Wynn, Design and Analysis of Computer Experiments, (with discussion), *Statistical Science*, **4**, 409–435. (1989).

P.J. Sanchez and A.H. Buss, A Model for Frequency Domain Experiments, *Proceedings of the 1987 Winter Simulation Conference,* A. Thesen, H. Grant, W.D. Kelton (Eds.), 424–427. Institute of Electrical and Electronics Engineers (1987).

P.J. Sanchez and S.M. Sanchez, Design of Frequency Domain Experiments for Discrete-Valued Factors. *Applied Mathematics and Computation* **42**(1): 1–21 (1991).

S.M. Sanchez. A Robust Design Tutorial. *Proceedings of the 1994 Winter Simulation Conference*, J.D. Tew, M.S. Mannivannan, D.A. Sadowski and A.F. Seila (Eds.), 106–113. Institute of Electrical and Electronics Engineers, Orlando, FL. (1994a).

S.M. Sanchez. Experiment Designs for System Assessment and Improvement when Noise Factors are Correlated. *Proceedings of the 1994 Winter Simulation Conference*, J.D. Tew, M.S. Manivannan, D.A. Sadowski and A.F. Seila (Eds.), 290–296. Institute of Electrical and Electronics Engineers, Orlando, FL. (1994b).

S.M. Sanchez, F. Moeeni and P.J. Sanchez, Assessing the Impact of Uncertainly on Kanban System Performance: a Frequency Domain Approach. *DSI Conference Proceedings*, November, 1782–1784 (1994).

S.M. Sanchez, J.S. Ramberg, J. Fiero and J.J. Pignatiello, Jr, Quality by Design. Chapter 10 in *Concurrent Engineering: Automation, Tools, and Techniques*, (ed. A. Kusiak), John Wiley and Sons, NY, 235–286 (1993).

L.W. Schruben and V.J. Cogliano, An Experimental Procedure for Simulation Response Surface Model Identification, *Communications of ACM*, **30**(8), 716–730 (1987).

L.W. Schruben and B.H. Margolin, Pseudo-Random Number Assignment in Statistically Designed Simulation and Distribution Sampling Experiments, *Journal of the American Statistical Association*, **73**, 504–525 (with discussions) (1978).

L.W. Schruben, S.M. Sanchez, P.J. Sanchez and V.A. Czitrom, Variance Reallocation in Taguchi's Robust Design Framework. *Proceedings of the 1992 Winter Simulation Conference*, J.J. Swain, D. Goldsman, R.C. Crain and J.R. Wilson (Eds.), 548–556. Institute of Electrical and Electronics Engineers, Washington, D.C. (1992).

L.W. Schruben and E. Yücesan, Complexity of Simulation Models: A Graph Theoretic Approach. *Proceedings of the 1993 Winter Simulation Conference*, C.W. Evans, M. Mollaghasemi, E.C. Russell and W.E. Biles (Eds.), Institute of Electrical and Electronics Engineers, Los Angeles, CA (1993).

G. Taguchi, *Introduction to Quality Engineering, UNIPUB/Krauss International*: White Plains, New York (1986).

G. Taguchi, *System of Experimental Design*, Vols. 1 and 2. UNIPUB/Krauss International:White Plains, New York (1987).

G. Taguchi and Y. Wu, *Introduction to Off-Line Quality Control.* Nagoya, Japan: Central Japan Quality Association (1980).

J.D. Tew, Using Central Composite Designs in Simulation Experiments. *Proceedings of the 1992 Winter Simulation Conference*, J.J. Swain, D. Goldsman, R.C. Crain and J.R. Wilson (Eds.), 529–538. Institute of Electrical and Electronic Engineers, Washington, D.C. (1992).

J.D. Tew and J.R. Wilson, Validation of Simulation Analysis Methods for the Schruben-Margolin Correlation-Induction Strategy. *Operations Research* **40**, 87–103 (1991).

J.D. Tew and J.R. Wilson, Estimating Simulation Metamodels Using Combined Correlation-Based Reduction Techniques, *IIE Transactions*, **26**(3), 2–16 (1994).

G.G. Vining and R.H. Myers, Combining Taguchi and Response Surface Philosophies: A Dual Response Approach, *Journal of Quality Technology*, **22**, 38–45 (1990).

W.J. Welch, T.K. Yu, S.M. Kang and J. Sacks, Computer Experiments for Quality Control by Robust Design, *Journal of Quality Technology*, **22**, 15–22 (1990).

R.H. Wild and J.J. Pignatiello, Jr., An Experimental Design Strategy for Designing Robust Systems Using Discrete-Event Simulation, *Simulation*, **57**(6), 358–368 (1991).

CHAPTER 13

Manufacturing System Simulation

CHARLES R. STANDRIDGE

Consultant, 4144 Golf Bag Lane, Terrehaute, FN 47802, USA

1. INTRODUCTION

Before a newly engineered manufacturing system is implemented, or significant changes to an existing manufacturing system are brought on-line, the performance of the system with respect to management goals and engineering design requirements must be assessed. The detailed, step-by-step behavior of the system resulting from operational control procedures must be determined. The response of the system to particular events, such as machine breakdowns, must be known.

Computer simulation is a technique widely used to perform the analyses necessary to meet the above requirements. A simulation study includes building a model of the manufacturing system. Simulation models emphasize the direct representation of the structure, logic, and relationships between entities of a manufacturing system as opposed to abstracting the system into a strictly mathematical and fixed form. Simulation models can incorporate stochastic elements, directly use data gathered from the manufacturing system, and estimate the values of performance variables peculiar to each system.

Such capabilities imply that simulation models are analytically intractable, that is new information in the form of exact values cannot be derived from the model by mathematical analysis. Instead, such inferencing is accomplished by experimental procedures that result in statistical estimates of the values of interest. These experiments must be designed as would any laboratory or field experiment. Proper statistical methods must be used in computing estimates and in interpreting experimental results.

Thus simulation generates the temporal behavior of the manufacturing system on a computer. This simulated temporal behavior is observed and the data required for decision making extracted. Such data can describe the behavior of individual entities such as parts, machines, and material handling devices. The data can be a time series of values, for example the state (e.g. busy, starved, blocked, or broken) of a machine or the level of a work in process inventory. The data can characterize both average behavior and system variability. Statistical quantities such as averages, histograms, and confidence intervals can be computed from the observations. Animation can be used to visualize system temporal behavior.

Computer simulation has been widely used over a significant period of time. This success can be attributed to several factors (Pritsker, 1989):

1. Computer simulation models conform both to manufacturing system structure and to available system data.
2. Simulation experimental results conform to individual manufacturing system requirements for information, including estimates of system variability.
3. The same simulation approach works for a wide variety of manufacturing systems.
4. Simulation models adapt easily over the course of a project.
5. Simulation supports experimentation with manufacturing systems at relatively low cost and at little risk.
6. Simulation is founded on an engineering heritage of problem solving.

This chapter presents a process for doing a simulation study. Within the context of this process, it discusses the perspectives for building models of manufacturing systems and the procedures for simulating them. Issues impacting the design of simulation experiments are presented. An example illustrates the simulation process and decision making based on simulation generated information.

2. THE SIMULATION PROCESS

Many authors have proposed processes for performing a simulation study (Pegden, Shannon, and Sadowski, 1990; Pritsker, 1987; Schriber, 1991; Shannon, 1975; Standridge and Brown-Standridge, 1995). One such process is shown in Figure 1 and views a simulation study as having five strategic phases.

Strategic Phase	*Tactics*
1. Define the Issues and Solution Objective	Identification of outputs: who uses them and their requirements as well as systems producing the outputs and the individuals responsible for those systems.
2. Build Models	Simulation model building, data acquisition, and data examination.
3. Identify Root Causes	Simulation model implementation, verification, validation, experimentation, and result analysis.
4. Propose and Assess Solutions	Simulation model modification, experimentation, and result analysis.
5. Implement Solutions and Evaluate	Designed process experimentation and process monitoring.

Fig. 1 A simulation process.

A simulation model should be built and experiments designed to address a particular, pre-determined set of manufacturing system issues. These issues help distinguish the significant system objects and relationships to include in the model from those that are secondary and thus may be eliminated or approximated. Therefore, the first strategic phase in the simulation process is the definition of the issues to be resolved and the characteristics of a solution to these issues. This requires identification of the manufacturing systems components and their outputs that are relevant to the problem as well as the users of the system outputs and their requirements.

The focus of the second phase is the construction of models of the manufacturing system as it currently exists or is designed. Simulation models are constructed using the techniques described in the next section. If necessary to aid in the construction of the simulation model, descriptive models such as flowcharts may be built.

Model building requires gathering and studying data from the system if it exists or the design of the system if it is proposed. The availability of system data influences the choice of simulation model parameters as well as which system entities and which of their attributes can be included in the model. Simulation model parameters are estimated using system data.

The third strategic phase involves identifying the manufacturing system operating parameters, control strategies, and organizational structure that impact the issues and solution objectives identified in the first phase. Cause and effect relationships between system components are uncovered. The most basic and fundamental factors effecting the issues of interest, the root causes, are discovered.

Information resulting from experimentation with the simulation model is essential to meeting these goals. Simulation experiments can be designed using standard design of experiments methods. At the same time, as many simulation experiments can be conducted as computer time and the limits on project duration allow. Thus, experiments can be replicated as needed for greater statistical precision, designed sequentially by basing future experiments on the results of preceding ones, and repeated to gather additional information needed for decision making.

The model defined in the previous strategic phase must be implemented as a computer program. Simulation software environments include model builders that provide the functionality and a user interface tailored to model building as well as automatically and transparently preparing the model for simulation. The process of assuring that the model implemented on the computer is the same as the one specified and built in the build model phase is called verification.

For the results of a simulation experiment to be usable for decision making, the simulation model and experiment need to be valid. This means that the model and experiment accurately represent the manufacturing system with respect to the issues and solution objectives determined in phase 1. This can be accomplished in a number of ways. The structure of the system represented in the model can be compared to the structure of the actual system or to the system design specification. System data and simulation estimates of the same quantity can be compared statistically. If no significant differences exist, the model can be said to be valid.

The fourth strategic phase seeks to identify and evaluate alternatives to the current system implementation or design as necessary to meet the solution objectives for the issues identified in the first phase. The simulation experiments defined in this phase are based on conclusions drawn by simulation analysts and manufacturing system experts from the system data, simulation model, and simulation experiment results of the preceding stage. The fourth phase may rely on the ability to design simulation experiments sequentially. Alternative solutions may be generated using formal ways for searching a solution space such as a response surface method. In addition, manufacturing system experts may suggest alternative strategies, for example alternative part routings based on the work-in-process inventory at work stations. Performing experiments involves modifications to the simulation model as well as using new model parameter values.

In the fifth phase, physical experiments using the actual manufacturing system or laboratory prototypes of the system may be performed to confirm the benefits of the selected system improvements. Improvements deemed worthy are implemented in the manufacturing system and the results monitored.

The simulation process is iterative. Information obtained at any phase may indicate that previous phases should be repeated. For example, the activities required to discover the root causes of the issues that gave rise to the simulation project may also give rise to new issues. In this case, the phases 1 and 2 of the simulation process should be repeated.

3. SIMULATION MODELING PERSPECTIVES

The construction of a simulation model is the goal of the second phase of the simulation process. There are three common approaches, or world views, employed in model building. The most flexible world view supports the exact representation of the unique details of any particular manufacturing system but requires the most time for model building. The least flexible world view provides standard manufacturing system modeling constructs such work stations, routings, and parts. This approach helps to build models quickly but the standard form may not match the reality of any particular manufacturing system. Current trends are toward a combined approach whereby as much of the manufacturing system as possible is modeled using pre-defined standard manufacturing system constructs and the remainder of the system is modeled using a more flexible world view. Simulation modeling languages are evolving to support this approach. Those that support only the more flexible world views are adding manufacturing specific constructs and those that began with more manufacturing specific constructs are providing the flexibility needed to deal with the peculiarities of individual manufacturing systems.

3.1. The Event World View

The event world view is the most flexible model building perspective. A manufacturing system is modeled in terms of state variables that define the status and conditions of the system at each point in time. An event changes the values of the state variables and tells when in simulated time other events or itself occurs. Each event can happen many times during one simulation. In total, events occur at a finite number of points in time. Thus, this method is often referred to as discrete event simulation. The values of the state variables in the model remain the same between events.

Event graphs (Schruben, 1983) are a diagramming technique for event world view models. An event graph consists of nodes and arcs. Nodes correspond to events. Arcs tell the relationships between events, that is the other events that each event causes to occur and the logical conditions that may restrict such occurrences. The logical conditions make use of the state variables.

For example, consider a work station where parts arrive, wait in a buffer, are processed one at a time by a single machine, and then depart. The event graph is shown in Figure 2. The state variables of interest are the number of parts in the buffer and the state of the machine, BUSY processing a part or IDLE. There are three events: part *arrives* that includes the part entering the buffer, *start service* that removes the part from the buffer and begins its processing as well as making the state of the work station BUSY, and *end service* that makes the state of the work station IDLE. The part arrives event causes itself to occur again, that is the next part to arrive, after a time interval specified by the time between arrivals. The part arrives event causes the start service event to begin processing the arriving part immediately if the work station is IDLE. The end service event follows the start service event and occurs after a time interval that is the part processing time. The end service event will initiate processing of the first part waiting in the buffer if there is one. The time between arrivals and the part processing time could be random variables.

This simple example, while illustrating the concepts of the event world view, does not sufficiently show its power. Consider the following possibilities.

1. An event models the random arrival of a material movement task for an AGV system. State variables show the location and state (BUSY, IDLE) of each AGV. The event determines the AGV to which the task is assigned based on minimum time to reach the pick-up point.

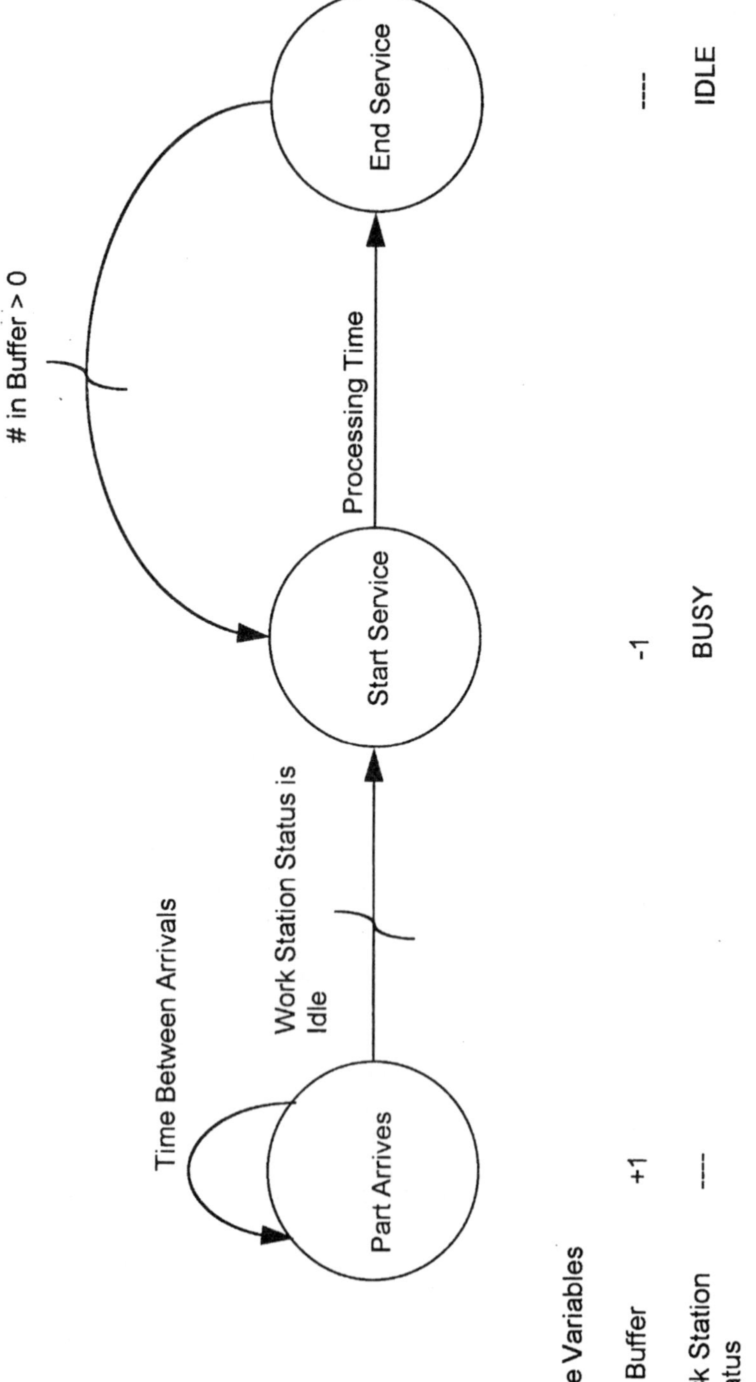

Fig. 2 Event graph for the single work station example

2. An event models the beginning of the day at a refinery making multiple products. Taking into account current production demands and the desire to minimize product switches due to significant down times for cleaning on each of several reactors, the event schedules specific products on individual reactors for the day.
3. An event determines the route of an automatic storage / retrieval machine through a warehouse to pick a particular order based on the current location of various products in the warehouse.
4. An event models the end of an operation by a robot and chooses between various waiting tasks that the robot could perform based on the prior sequence of tasks performed and the number of parts waiting for each task.

3.2. The Process World View

The process world view focuses on describing the processing steps for a part or whatever other entity is transformed by a manufacturing system. The generic name for these entities is transactions. Each transaction can be distinguished from all the others by quantities called attributes. Typical attributes are part type, time of arrival to the manufacturing system, and location on route. The availability of machines, material handling devices, workers, raw material, and the like constrains the processing of the part transactions, causing them to wait in buffers. Resource is the generic name for an entity for which transactions wait in buffers.

Each process world view modeling language provides a standard set of graphical symbols for describing the movement of transactions through a process constrained by the availability of resources. To increase modeling flexibility, most such languages allow a process model to be embellished with events.

For example, consider the GPSS model of the single work station shown in Figure 3. The model consists of a set of blocks connected by lines that order the steps of the process. The GENERATE block creates transactions and models the arrival of parts over time using the time between arrivals parameter. The SEIZE block causes each part transaction to wait until the WORK STATION resource can process it. The time delay for processing is specified in the ADVANCE block. The RELEASE block specifies the end of the use of the WORK STATION by the transaction. The TERMINATE block models the part leaving the work station.

Notice that some blocks occur in pairs. The GENERATE block models the arrival of parts and the TERMINATE block their departure. The SEIZE block models the acquisition of the WORK STATION resource by a part and the RELEASE block models the end of the processing of the transaction by the WORK STATION.

The same state variable values computed explicitly in the event world view model are computed transparently in the process world view model. The SEIZE block adds one to the number in the buffer when a transaction arrives. When a resource begins processing a transaction, the SEIZE block subtracts one from the number in the buffer and makes the state of the resource BUSY. The RELEASE block makes the state of the resource IDLE and looks in the buffer for another transaction for the resource to process. Thus, the number of transactions in the buffer and the state of the resource can be used as a part of the conditional logic of the model

Typically, process world view languages provide a variety of modeling constructs relevant to manufacturing systems such as:

1. The assembling of transactions into batches based on attribute values such as the type of part.
2. Conditional logic for modeling alternative routes through manufacturing processes. Conditions can be based on attributes and state variable values.
3. Interfaces with events for computing non-standard decisions, for example choosing between the various tasks awaiting a robot upon completion of its current task.
4. Material handling operations by conveyors, AGV's, and AS/RS machines.
5. Matching transactions that represent different part types to be assembled.
6. Pre-emption of a working resource to perform a more important task.

3.3. Simulators

A simulator supports model building by description of the entities comprising a system, their attributes, and the logic that relates them. A simulator is domain specific and has built-in knowledge of how systems in that domain operate. More sophisticated simulators have built in operation decision rules for complex situations. The domain specific information does not need to be specified for each simulation model. This speeds the model building process. Most current simulators can be enhanced by event or process world view models to supply details needed for a particular system.

Manufacturing simulators typically provide model building constructs that correspond one to one with the entities that make up a manufacturing system such as work stations, buffers, parts, the routes the parts follow through the work stations, material handling devices, standard decision rules, and the like. Model building consists of describing each entity in the manufacturing system by giving values to the attributes of the corresponding standard construct.

Consider a small manufacturing system that processes two types of parts using three work stations. Part type A is processed by stations 1 and 2. Part type B is processed by stations 1, 2, and 3. A conveyor moves parts between the work stations. Figure 4 shows a possible manufacturing simulator model of the shop. The three work stations are connected by conveyors. Each work station is assumed to have its own buffer that is modeled as having infinite capacity. This means that the work station buffer size is not modeled as a constraint on the operation of the system. The processing times for each part type at each work station are given. The route of each part through the system is specified. The time between arrivals for each part type is included.

The simulator must have built-in knowledge of the operation of the manufacturing system components, for example:

1. How a work station processes parts, that is the logic of the discrete event model of section 3.1. is contained within the simulator and is applied transparently to the modeler.
2. How a conveyor transports parts between work stations.
3. How a part follows its route.
4. The arrival process for parts.

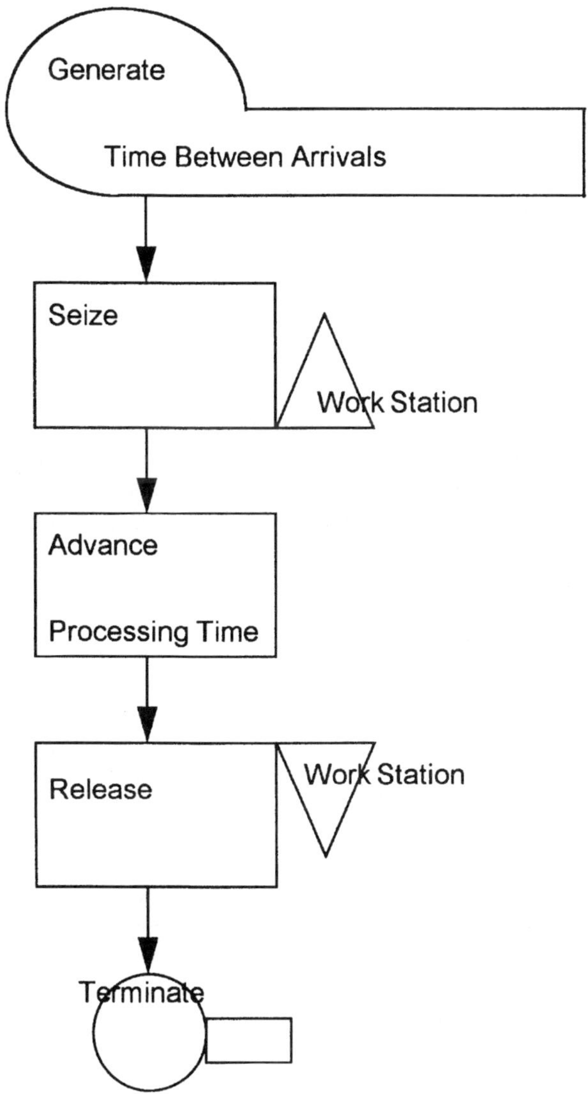

Fig. 3 GPSS process model of the single work station.

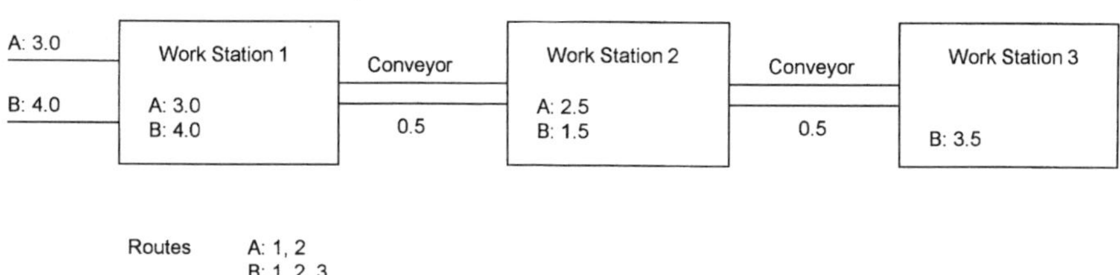

Fig. 4 Manufacturing simulator model of a serial system.

This information is not supplied by the modeler as it would need to be if either of the other two world views was used.

4. THE SIMULATION ENGINE

A model is analyzed to derive additional information such as the values of quantities of interest, sometimes called performance measures. If the model is expressed in a mathematical form, a closed form mathematical result is sought to give the exact values of the performance measures. However, it may not be possible to derive a closed form mathematical result for several reasons including the following:

1. The stochastic elements of the model do not meet the distribution requirements need for a closed form solution, for example the arrival of parts could be modeled using the exact arrival times from historical data gathered from the manufacturing system.
2. Averaging, for example processing times at a station across part types, removes variance from the model and may underestimate performance measures such as the number of parts in a buffer.
3. Model parameter values, such as part arrival rates to the system, may vary in time.
4. The model includes conditional logic that cannot be expressed in a strictly mathematical form, for example the logic that chooses among the waiting tasks for a robot just completing its current task.
5. The computation from the model of time series of values or distributions of values of performance measures, as well as performance measures specific to a particular system, may be analytically intractable.

When analytic methods fail, simulation experimental procedures must be used. All simulation experiments can be conducted from the event world view. The modeling constructs used in a process world view model correspond to a sequence of events. This correspondence is transparent to the modeler. In the same way, the modeling constructs used in a simulator correspond to a set of events.

Simulation experimental procedures replicate the temporal behavior of the manufacturing system as described in the model and observe values of the performance measures of interest. Each change in each state variable value, transactions, or resource must be generated and managed. Meeting these requirements involves procedures for random sampling, event list management, and state variable computation and observation. For illustration, simulation procedures will be discussed in the context of the event world view single work station model presented in Section 3.1.

4.1. Random Sampling

The part arrives event must compute a specific time till the next arrival value each time the event occurs. Similarly, the start service event must compute each time it occurs an individual processing time for a part and as a result the time the end service event occurs for that part.

Suppose for illustration that the time between arrivals is exponentially distributed with mean 3.0 minutes. This implies that all of the time between arrival values taken together must not have a statistically significant difference from the exponential distribution with mean 3.0. In addition, suppose that the processing time is uniformly distributed between 1.5 and 3.5 minutes. In the same way, this implies that all of the processing time values taken together must not have a statistically significant difference from the uniform distribution between 1.5 and 3.5.

Achieving this goal requires computing each time between arrivals value as a random sample from the exponential distribution with mean 3.0 and the processing time values as a random sample from the uniform distribution between 1.5 and 3.5. These computations are performed using the inverse transformation method as follows.

The time between arrivals to the work station has the following cumulative distribution function:

$$Y = F(X) = 1 - e^{-X/3}$$

and therefore

$$X = -3\ln(1 - Y).$$

In the same way, the processing time has the following cumulative distribution function:

$$Y = F(X) = (X - 1.5)/(3.5 - 1.5)$$

and therefore

$$X = Y^{*}(3.5 - 1.5) + 1.5$$

Notice in each case the inverse of the cumulative distribution is taken, hence the name inverse transformation method. This reduces each case to the same problem, determining the value of y. Since y is a probability and a cumulative distribution, it ranges uniformly between 0 and 1. Good experimental procedure requires a random sample and so a random sample of y must be chosen.

Fortunately, there are well known algorithms for generating a sequence of random samples uniformly distributed in the range (0,1). Each sample value is called a pseudo – random number and a sequence of pseudo – random numbers is called a stream. The algorithm that generates a stream is deterministic but the properties of the numbers in the stream make them look random. These properties include the following:

1. The numbers show no statistically significant auto-correlation.
2. There are many, many numbers in sequence before any number repeats.
3. The numbers are uniformly distributed between 0 and 1.

Thus, generating a random sample of the time between arrivals or the processing time involves generating a pseudo-random number r and using the inverse cumulative distribution function as follows:

Time Between Arrivals: $X = -3ln\,(1 - r)$
Processing Time: $X = r^*\,(3.5 - 1.5) + 1.5$

Other methods of generating random samples must be used when the cumulative distribution function does not have a closed form inverse as discussed in Law and Kelton (1991).

4.2. Event List Management

The simulation engine maintains a list of events that have been scheduled to occur at future times, or possibly, at the current time. This list is sequenced by the time each event is schedule to occur, earliest time first. The simulation proceeds by processing the first event on the list. This event could place other events on the list at any future time or the current time.

Consider the event world view work station model of section 3.1. At any point in time, the event list could contain up to three entries:

1. The part arrival event at a future point in time.
2. The end service event at a future point in time.
3. The start service event at the same time as the part arrival event or the end service event.

To illustrate, consider the event list at time 2.0, a point in time when no events occur.

Event	*Time of Occurrence*
Part Arrives	2.5
End Service	3.5

The next part will arrive at time 2.5 and the end of service for the part currently being processed will occur at time 3.5.

Consider the event list at time 14.0, another point in time when no event occurs.

Event	*Time of Occurrence*
Part Arrives	14.5

The next part will arrive at time 14.5 and no part is currently being processed since no end service event is scheduled. The part arrives event will schedule a start service event at time 14.5 and the next part arrives event sometime in the future, say at time 17. Upon completion of the part arrives event, the event list will be as follows:

Event	*Time of Occurrence*
Start Service	14.5
Part Arrives	17.0

The start service event will schedule the end service event, say at time 19. Thus, after the start service event is processed, the event list will be as follows:

Event	Time of Occurrence
Part Arrives	17.0
End Service	19.0

4.3. State Variable Computations

Table 1 shows one possible way that the simulation of the event world view model of the single work station could happen. Notice that at least one of the two state variables, number in the buffer and status of the work station, changes value at each event. The time each possible future event, part arrival and end service, is shown.

TABLE 1
Simulation of the event world view model of the single work station.

Current Simulation Time	Event	Simulation Time of Next Part Arrival Event	Simulation Time of Next End Service Event	Number in Buffer	Status of Work Station
0.0	Initial Conditions	0.0	N/A	0	IDLE
0.0	Part Arrives	1.0	N/A	1	IDLE
0.0	Start Service	1.0	2.0	0	BUSY
1.0	Part Arrives	3.7	2.0	1	BUSY
2.0	End Service	3.7	N/A	1	IDLE
2.0	Start Service	3.7	4.5	0	BUSY
3.7	Part Arrives	4.2	4.5	1	BUSY
4.2	Part Arrives	9.3	4.5	2	BUSY
4.5	End Service	9.3	N/A	2	IDLE
4.5	Start Service	9.3	7.6	1	BUSY
7.6	End Service	9.3	N/A	1	IDLE
7.6	Start Service	9.3	9.5	0	BUSY

The start service event can occur only at the same time as a part arrival event or an end service event.

Initially, there are no parts in the buffer and the work station is IDLE. The first part arrives at the start of the simulation, time 0, and starts processing at that time. The second part arrives at time 1 and must wait in the buffer for the first part to complete processing which occurs at time 2. The processing of the second part starts at the same time. Two more parts arrive before the processing of the second part is completed. These are processed when the work station completes the proceeding part.

5. DESIGN AND ANALYSIS OF SIMULATION EXPERIMENTS

Simulation experiments can be designed and analyzed as would any other laboratory or field experiment, using proper statistical experimental design techniques and data analysis methods. These are widely documented. In this section, some of the unique issues in the design of simulation experiments are discussed.

5.1. Length of the Simulation

What event ends the simulation is a fundamental question in simulation experimental design. There are two basic answers to this question:

i. The end of the planning period, the completion of all production requirements, or the like occurs: a predetermined point in time or condition that happens in the normal operation of the manufacturing system. This is called a terminating simulation.
ii. The long term steady-state potential of the manufacturing system is reached, as measured by the estimate of some statistic concerning a performance measure.

In most manufacturing systems, the former approach for setting the simulation run length is used. This avoids complex procedures for estimating the point in simulated time at which steady-state is achieved for estimates concerning multiple performance measures. Furthermore, the operating conditions of most manufacturing systems change over time so that steady-state is a condition that never occurs in reality. The production requirements for the system change in both volume and mix. The human and other resources available to the system may be dynamic in time. The need to examine transient, non-steady state behavior, such as the changes in work-in-process inventory due to a machine failure, is significant.

5.2. Independence of Observations

Most statistical analysis procedures require independent observations of performance measure values. However, the observations in a simulation are typically dependent. To illustrate, consider the time the nth part processed would spend at a work station:

$$\text{Time at Work Station}_n = \text{Time in Buffer}_n + \text{Operation Time}_n$$

The Time in Buffer for the nth part include the operation times for the parts that preceded it in processing at the work station while the nth part was in the buffer. For example, in Table 1 the fourth part to arrive does so while the second part to arrive is being processed. So the time the fourth part spends in the buffer is equal to part of the operation time for the second part and the entire operation time for the third part:

$$\text{Time at Work Station}_4 = f(\text{Operation time}_2, \text{Operation time}_3) + \text{Operation Time}_4$$

Thus, the time spent at the work station by the fourth part is correlated with the time spent by the second and the third parts.

Therefore, rather than using performance measure observations directly in statistical analysis computations, independent observations must be constructed. How this construction is done depends on whether a terminating or a steady-state simulation is run.

Replication is used for a terminating simulation. One replicate of a simulation varies from another only in the random number streams used. Thus, differences in replicates are due to random variation only and therefore replicates are independent of each other. Each replicate yields one independent estimate of each performance measure, such as the average time in the system for parts, the average work in process inventory in each buffer, and the percent of time in the BUSY state for each machine. Standard statistical analysis procedures can be applied to these independent replicates.

Replication can also be used for steady-state simulations. Another method for creating independent observations is that of batch means. The simulation after the point that steady-state is reached is divided into n parts each with an equal number of observations of the performance measure of interest or each of equal length in simulated time. Each of these batches is considered to be independent of the others and each like a replicate yields one independent estimate of each performance measure. Statistical tests of correlation can be performed to determine if the observations from the batches are truly statistically independent of each other.

5.3. Initial Conditions

In order to start a simulation, the initial values of the state variables in the simulation model must be given. These include the state of each machine or other manufacturing system resources as well as the number of parts in each buffer.

Setting the initial conditions requires an understand of how the manufacturing system operates. The initial conditions may be the actual start-up state of the manufacturing system at the beginning of the production period being modeled. For example, the initial conditions could be the set of orders to process in the next production period and the reality that the production starts with all machines IDLE and all buffers empty. If the system context is such that no well defined start-up state exists, the initial conditions should reflect a combination of state variable values that can occur in the manufacturing system. Wilson and Pritsker (1979) suggest that the combination of system state variables values that occurs the most often in reality is the best choice with respect to achieving steady-state in the shortest period of simulated time.

Failure to set the initial conditions in the manner suggested above can lead to biased observations of system performance measures. A biased observation is a value that could not occur in the actual manufacturing system. Using biased observations can lead to biased estimates of performance measure statistics and inaccurate conclusions about manufacturing system operations and possible improvements. As an example of a biased observation, consider again the part time at the work station discussed in the previous section.

Time at Work Station = Time in Buffer + Operation Time

Suppose the system buffers are never empty, but the initial conditions given for a simulation of the system included empty buffers. In this case, the time at the work station for the first part processed would be the operation time and time in the buffer would be zero. However, in the actual system the time in the buffer would never be zero. Thus, the time at the work station is under estimated for the first part.

6. EXAMPLE DESIGN PROBLEM

The following example illustrates the simulation process through the study of an automated work cell. A process world view model is constructed. Modeling constructs for batching parts and setting up machines are employed. Selection rules for choosing between tasks waiting for the same robot are incorporated. System buffer size requirements are estimated. New control logic is simulated and compared to the original system to determine if performance in terms of part time in the system is improved.

6.1. Define the Problem and the Improvement Objectives

A manufacturing process requires three operations in sequence: drill, deburr, and polish. These operations are performed by three machines in a U-shaped configuration as shown in Figure 5. Parts wait for machines in a common waiting area. A single robot serves the three machines by moving parts to and from the common waiting area as well as loading and unloading parts on machines.

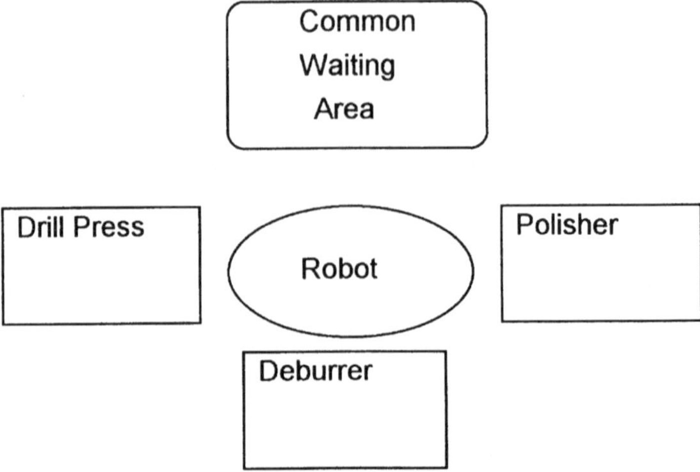

Fig. 5 Automated Manufacturing Process Layout.

Currently, the manufacturing process serves three types of parts which arrive in a random sequence with a constant time between arrivals of 9.6 minutes. The proportion of each type of part is as follows: Type 1, 24%; Type 2, 44%; and Type 3, 32%. The drill press only uses different tooling for each part type. Thus, a setup task is required on the drill press if the type of the current part is different from the type of the preceding part. As a completely automated machine, the drill press can perform the change of tooling on its own. Each part is served by the drill press, deburrer, and polisher in sequence. The polisher works on two parts of the same type concurrently. The operation time on each machine varies by part type. Drill press setup time is a constant.

6.2. Build the Model

First consider that the robot has six different tasks to perform, moving parts to and loading as well as unloading and moving parts from each of three work stations. Often parts requiring different tasks will be waiting for the robot. The relative priority of these waiting tasks is important. One priority scheme for the six tasks is as follows. Removing a completed part from a work station is prerequisite for moving a new part to the work station. Thus, unload and move from the work station tasks should precede move to the work station and load tasks. It is important to complete work on each part as soon as possible. Thus, priority is given to polisher station tasks over deburrer station tasks over drill press tasks. In summary, this priority scheme yields the following:

1. Unload and move a part from the polisher.
2. Unload and move a part from the deburrer.
3. Unload and move a part from the drill press.
4. Move a part to the polisher and load.
5. Move a part to the deburrer and load.
6. Move a part to the drill press and load.

Next, consider the arrival process as follows:

1. A part arrives with time of arrival assigned to attribute A1 and part type assigned to attribute A2. The part type value is a random sample from the discrete distribution of part types computed using the inverse transformation method discussed in Section 4.1.
2. The part enters the common storage area and one is added to the number of parts in the common storage area.

The processing time at each machine varies by part type as shown in the operation time matrix in Table 2.

The process model for the drill press is as follows:

1. If the next part to be processed by the drill press is of a different type than its predecessor, the setup operation is performed automatically by the drill press in 1.0 minute.
2. The robot moves a part from the common storage area to the drill press and loads the part on the machine in 1.5 minutes. The number of parts in the common storage area is decreased by 1.

3. A part is processed by the drill press in constant time that depends on the part type as given in the operation time matrix in Table 2.
4. The robot unloads and moves the completed part from the drill press to the common storage area. The number of parts in the common storage area is increased by 1.

Note again that the drill press station must have an attribute of the type of the predecessor part.

The process for the deburrer station is the same except that the setup step is omitted. The process for the polisher station must include batching two parts of the same type together as follows:

1. A part completes the deburr operation and is returned to the common storage area. The number of parts in the common storage area is increased by 1
2. The number of parts of that type in the forming batch is incremented by 1.
3. If 2 parts of that type have accumulated, the pair of parts may be processed on the polisher.
4. Each of the two parts is moved in turn from the common storage area to the polisher and loaded by the robot in 3.0 minutes for both. The number of parts in the common storage area is decreased by 2.
5. A pair of parts is processed by the polisher in the operation time given in the operation time matrix in Table 2.
6. Each of the two parts is unloaded and moved in turn from the polisher to an output buffer by the robot in 3.0 minutes for both.

6.3. Find the Root Cause of the Problem – Experimental Design

Part time in the automated work cell and the number of parts waiting in the common storage area are key performance measures. Since the size of the common storage area is of interest, the statistic concerning the number of parts waiting in the common storage area should measure the maximum size. The 99% point of the distribution of the number of parts in the common storage area was chosen.

Management has a policy of working over time to complete all parts by the end of the month. Thus, the initial conditions for starting a month can be empty and idle.

TABLE 2
Operation time matrix.

Part Type	Drill Press	Deburrer	Polisher
1	4.5	6.0	4.5
2	5.5	7.5	4.5
3	11.5	15.0	12.0

There is one random variable in the model, the type of part. Replication will be used to obtain independent estimates of the performance measures. Thirty replicates will be simulated.

In summary, the experimental design is as follows:

Performance measures:	Part time in the system. Number of parts in the common storage area.
Initial Conditions:	Empty and idle.
Simulated Time Interval:	0–184 hours.
Streams:	Part type.
Replicates:	30

6.4. Find the Root Cause of the Problem – Conclusion Drawing

Table 3 shows the average part time in the system and the 99% point of the distribution of the number of parts in the common storage area. The average time parts spend in the system slightly exceeds one hour. The 95% confidence interval for the 99% point of the number of parts in the common storage area is between 7 and 8.

6.5. Propose and Assess Solution Alternatives

Next, the simulation project team reviews the model, experiments, and simulation results produced so far. Based on this information, new alternatives to consider are often identified. As a part of this discussion, someone pointed out that it seemed inefficient for the robot to move a part from the drill press back to the storage area and then immediately from the storage area to the deburrer if the deburrer was not in use. The robot should move the part from the drill press directly to the deburrer.

TABLE 3
Simulation results-initial evaluation.

Statistic	*Average Part Time in System (Minutes)*	*Number of Parts in the Common Storage Area (99% point)*
Average	65.76	7.47
Variance	7.65	0.60
Standard Error	0.50	0.14
Upper 95% Confidence Limit	66.80	7.76
Lower 95% Confidence Limit	64.73	7.18

This logic was incorporated in the process model for the drill press as shown below.

1. If the next part to be processed by the drill press is of a different type than its predecessor, the setup operation is performed automatically by the drill press in 1.0 minute.
2. The robot moves a part from the common storage area to the drill press and loads the part on the machine in 1.5 minutes. The number of parts in the common storage area is decreased by 1.
3. A part is processed by the drill press in constant time that depends on the part type as given in the operation time matrix in Table 2.
4. If the deburr machine is in use, the robot unloads and moves the completed part from the drill press to the common storage area. The number of parts in the common storage area is increased by 1.
5. If the deburr machine is not in use, the robot unloads and moves the completed part from the drill press to the deburr machine.

The experimental design is the same as in section 6.3. The simulation results for both the new and the old control strategy are shown in Table 4.

TABLE 4
Simulation results-old versus new control strategy.

Statistic	*Average Part Time in System (Minutes)*	*Number of Parts in the Common Storage Area (99% point)*	*Average Part Time in System (Minutes)*	*Number of Parts in the Common Storage Area (99% point)*
Average	65.76	7.47	62.94	7.17
Variance	7.65	0.60	7.22	0.70
Standard Error	0.50	0.14	0.49	0.15
Upper 95% Confidence Limit	66.80	7.76	63.95	7.48
Lower 95% Confidence Limit	64.73	7.18	61.94	6.86

Note that the confidence intervals for the average part time in the system do not overlap. This indicates that the new control strategy produces a statistically significant difference in the average part time in the system. Thus, this strategy was adopted.

The confidence intervals for the 99% point of the buffer size overlap indicating no statistically significant difference. The upper bound of both was between 7 and 8 so a common storage area size of 8 parts was selected.

6.6. Implement the Selected Solution and Evaluate

The automated work cell was implemented with direct movement of parts from the drill press to the robot when possible and with a common storage area capacity of 8 parts.

References

R.G. Askin and C.R. Standridge *Modeling and Analysis of Manufacturing Systems*, John Wiley and Sons, New York (1993).

R. Godziela, Simulation of a flexible manufacturing cell,“ in *Proceedings of the 1986 Winter Simulation Conference*, (eds. J.R. Wilson, J.O. Henriksen, and S.D. Roberts), Association for Computing Machinery, New York, pp. 641–648 (1986).

F.H. Grant, Simulation in designing and scheduling manufacturing systems in *Design and Analysis of Integrated Manufacturing Systems*, (ed. W.D. Compton), National Academy Press, Washington, D. C., pp. 134–147 (1988).

W. Hancock, R. Dissen and A. Merten, An example of simulation to improve plant productivity. *AIIE Transactions*, **9**, pp. 2–10 (1977).

D.S. Hira and P.C. Pandey, A computer simulation study of manual flow lines. *Journal of Manufacturing Systems*, **2**(2), p. 117 (1983).

P.Y. Huang, E.R. Clayton and L.J. Moore, Analysis of material and capacity requirements with Q-GERT. *International Journal of Production Research*, **20**(6), pp. 701–713 (1982).

P.Y. Huang, L.P. Lees and B.W. Taylor III, A simulation analysis of Japanese just-in-time technique (with kanban) for the multiline, multistage production system. *Decision Sciences*, **14**, pp. 326–344 (1983).

O. Kimura and H. Terada, Design and analysis of pull systems, a method of multi-stage production control, “*International Journal of Production Research*, **19**, pp. 241–253 (1981).

J.P.C. Kleijnen, *Statistical Tools for Simulation Practitioners*, Marcel-Dekker, Amsterdam (1986).

J.P.C. Kleijnen and C.R. Standridge, Experimental design and regression analysis in simulation: an FMS case study. *European Journal of Operations Research*, **33**, pp. 257–261 (1988).

A.M. Law and W.D. Kelton, *Simulation Modeling and Analysis, 2nd ed.*, McGraw-Hill, New York (1991).

P. L'Ecuyer, Random numbers for simulation. *Communications of the ACM*, **33** (1990).

L. Lin and J.K. Cochran, Optimization of a complex flow line for printed circuit board fabrication by computer simulation. *Journal of Manufacturing Systems*, **6**(1), pp. 47–57.

M.C.Miles, Using group technology, simulation and analytic modeling in the design of a cellular manufacturing facility, “in *Proceedings of the 1986 Winter Simulation Conference*, (eds. J.R. Wilson, J.O. Henriksen and S.D. Roberts), Association for Computing Machinery, New York, pp. 657–660 (1986).

R.J. Miner, D.B. Wortman and D. Cascio, Improving the throughput of a chemical plant. *Simulation*, **35**, pp. 125–132 (1980).

S. Mittal and H.-P. Wang, Simulation of JIT production to determine number of kanbans. *International Journal of Advanced Manufacturing Technology*, **7**, pp. 292–308 (1992).

K.J. Musselman, Computer simulation: a design tool for FMS. *Manufacturing Engineering*, **93**, pp. 115–120 (1984).

C.D. Pegden, R.E. Shannon and R.P. Sadowski, *Introduction to Simulation Using SIMAN*, McGraw-Hill, New York, N.Y (1990).

A.A.B. Pritsker, *Introduction to Simulation and SLAM II*, 3rd ed., Halsted, New York (1987).

A.A.B. Pritsker, C. E. Sigal and R.D.J. Hammesfahr, *SLAM II Network Models for Decision Support*, The Scientific Press, Palo Alto, CA (1994).

A.A.B. Pritsker, Why simulation works, in *Proceedings of the 1989 Winter Simulation Conference*, (eds. E.A. MacNair, K.J. Musselman and P. Heidelberger), Association for Computing Machinery, New York, pp. 1–8 (1989).

A.A.B. Pritsker, Applications of SLAM. *IIE Transactions*, **14**, pp. 70–77 (1982).

P. Rogers and R.J. Gordon, Simulation for real-time decision making in manufacturing systems, in *Proceedings of the 1993 Winter Simulation Conference*, (eds. G.W. Evans, M. Mallaghasemi, E. C. Russel and W.E. Biles), Association for Computing Machinery, New York, pp. 886–874 (1993).

R.G. Sargent, Validation and verification of simulation models, in *Proceedings of the 1992 Winter Simulation Conference*, (eds. J.J. Swain, D. Goldsman, R.C. Crain and J.R. Wilson), Association for Computing Machinery, New York, pp. 104–114 (1992).

T.J. Schriber, *An Introduction to Simulation Using GPSS/H*, John Wiley and Sons, New York (1991).

B.J. Schroer, J.T. Black and S.X. Zhang, Just in time (JIT) with kanban. *Manufacturing System Simulation on Microcomputers*. Society for Computer Simulation, La Jolla, CA (1985).

L. Schruben, Simulation modeling with event graphs. *Communications of the ACM*, **26**(11), pp. 957–963 (1983).

L.B. Schwarz, S.C. Graves and W.H. Hausman, Scheduling policies for automatic warehousing systems: simulation results. *IIE Transactions*, **10**(3), pp. 260–270 (1978).

R.E. Shannon, *System Simulation: The Art and Science*, Prentice-Hall, N.J. Englewood Cliffs, (1975).

C.R. Standridge and M.D. Brown-Standridge, Combining total quality management and simulation with application to family therapy process design. *Journal of the Society for Health Systems*, **5**(1), pp. 23–40 (1994).

J. Talavage and R.G. Hannam, *Flexible Manufacturing Systems in Practice: Applications, Design and Simulation*, Marcel-Dekker, New York (1988).

B.W. Taylor III, L.M. Moore and R.D. Hammesfahr, Global analysis of a multi-product, multi-line production system using Q-GERT modeling and simulation. *AIIE Transactions*, June, pp. 145–155 (1980).

A.R. Tenner and I.J. DeToro, *Total Quality Management: Three Steps to Continuous Improvement*, Addison-Wesley, New York (1992).

J.R. Wilson and A.A.B. Pritsker, A procedure for evaluating startup policies in simulation experiments. *Simulation*, **31**, pp. 79–89 (1978).

D.B. Wortman and J.R. Wilson, Optimizing a manufacturing plant by computer simulation. *Computer-aided Engineering*, **3**, pp. 48–54 (1984).

S.-Y.D. Wu. and R.A. Wysk, An application of discrete-event simulation to on-line control and scheduling in flexible manufacturing. *International Journal of Production Research*, **27**, pp. 1603–1623 (1989).

CHAPTER 14

Interactive Geometric Modeling and Simulation of Manufacturing Systems

YONG FANG and F. W. LIOU

Department of Mechanical and Aerospace Engineering and Engineering Mechanics/ Intelligent Systems Center, University of Missouri-Rolla, Rolla, Missouri 65409-1350, USA

1. INTRODUCTION

In engineering applications, there is often a need to examine an entire manufacturing system rather than an individual component itself. Conventionally, if a designer wants to inspect the dynamic or kinematic performance of a manufacturing system, a model for the analysis has to be generated from the manufacturing system drawing. The mating conditions between components will have to be specified as idealized joints, such as revolute, prismatic, spherical, gear joint, etc. Although there are good reasons for using these standard joints, the following problems may occur when joints are assumed to be perfect:

1. The dynamic analysis of a manufacturing system may not be accurate as clearances always exist between the components in the real world. The clearances will cause the components to impact with each other and the geometric configuration of the system to change during the movement.
2. Many manufacturing systems, such as the part-feeding systems and the material handling systems, can not be modeled using idealized joints. To analyze such

kind of manufacturing systems using traditional approach, a great deal of simplifications must be made.
3. If a designer is forced to use the idealized traditional joints, the creativity of the designer is limited.

Computer simulation of manufacturing systems which cannot be described by idealized traditional joints due to changes of the geometric configurations is a field that has seen only moderate development. Such systems are commonly encountered in manufacturing and material handling equipment and must be carefully analyzed during the design process. Smith (1971), Paul (1975), Haug *et al.* (1986), and Wehage and Haug (1982) presented a computer strategy which can automatically handle mating joints addition and deletion. However, the time and duration of a contact and the resulting kinematic constraint due to the collision must be handled by the user. This can be a formidable task when the mating joints are dynamically changing as a function of the system response. Haug (1989), Nikravesh (1988), and Wittenburg (1977) developed simulation algorithms for hinged mechanical systems. These algorithms have an event handler for changing mating conditions; however, the mating condition changes still need be manipulated by the user. Therefore, these approaches are suitable for systems with small relative distances between contacting bodies or for straight forward predictable motions. Gilmore and Cipra (1991a, 1991b) and Han *et al.* (1993) presented an algorithm that automatically predicts and detects the changes in the mating conditions and then reformulates the equations of motion for planar dynamic mechanical systems with changing geometric configurations. The force closure concept and point-to-line constraint is adopted to facilitate the reformulation of the governing equations of motion.

Presented in this chapter is an advanced modeling system approach for general three-dimensional manufacturing systems. A designer can interactively create an assembly of components in a simulated physical environment, and each component will respond according to basic physical laws. Therefore, without specifying the mating conditions between the individual components, the modeling system can automatically simulate the kinematic and dynamic response of the manufacturing system.

2. FORCE CLOSURE FORMULATION

The modeling system is developed based on the force closed kinematic joint (Reuleaux, 1876) which maintains the kinematic constraint between two bodies by the use of a closing force. This is in contrast to a form closed kinematic joint which is constructed such that the kinematic constraint is physically provided by the bodies. The closing force in the force closed joint acts in the same direction as the reaction force would in an equivalent form closed joint.

From a kinematic viewpoint, if a point on a rigid body is constrained to remain in contact with a surface of another rigid body, the point is restricted to move in the direction tangent to the surface. Hence, one degree of freedom is lost. In addition, a reaction force exists at each contact point. When considering the kinematic constraint in this manner, it becomes possible to consider a kinematic joint which may change its degrees of freedom at any time depending on its contact force condition.

2.1. Equivalent Point Contacts for Three-Dimensional Components

The location of a constrained point on a surface is instrumental in allowing the proper motion and in modeling a manufacturing system with changing geometric configurations. Considering spatial motion, each component has several boundary surfaces (faces), each face has several boundary lines (edges), and each edge has two endpoints (vertices). The vertices of one body may come into contact with a face of any body thus establishing a point-to-surface contact. When an edge of a body collides with a face of another body, the bodies are in line contact. A line contact is kinematically equivalent to two point-to-surface contacts. In a similar way, a surface-to-surface contact can be substituted by three point-to-surface contacts. Figure 1 shows the spatial kinematic joints of various degrees of freedom in their form closed and force closed versions.

Force Closed Constraints	Form Closed Equivalents
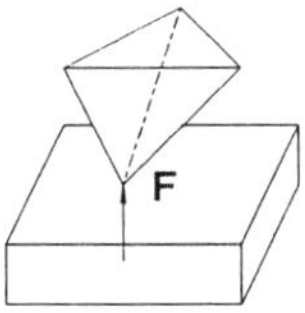 F	NO
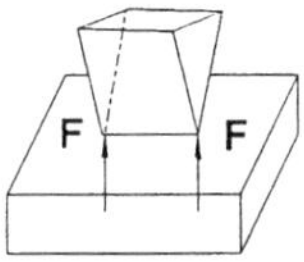 F F	NO
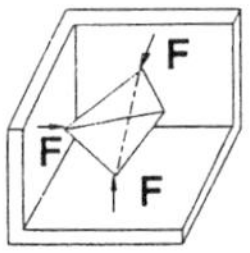 F F F	Spherical Joint
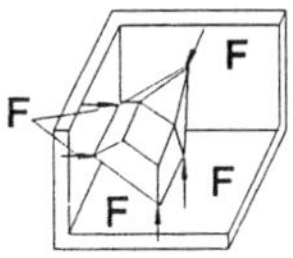 F F F F	Revolute Joint
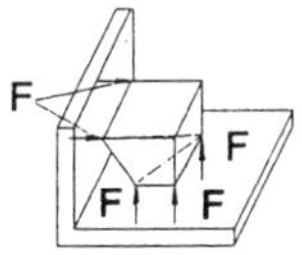 F F F F	Slider Joint

Fig. 1 Forced closed constraints and their equivalents.

The point-to-surface contact and force closure concept lend a great deal of flexibility to this approach. They allow a manufacturing system being modeled to change its geometric configuration by changing the number of point-to-surface contacts. The point-to-surface force-closed constraint joint method simply adds or deletes point-to-surface constraint equations, eliminating the need to determine the type of joint between the components and to maintain a joint constraint library.

2.2. Point-to-Surface Constraint Joint

A general point-to-surface kinematic constraint joint is shown in Figure 2. The reaction force F between two bodies results in a force-closed kinematic constraint joint at point C_i. The (x, y, z) coordinate system is a global inertial coordinate system while the local coordinate system $(\xi, \eta, \zeta)_i$ is fixed to each body. Point C_i of body i is instantaneously in contact with point C_j on a surface of body j. While the contact is maintained, point C_i is constrained such that it does not separate from the surface of body j but is able to slide along the surface or rotate about it. Thus the two-body system has five relative degrees of freedom. This point-to-surface constraint joint can

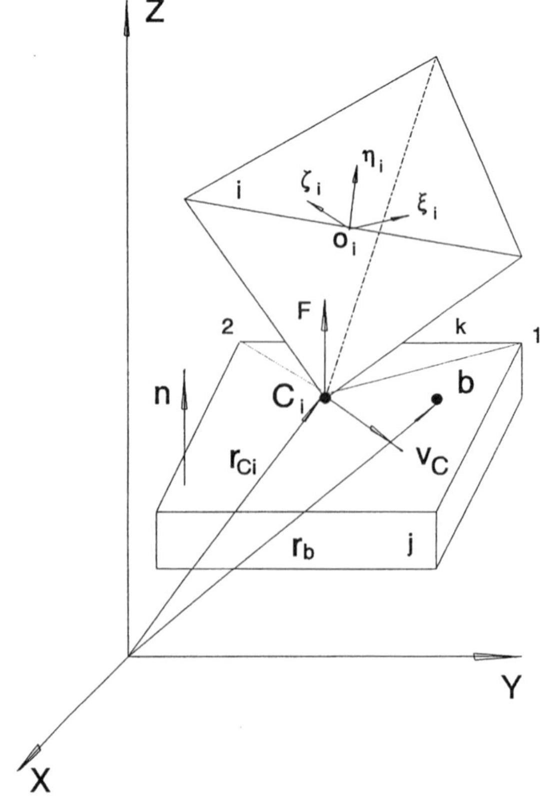

Fig. 2 A general point-to-surface constraint.

be expressed as

$$\phi(q,t) = \vec{n} \cdot (\vec{r}_{C_i} - \vec{r}_b) = 0 \tag{1}$$

where q is the generalized coordinates of the system with

$$q_i = (x, y, z, e_0, e_1, e_2, e_3)_i = (r, e)_i \tag{2}$$

and (e_0, e_1, e_2, e_3) are the Euler parameters which define the orientation of body i. t denotes time, $\vec{n}$ is the outward normal of the contact surface, $\vec{r}_{C_i}$ and $\vec{r}_b$ represent the position vectors of point C_i and b respectively, $\vec{r}_{C_i} - \vec{r}_b$ is a vector on the contact surface, and b is an arbitrary point on the contact surface. All the vectors are measured with respect to the global coordinate system.

Equation (1) can be differentiated with respect to time to yield the velocity constraint

$$\dot{n} \cdot (\vec{r}_{C_i} - \vec{r}_b) + \vec{n} \cdot (\dot{r}_{C_i} - \dot{r}_b) = 0 \tag{3}$$

where the upper dot denotes differentiation with respect to time. From Figure 2, it can be seen that

$$\vec{r}_{C_i} = \vec{r}_{oi} + A_i \vec{S}_{C_i} = \vec{r}_{oi} + \vec{S}_{C_i} \tag{4}$$

$$\vec{r}_b = \vec{r}_{oj} + A_j \vec{S}'_b = \vec{r}_{oj} + \vec{S}_b \tag{5}$$

where $\vec{r}_{oi}$ and $\vec{r}_{oj}$ are the locations of the origin of the body fixed coordinate system for bodies i and j, $\vec{S}'_{C_i}$ and $\vec{S}'_b$are the coordinates of the points C_i and b with respect to their local coordinate systems, prime denotes the coordinates with respect to local coordinate system, and A_i and A_j are the coordinate transformation matrices for bodies i and j. Substituting equations (4) and (5) into equation (3), yields

$$\dot{n} \cdot (\vec{r}_{C_i} - \vec{r}_b) + \vec{n} \cdot (\vec{v}_i + \dot{S}_{C_i} - \vec{v}_j - \dot{S}_b) = 0 \tag{6}$$

where

$$\dot{n} = \vec{\omega}_j \times \vec{n} = A_j (\vec{\omega}'_j \times \vec{n}') \tag{7}$$

$$\dot{S}_{C_i} = \vec{\omega}_i \times \vec{S}_{C_i} = A_i (\vec{\omega}'_i \times \vec{S}'_{C_i}) \tag{8}$$

$$\dot{S}_b = \vec{\omega}_j \times \vec{S}_b = A_j(\vec{\omega}'_j \times \vec{S}'_b) \tag{9}$$

where $\vec{V}_i$ and $\vec{V}_j$ are the translational velocities, and $\vec{\omega}_i$ and $\vec{\omega}_j$ are the angular velocities of the bodies i and j respectively. Substituting equations (7), (8), and (9) into equation (6), yields

$$(\vec{\omega}_j \times \vec{n}) \cdot (\vec{r}_{C_i} - \vec{r}_b) + \vec{n} \cdot (\vec{v}_i + \vec{\omega}_i \times \vec{S}_{C_i} - \vec{v}_j - \vec{\omega}_j \times \vec{S}_b) = 0 \tag{10}$$

Similarly, differentiating equation (10) can yield the acceleration constraint equation

$$\begin{aligned} &\vec{n}\cdot(\vec{a}_i-\vec{a}_j)+(\vec{\alpha}_j\times\vec{n})\cdot(\vec{r}_{C_i}-\vec{r}_b)+\vec{n}\cdot(\vec{\alpha}_i\times\vec{S}_{C_i}-\vec{\alpha}_j\times\vec{S}_b)= \\ &-\vec{\omega}_j\times(\vec{\omega}_j\times\vec{n})\cdot(\vec{r}_{C_i}-\vec{r}_b)-2(\vec{\omega}_j\times\vec{n})\cdot(\vec{\omega}_i\times\vec{S}_{C_i}+\vec{v}_i-\vec{\omega}_j\times\vec{S}_b-\vec{v}_j) \\ &-\vec{n}\cdot[\vec{\omega}_i\times(\vec{\omega}_i\times\vec{S}_{C_i})-\vec{\omega}_j\times(\vec{\omega}_j\times\vec{S}_b)]=\gamma \end{aligned} \tag{11}$$

where $\vec{a}_i$ and $\vec{a}_j$ are the translational, and $\vec{\alpha}_i$ and $\vec{\alpha}_j$ are the angular accelerations of bodies i and j respectively. Rearranging equation (11), one obtains

$$\vec{n}\cdot\vec{a}_i+(\vec{S}_{C_i}\times\vec{n})\cdot\vec{\alpha}_i-\vec{n}\cdot\vec{a}_j+[\vec{n}\times(\vec{r}_{C_i}-\vec{r}_b)-\vec{S}_b\times\vec{n}]\cdot\vec{\alpha}_j=\gamma(\vec{\omega}) \tag{12}$$

From the geometric analysis (Shabana, 1989), the following relations can be obtained

$$\vec{\omega}_k=2L_i\dot{e} \qquad k=i,j \tag{13}$$

$$\vec{\alpha}_k=2L_k\ddot{e} \qquad k=i,j \tag{14}$$

where L_i is defined as

$$L_i=\begin{pmatrix} -e_1 & e_0 & -e_3 & e_2 \\ -e_2 & e_3 & e_0 & e_1 \\ -e_3 & -e_2 & e_1 & e_0 \end{pmatrix} \tag{15}$$

Substituting equations (13) and (14) into equation (12), yields

$$\phi_q\ddot{q}=\begin{pmatrix} \vec{n} \\ 2L_i^T(\vec{S}_{a_i}\times\vec{n}) \\ -\vec{n} \\ 2L_j^T[\vec{n}\times(\vec{r}_{a_i}-\vec{r}_b)-\vec{S}_b\times\vec{n}] \end{pmatrix}^T \begin{pmatrix} \ddot{r}_i \\ \ddot{e}_i \\ \ddot{r}_j \\ \ddot{e}_j \end{pmatrix}=\gamma(\dot{e}) \tag{16}$$

where

$$\phi_q=\begin{pmatrix} \vec{n} \\ 2L_i^T(\vec{S}_{a_i}\times\vec{n}) \\ -\vec{n} \\ 2L_j^T\,[\vec{n}\times(\vec{r}_{a_i}-\vec{r}_b-\vec{S}_b\times\vec{n}] \end{pmatrix}^T \tag{17}$$

is called the constraint Jacobin and plays a critical role in the governing equations of motion of the system.

Sometimes two edges can collide with each other and form an edge-to-edge kinematic constraint joint. This edge-to-edge constraint joint is a special kind of point-to-surface constraint. Where the point is the intersection point between two edges, the surface is defined by two contact edges and the outward normal can be calculated using the following equation as shown in Figure 3.

$$\vec{n} = \vec{AB} \times \vec{CD} \tag{18}$$

where $\vec{AB}$ and $\vec{CD}$ are two contacting edges. In this instance, the contact point is changing with time due to the relative movement of the two edges.

3. COLLISION DETECTION

Collision detection involves determining when one body contacts with another, and the corresponding contact point(s). It is a computationally expensive process, especially when large number of components are involved and the components have complex shapes. For simplicity, only convex polyhedrons are considered in this chapter. However, complex components can be decomposed into a set of convex

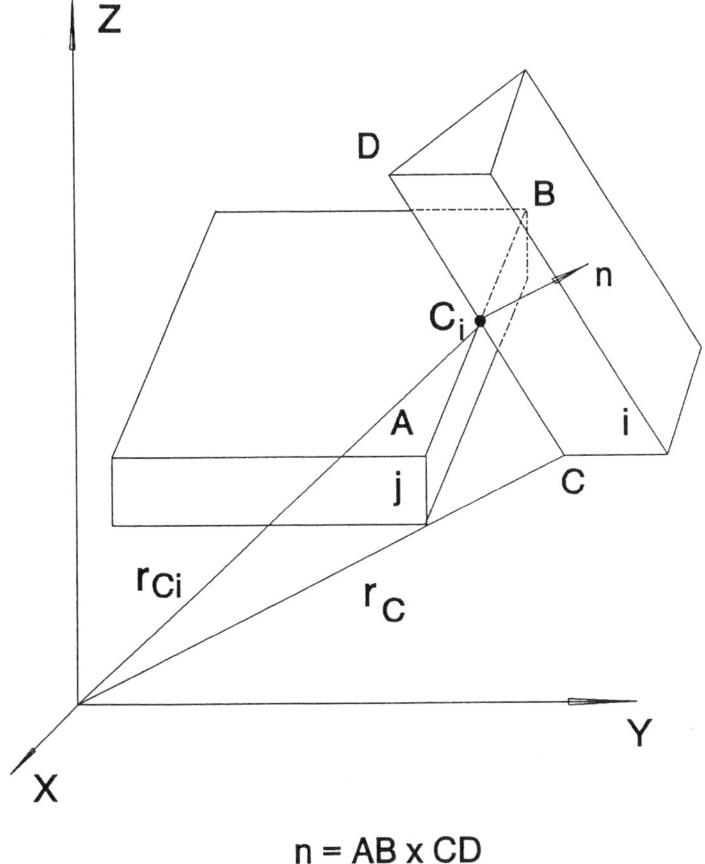

Fig. 3 An edge-to-edge constraint.

polyhedrons. Based on the direct minimization approach, Bobrow (1989) presented an optimization method for calculating the minimum distance between two bodies. When the minimum distance is less than a predefined threshold, two bodies are considered to be in contact and the corresponding two minimum points are selected as the contact points.

3.1. Gradient Projection Method

Let O_1 and O_2 be two convex polyhedron bodies defined respectively by l_1 and l_2 boundary surfaces. Object O_i can be designated as the collection of the points which satisfy

$$\langle (\boldsymbol{x} - \boldsymbol{x}_{bj}),\ \boldsymbol{n}_{ij} \rangle \quad \leq 0 \qquad j = 1, 2, \ldots, l_i \tag{19}$$

where $\langle\, , \rangle$ represents an inner product operator, $\mathbf{x}$ is an arbitrary point in R^3 space, $\mathbf{x}_{bj}$ is an arbitrary point on the boundary surface j, and $\mathbf{n}_{ij}$ is the outward normal of the surface j. Object i can also be represented as

$$O_i = \bigcap_{j=1}^{l_i} \{\boldsymbol{x} \mid \langle (\boldsymbol{x} - \boldsymbol{x}_{bj}), \boldsymbol{n}_{ij} \rangle \leq 0\} \tag{20}$$

Therefore, the minimum distance between two objects can be stated as follows:

Find $\mathbf{x}_1$ of O_1 and $\mathbf{x}_2$ of O_2 such that the objective function

$$f(\boldsymbol{x}_1, \boldsymbol{x}_2) = \frac{\|\boldsymbol{x}_1 - \boldsymbol{x}_2\|^2}{2} \tag{21}$$

is minimized while subjected to the linear constraints:

$$\begin{aligned} g_{ij}(\boldsymbol{x}_i) &= \langle (\boldsymbol{x}_i - \boldsymbol{x}_{bj}), \boldsymbol{n}_{ij} \rangle \leq 0 \\ j &= 1, 2, \ldots, l_i \qquad i = 1, 2 \end{aligned} \tag{22}$$

The above formulation defined by equations (21) and (22) is a nonlinear optimization problem. However, since the constraint equations are linear, the problem is solved by adopting the gradient projection method as described in the following.

For iteration $k+1$, the point $\mathbf{x}_{i,k+1}$ on each body is given by

$$\boldsymbol{x}_{i,k+1} = \boldsymbol{x}_{i,k} + \beta_i \boldsymbol{S}_{i,k} \tag{23}$$

where $\mathbf{x}_{i,k}$ is the starting point for the $(k+1)^{\text{th}}$ iteration, $i = 1$ and 2 for body 1 and body 2, respectively, β_i is the step length and $\mathbf{S}_{i,k}$ is the search direction.

The search direction is determined by projecting the gradient of the objective function onto the active constraints ($g_{ij} = 0$) at $\mathbf{x}_{i,k}$.

The necessary and sufficient conditions for $\mathbf{x}_{i,k}$ to be a global minimum is given by the Kuhn-Tucker Conditions (KTC) (Fox, 1971).

$$-\nabla f = \sum_{j=1}^{m_i} \alpha_{ij} \nabla g_{ij} \quad \alpha_{ij} \geq 0 \quad i = 1, 2 \tag{24}$$

where m_i is the number of active constraints, ∇f and $\nabla g_{i,j}$ are the gradients of f and g_{ij}, respectively, and α_{ij} is unknown. Furthermore, for polyhedrons,

$$\nabla g_{ij} \quad = \quad \boldsymbol{n}_{ij} \tag{25}$$

Geometrically, the KTC represent the case where the gradient line lies within the region defined by the surfaces which are normal to the boundary of the bodies as shown in Figure 4. For the specified objective function (21) and constraint equations (22), equation (25) can be simplified to

$$\boldsymbol{x}_{2,k} - \boldsymbol{x}_{1,k} = \sum_{j=1}^{m_1} \alpha_{1j} \boldsymbol{n}_{1j} \qquad \alpha_{1j} \geq 0 \tag{26}$$

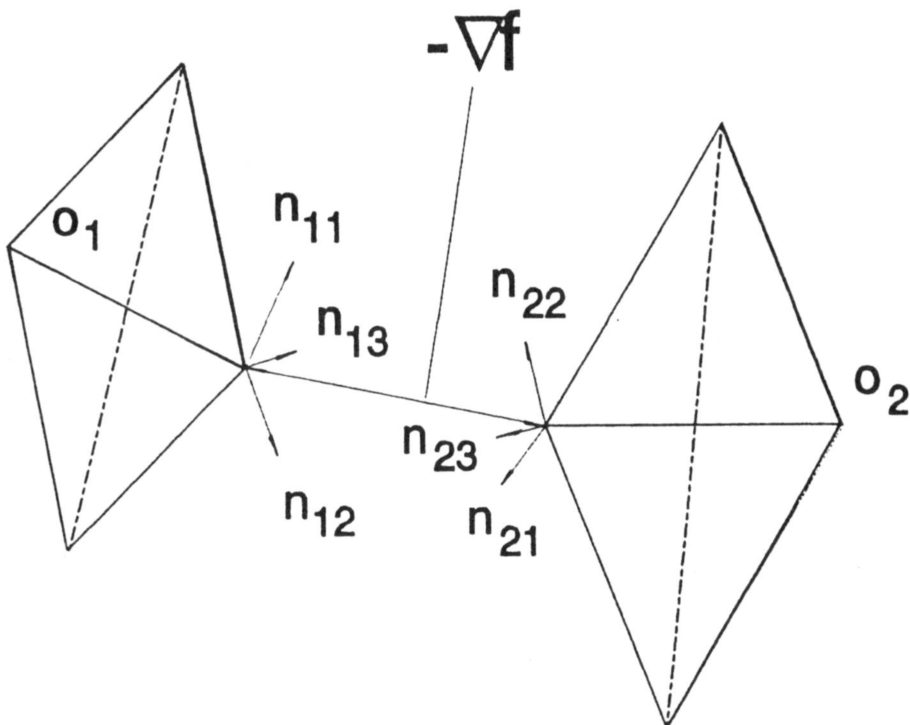

Fig. 4 Kuhn-Tucker conditions at a global minimum.

$$\boldsymbol{x}_{1,k} - \boldsymbol{x}_{2,k} = \sum_{j=1}^{m_2} \alpha_{2j} \boldsymbol{n}_{2j} \qquad \alpha_{2j} \geq 0 \tag{27}$$

3.2. Intersection Detection

In iteration k, a pair of points $(\mathbf{x}_{1,k}, \mathbf{x}_{2,k})$ which make up the minimum distance between two bodies is given. It is assumed that the two bodies are separate with each other at step k.

For iteration $k+1$, the pair $(\mathbf{x}_{1,k}, \mathbf{x}_{2,k})$ are used as the initial points. Before the new pair of points $(\mathbf{x}_{1,k+1}, \mathbf{x}_{2,k+1})$ are searched, the KTC conditions, equations (26) and (27) are used to see whether the two bodies are still separate or just contact each other.

If all the calculated coefficients, α_s, are less than zero, then two components are intersected with each other. The time step should be reduced until the two bodies are separated or contact with each other.

3.3. Algorithm for Minimum Distance Calculation

The steps used to implement the above algorithm are:

1. Find initial points $\mathbf{x}_{1,0}$ and $\mathbf{x}_{2,0}$
2. Determine the active constraints at $\mathbf{x}_{1,k}$ and $\mathbf{x}_{2,k}$.
3. Evaluate the KTC. If they are satisfied for both objects, the iteration is complete.
4. Check for intersection. If it is intersected, then set $k = k - 1$, $(\mathbf{x}_{1,k}, \mathbf{x}_{2,k}) = (\mathbf{x}_{1,k-1}, \mathbf{x}_{2,k-1})$ and time step $=$ time step/2. Go to step 2.
5. Compute new feasible search direction on each object by gradient projection if the KTC are not satisfied for both objects.
6. Determine the new minimum points along the new search directions.
7. Reset $(\mathbf{x}_{1,k}, \mathbf{x}_{2,k})$ to this new minimum and return to step 2.

4. COLLISION RESPONSE

In an advanced modeling system, the system itself must respond to a collision by determining new linear and angular velocities for the colliding components. These new velocities must conserve linear and angular momentums. The elasticity of the contacting surfaces must also be taken into account, as this determines how much the kinematic energy is lost in a collision.

Consider a rigid body system under an impact condition as shown in Figure 5. Point **a** of body i will collide with point $\mathbf{a}'$ on the surface of object j. $\mathbf{v_r}$ is the relative velocity of point **a** with respect to point $\mathbf{a}'$ and **n** is the outward normal of the surface. Let $x'y'z'$ be a surface fixed coordinate system with z' coincident with **n** and x' coincident with tangent relative velocity $\mathbf{v}_t$. The normal and relative tangent velocities can be calculated as

$$\boldsymbol{v}\boldsymbol{n}_n = (\boldsymbol{v}_r \cdot \boldsymbol{n})\boldsymbol{n} \tag{28}$$

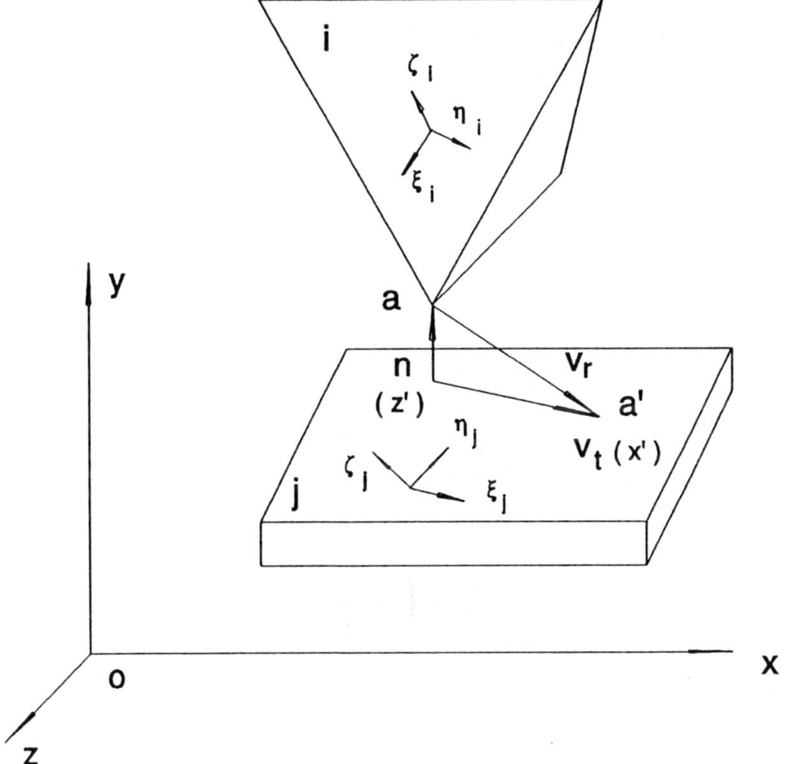

Fig. 5 Three-dimensional rigid bodies in collision.

$$\boldsymbol{v}_t = \boldsymbol{v}_r - (\boldsymbol{v}_r \cdot \boldsymbol{n})\boldsymbol{n} \tag{29}$$

As shown in Figure 4, the collision impulse are $R_{x'}$, $R_{y'}$ and $R_{z'}$,. Where $R_{z'}$ is the normal impulse and $R_{x'}$ and $R_{y'}$ are the tangential impulses. Using Coulomb's law, one obtains

$$R_{x'} = \mu \cos\alpha\, R_{z'} = \mu\, R_{z'} \tag{30}$$

$$R_{y'} = \mu \sin\alpha\, R_{z'} = 0 \tag{31}$$

where μ is the coefficient of the friction and α is the angle between $\mathbf{v_t}$ and x', which is zero in this case.

Let B represents the transformation matrix which transform the point form surface coordinate system $x'y'z'$ to global coordinate system xyz. Then the collision impulses along the global x, y, and z axis can be written as:

$$\begin{pmatrix} R_x \\ R_y \\ R_z \end{pmatrix} = R_{z'} B \begin{pmatrix} \mu \\ 0 \\ 1 \end{pmatrix} = R_{z'} \begin{pmatrix} r_x \\ r_y \\ r_z \end{pmatrix} \tag{32}$$

Let $\xi\eta\zeta$ represents the object fixed coordinate system and A represents the transformation matrix from the object fixed system to the global system. Then the impulses along the ξ, η, ζ axis of object j are

$$\begin{pmatrix} R_\xi \\ R_\eta \\ R_\zeta \end{pmatrix}_j = R_{z'} A_j^T B \begin{pmatrix} \mu \\ 0 \\ 1 \end{pmatrix} = R_{z'} \begin{pmatrix} r_\xi \\ r_\eta \\ r_\zeta \end{pmatrix}_j \tag{33}$$

For the object i, one can similarly obtain

$$\begin{pmatrix} R_\xi \\ R_\eta \\ R_\zeta \end{pmatrix}_i = -R_{z'} A_i^T B \begin{pmatrix} \mu \\ 0 \\ 1 \end{pmatrix} = R_{z'} \begin{pmatrix} r_\xi \\ r_\eta \\ r_\zeta \end{pmatrix}_i \tag{34}$$

The impulsive moments with respect to the object fixed coordinate system, $\xi\eta\zeta$, can be written as

$$\begin{pmatrix} M_\xi \\ M_\eta \\ M_\zeta \end{pmatrix}_i = \tilde{r}_{a_i} \times \begin{pmatrix} R_\xi \\ R_\eta \\ R_\zeta \end{pmatrix}_i = R_{z'} \begin{pmatrix} m_\xi \\ m_\eta \\ m_\zeta \end{pmatrix}_i \tag{35}$$

and

$$\begin{pmatrix} M_\xi \\ M_\eta \\ M_\zeta \end{pmatrix}_j = \tilde{r}_{a'_j} \times \begin{pmatrix} R_\xi \\ R_\eta \\ R_\zeta \end{pmatrix}_j = R_{z'} \begin{pmatrix} m_\xi \\ m_\eta \\ m_\zeta \end{pmatrix}_j \tag{36}$$

where

$$\tilde{r}_{a_i} = \begin{pmatrix} 0 & -a_\zeta & a_\eta \\ a_\zeta & 0 & -a_\xi \\ -a_\eta & a_\xi & 0 \end{pmatrix}_i \tag{37}$$

$$\tilde{r}_{a'_j} = \begin{pmatrix} 0 & -a'_\zeta & a'_\eta \\ a'_\zeta & 0 & -a'_\xi \\ -a'_\eta & a'_\xi & 0 \end{pmatrix}_j \tag{38}$$

If Euler parameters are selected as the generalized coordinate system, the generalized impulsive moments can be written as

$$\begin{pmatrix} M_{e_0} \\ M_{e_1} \\ M_{e_2} \\ M_{e_3} \end{pmatrix}_k = 2l_k^T \begin{pmatrix} M_\xi \\ M_\eta \\ M_\zeta \end{pmatrix}_k = R_{z'} \begin{pmatrix} m_{e0} \\ m_{e1} \\ m_{e2} \\ m_{e3} \end{pmatrix}_k \quad k = i,j \tag{39}$$

where l_k, a 3 x 4 transformation matrix (Shabana, 1989), is

$$l_k = \begin{pmatrix} -e_1 & e_o & -e_3 & e_2 \\ -e_2 & e_3 & e_0 & -e_1 \\ -e_3 & -e_2 & e_1 & e_0 \end{pmatrix}_k \quad k = i,j \tag{40}$$

It can be seen that the generalized impulsive forces and moments are only the function of the normal impulsive force $R_z{}'$. Therefore, the generalized impulsive forces can be written as

$$Q_i^I = R_{z'} d_i \text{ and } Q_j^I = R_{z'} d_j \tag{41}$$

where d_i and d_j are respectively

$$d_i^T = (-r_x, -r_y, -r_z, m_{e0}, m_{e1}, m_{e2}, m_{e3})^T \tag{42}$$

$$d_j^T = (r_x, r_y, r_z, m_{e0}, m_{e1}, m_{e2}, m_{e3})^T \tag{43}$$

Using the equation derived by Fang and Liou (1993), the velocity jumps can be written as

$$\begin{pmatrix} K_i & \phi_{qi}^T \\ \phi_{qi} & 0 \end{pmatrix} \begin{pmatrix} \dot{q}_i(t_I^+) \\ p^\lambda \end{pmatrix} = \begin{pmatrix} Q_i^I \\ 0 \end{pmatrix} + \begin{pmatrix} K_i & \phi_{qi}^T \\ \phi_{qi} & 0 \end{pmatrix} \begin{pmatrix} \dot{q}_i(t_I^-) \\ 0 \end{pmatrix} \tag{44}$$

where K_i is the mass matrix, ϕ_{q_i} is the constraint Jacobin, t_I^- and t_I^+ are the time just before and after the collision.

The constraint equation for the collision response, considering the experimentally obtained coefficient of restitution e, can be written as

$$(\boldsymbol{v}_i + \omega_i \times \boldsymbol{r_a} - \boldsymbol{v}_j - \omega_j \times \boldsymbol{r}_{a'}) \cdot \boldsymbol{n}|_{t=t_I^+} = -e(\boldsymbol{v}_i + \omega_i \times \boldsymbol{r}_a - v_j - \omega_j \times \boldsymbol{r}_{a'}) \cdot \boldsymbol{n}|_{t=t_I^-} \tag{45}$$

where $\mathbf{v}_i$ are the velocities of the mass center of the object i and

$$\omega_i = 2l_i \begin{pmatrix} \dot{e}_0 \\ \dot{e}_1 \\ \dot{e}_2 \\ \dot{e}_3 \end{pmatrix}_i \tag{46}$$

are the angular velocities of the object i. Substituting equation (46) into equation (45) and rearranging (45), one obtains

$$c\dot{q}(t_I^+) = -ec\dot{q}(t_I^-) \tag{47}$$

where c is the function of (e).

Combining equations (44) and (47), one obtains the equation to govern the collision response of the system

$$\begin{pmatrix} K_i & \phi_{q_i}^T \\ \phi_{q_i} & 0 \\ c & 0 \end{pmatrix} \begin{pmatrix} \dot{q}(t_I^+) \\ P^\lambda \end{pmatrix} = R_{z'} \begin{pmatrix} d \\ 0 \\ 0 \end{pmatrix} + \begin{pmatrix} K_i \\ \phi_{q_i} \\ -ec \end{pmatrix} \dot{q}(t_I^-) \tag{48}$$

Moving the first term in the right side of equation (48) to the left side and rearranging the equation, one obtains

$$\begin{pmatrix} K_i & \phi_{q_i}^T & d \\ \phi_{q_i} & 0 & 0 \\ c & 0 & 0 \end{pmatrix} \begin{pmatrix} \dot{q}(t_I^+) \\ p^\lambda \\ -R_{z'} \end{pmatrix} = \begin{pmatrix} K_i \\ \phi_{q_i} \\ -ec \end{pmatrix} \dot{q}(t_I^-) \tag{49}$$

5. PROTOTYPE SYSTEM DEVELOPMENT

5.1. Equations of Motion

Let q_i represent the generalized coordinates of the component i. q_i is defined as

$$q_i = (x, y, z, e_0, e_1, e_2, e_3)_i^T \tag{50}$$

where e_0, e_1, e_2, and e_3 are four Euler parameters which are used to represent the orientation of each component. Then the configuration of a manufacturing system can be represented by a set of such generalized coordinates $q = (q_1, q_2, \ldots, q_n)^T$ where n is the number of the components in the system.

The inertial characteristics of the system can be defined by the kinetic energy of the system

$$T = \sum \frac{1}{2} v_i^T m_i v_i + \sum \frac{1}{2} \omega_i^T I_i \omega_i \tag{51}$$

where the first term of the right side represents the translational kinetic energy and the second term represents the rotational energy of the system. $v_i = (\dot{x}, \dot{y}, \dot{z})_i^T$ and $\omega_i = (\omega_x, \omega_y, \omega_{z'})_i^T$ are, respectively, the translation and angular velocities of component i. m_i is the mass matrix of component i and I_i is the moment of inertia matrix of component i with respect to the local coordinate system. I_i is defined as

$$I_i = \begin{pmatrix} I_{\xi\xi} & I_{\xi\eta} & I_{\xi\zeta} \\ I_{\eta\xi} & I_{\eta\eta} & I_{\eta\zeta} \\ I_{\zeta\xi} & I_{\zeta\eta} & I_{\zeta\zeta} \end{pmatrix} \tag{52}$$

From kinematic analysis (Shabana, 1989), the angular velocity of component i, ω_i, can be represented by

$$\omega_i = 2\, L_i\, \dot{e}_i \tag{53}$$

where $\dot{e}_i = (\dot{e}_0, \dot{e}_1, \dot{e}_2, \dot{e}_3)_i$ and

$$L_i = \begin{pmatrix} -e_1 & e_0 & e_3 & -e_2 \\ -e_2 & -e_3 & e_0 & e_1 \\ -e_3 & e_2 & -e_1 & e_0 \end{pmatrix} \tag{54}$$

Substituting equation (53) into equation (51), yields

$$T = \sum \frac{1}{2} v_i^T m_i v_i + \sum \frac{1}{2} \dot{e}_i^T (4L_i^T I_i L_i) \dot{e}_i = \sum \frac{1}{2} \dot{q}_i^T M_i \dot{q}_i \tag{55}$$

where

$$M_i = \begin{pmatrix} m_i & 0 \\ 0 & 4L_i^T I_i L_i \end{pmatrix} \tag{56}$$

is a 7×7 matrix. It can be seen that the matrix M_i is the function of the generalized coordinate q_i even though the terms m_i and I_i are constant matrices. If the summation notation is neglected, equation (56) becomes

$$T = \frac{1}{2} \dot{q}^T M \dot{q} \tag{57}$$

where M is the mass matrix of the system and $\dot{q}$ is defined as

$$\dot{q} = (\dot{q}_1, \dot{q}_2, \ldots, \dot{q}_n)^T \tag{58}$$

Let us consider the case in which there are p external forces acting upon the manufacturing system. The virtual work done by the external forces are

$$\delta W = \sum_{j=1}^{p} F_j \delta r_j \tag{59}$$

where F_j is the external force and δr_j is the virtual displacement of the point the external force acts upon. The virtual displacement can be displayed in the form

$$\delta r_j = \sum_{k=1}^{n} \frac{\partial r_j}{\partial q_k} \delta q_k \tag{60}$$

Substituting equation (60) into equation (59), yields

$$\delta W = \sum_{k=1}^{n} \left(\sum_{j=1}^{p} F_j \frac{\partial r_j}{\partial q_k} \right) \delta q_k = \sum_{k=1}^{n} Q_k^e \delta q_k \tag{61}$$

where

$$Q_k^e = \sum_{j=1}^{p} F_j \frac{\partial r_j}{\partial q_k} \tag{62}$$

is the generalized external force.

The explicit form of the generalized external forces are (Shabana, 1989)

$$Q_i^e = \begin{pmatrix} F_i \\ 2L_i^T N_I \end{pmatrix} \tag{63}$$

where F_i is the total external force and N_i is the total external torque measured with respect to the local coordinate system acting upon body i.

In manufacturing systems of practical interest, components do not move independently, but are connected by form closed or force closed joints. Such constraint joints are defined by a set of algebraic constraint equations of the form

$$g_l(q_1, q_2, \ldots, q_n, t) = 0 \quad l = 1, 2, \ldots, m \tag{64}$$

It is assumed that each component $g_l(q)$ has two continuous derivatives with respect to its arguments.

Differentiating equation (64) obtains the velocity constraint equation as

$$\frac{\partial g_l}{\partial q_1}\dot{q}_1 + \frac{\partial g_l}{\partial q_2}\dot{q}_2 + \cdots + \frac{\partial g_l}{\partial q_n}\dot{q}_n + \frac{\partial g_l}{\partial t} = 0 \qquad l = 1, 2, \ldots, m \tag{65}$$

Here only holonomic catastatic constraints are considered. Equation (65) can be written in a matrix format as

$$g_q \dot{q} = -g_t \tag{66}$$

where g_q is a m by n matrix.

Differentiating equation (66) again leads to the acceleration constraint equation

$$g_q \ddot{q} = -((g_q \dot{q})_q \dot{q} + 2g_{tq} + g_{tt}) = \gamma \tag{67}$$

Equations (64), (66), and (67) comprise the system of kinematic equations of the manufacturing system.

The equations of motion of the system can be written using a Lagrangian formulation, with Lagrangian multiplier λ to account for constraint, in the form

$$\frac{d}{dt}\left(\frac{\partial T}{\partial \dot{q}}\right) - \frac{\partial T}{\partial q} + g_q^T \lambda = Q^e \tag{68}$$

where

$$\frac{\partial T}{\partial \dot{q}} = M\dot{q} \tag{69}$$

$$\frac{d}{dt}\left(\frac{\partial T}{\partial \dot{q}}\right) = \frac{dM}{dt}\dot{q} + M\ddot{q} = \dot{q}^T \frac{\partial M}{\partial q}\dot{q} + M\ddot{q} \tag{70}$$

$$\frac{\partial T}{\partial q} = \frac{1}{2}\dot{q}^T \frac{\partial M}{\partial q}\dot{q} \tag{71}$$

Substituting equations (69), (70), and (71) into (68), yields

$$M\ddot{q} + \dot{q}^T \frac{\partial M}{\partial q}\dot{q} - \frac{1}{2}\dot{q}^T \frac{\partial M}{\partial q}\dot{q} + g_q\lambda = Q^e \tag{72}$$

$$M\ddot{q} + g_q^T\lambda = Q^e - \frac{1}{2}\dot{q}^T \frac{\partial M}{\partial q}\dot{q} = Q^e + Q^* \tag{73}$$

where

$$Q^* = -\frac{1}{2}\dot{q}^T \frac{\partial M}{\partial q}\dot{q} \tag{74}$$

is the quadratic in terms of velocities whose coefficients are dependent on the generalized coordinates.

Combining equations (67) and (73), yields the governing equations of motion of the system

$$\begin{pmatrix} M & g_q^T \\ g_q & 0 \end{pmatrix}\begin{pmatrix} \ddot{q} \\ \lambda \end{pmatrix} = \begin{pmatrix} Q^e + Q^* \\ \gamma \end{pmatrix} \tag{75}$$

It is known that the Lagrangian multipliers, λ_s, in the above equation determine the reaction forces that act in the kinematic constraints of the system. For the specific point-to-surface constraint, which will be described later, λ represents reaction forces which are normal to the contact surfaces.

5.2. Friction Force

Recall that the Lagrangian multiplier, λ_i, in equation (75) represents the normal reaction force of the kinematic constraint g_1 as shown in Figure 6. Where (x', y', z') is a surface fixed coordinate system with the restriction that the z' axis must be coincident with the outward normal of the surface, and β is the angle between the x' axis and the relative velocity vector V_C of point C_i with respect to point C_j. If x' axis is chosen to coincide with the relative velocity vector, β is zero; otherwise, it is arbitrary.

From Coulomb's law, the frictional force can be expressed as

$$\begin{aligned} f_{x'} &= -\lambda\mu\ \cos\beta, \quad \text{and} \\ f_{y'} &= -\lambda\mu\ \sin\beta \end{aligned} \tag{76}$$

where μ is the coefficient of friction. Projecting the frictional forces into global coordinates yields

$$f = \begin{pmatrix} f_x \\ f_y \\ f_z \end{pmatrix} = B\begin{pmatrix} f_{x'} \\ f_{y'} \\ 0 \end{pmatrix} = \lambda\begin{pmatrix} f'_x \\ f'_y \\ f'_z \end{pmatrix} = \lambda f' \tag{77}$$

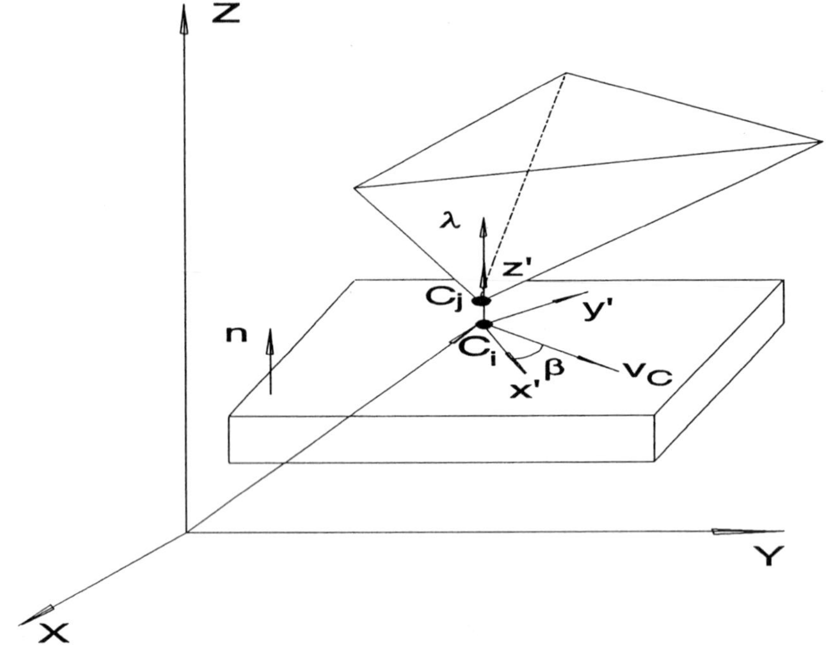

Fig. 6 A surface fixed coordinate system for friction analysis.

where

$$f'_k = -B_{k_1}\,\mu\cos\beta - B_{k_2}\mu\,\sin\beta \quad k = x, y, z \tag{78}$$

B is the transformation matrix from the surface fixed coordinate system (x', y', z') to the global coordinate system. The torques of the frictional forces, M_ξ, M_η, and M_ζ, can be calculated using the following equations

$$\begin{aligned}\begin{pmatrix} M_\xi \\ M_\eta \\ M_\zeta \end{pmatrix}_i &= \vec{r}'_{C_i} \times A_i^T f = \lambda \vec{r}\,'_{C_i} \times A_i^T f' \\ &= \lambda \begin{pmatrix} m_\xi \\ m_\eta \\ m_\zeta \end{pmatrix}_i = \lambda m'_i \end{aligned} \tag{79}$$

$$\begin{aligned}\begin{pmatrix} M_\xi \\ M_\eta \\ M_\zeta \end{pmatrix}_j &= \vec{r}'_{C_j} \times A_i^T f = \lambda \vec{r}\,'_{C_j} \times A_j^T f' \\ &= \lambda \begin{pmatrix} m_\xi \\ m_\eta \\ m_\zeta \end{pmatrix}_j = \lambda m'_j \end{aligned} \tag{80}$$

where $\vec{r}'_{C_i}$ and $\vec{r}'_{C_j}$ are the position vectors of the point C_i and C_j with respect to the bodies i and j respectively and

Thus the corresponding generalized frictional forces are

$$\begin{pmatrix} m_\xi \\ m_\eta \\ m_\zeta \end{pmatrix}_i = \vec{r}'_{C_i} \times A_i^T f' \tag{81}$$

$$Q_i^f = \begin{pmatrix} f \\ 2L_i^T M_i \end{pmatrix} = \lambda \begin{pmatrix} f' \\ 2L_i^T m_i \end{pmatrix} = \lambda d_i \tag{82}$$

$$Q_j^f = \begin{pmatrix} f \\ 2L_j^T M_j \end{pmatrix} = \lambda \begin{pmatrix} -f' \\ 2L_j^T m_j \end{pmatrix} = \lambda d_j \tag{83}$$

where

$$d_i = \begin{pmatrix} f' \\ 2L_i^T m'_i \end{pmatrix} \tag{84}$$

$$d_j = \begin{pmatrix} -f' \\ 2L_j^T m'_j \end{pmatrix} \tag{85}$$

The generalized frictional forces for the point-to-surface constraint g_l is

$$d_l = (0, \cdots, d_i, \cdots, d_j, \cdots, 0)^T \tag{86}$$

Substituting equation (83) into the equation (73) as part of Q^e, yields

$$M\ddot{q} + g_q^T \lambda = Q^e + Q^* + Q^f \tag{87}$$

$$M\ddot{q} + g_q^T \lambda = Q^e + Q^* + d\lambda \tag{88}$$

$$M\ddot{q} + (g_q^T - d)\lambda = Q^e + Q^* \tag{89}$$

where

$$d = (d_1, d_2, \cdots, d_m)^T \tag{90}$$

is an $n \times m$ matrix. Combining equation (67) and equation (89), yields

$$\begin{pmatrix} M & g_q^T - d \\ g_q & 0 \end{pmatrix} \begin{pmatrix} \ddot{q} \\ \lambda \end{pmatrix} = \begin{pmatrix} Q^e + Q^* \\ \gamma \end{pmatrix} \tag{91}$$

It can be seen that the coefficient matrix is not symmetric due to the existence of the frictional forces. The above equation is the governing equation of the system when there is a sliding condition between each point-to-surface constraint. If at time t_s, the relative velocity of a point-to-surface constraint goes to zero, a phenomenon called stiction occurs. During the stiction period, the friction forces may prevent the point to move tangen-tially. Therefore, the original point-to-surface constraint changes to a spherical constraint which can be represented by three point-to-surface constraints, i.e., the original constraint g_k can be represented by three new constraints g_{k1}, g_{k2}, and g_{k3}. Where the function of g_{k1} is the same as the original constraint, and the function of g_{k2} and g_{k3} is to prevent the point to move along x' axis and y' axis. After time t_s, the new system is integrated and $\lambda_{k1}, \lambda_{k2}$, and λ_{k3} are calculated. If the following equation can be satisfied,

$$\lambda_{k1} \geq \mu\sqrt{\lambda_{k2}^2 + \lambda_{k3}^2} \tag{92}$$

then the stiction is holding, otherwise, the stiction is destroyed and the original constraint system is used again.

The computational algorithm for Coulomb's friction and stiction is, therefore, to solve equation (91) as long as the relative velocity of each point-to-surface constraint is not zero. When the relative velocity goes to zero at time t_c, the original constraint g_k is represented by three new constraints g_{k1}, g_{k2}, and g_{k3}, and the system equation is integrated again. A check is made using equation (92) to determine whether or not stiction is holding or destroyed. If it is destroyed, then the old constraint g_k is used again.

5.3. Automatic Constraint Addition and Deletion

To easily handle the changing configuration of a general manufacturing assembly, the force closed point-to-surface constraint joint (Fang and Liou, 1993) is adopted. A general point-to-surface constraint joint can be expressed as

$$g_i(q,t) = n_x(x_C - x_b) + n_y(y_C - y_b) + n_z(z_C - z_b) = 0 \tag{93}$$

where (x_C, y_C, z_C) is the coordinate of point C with respect to the global coordinate system; (x_b, y_b, z_b) is the coordinate of an arbitrary point on the contact surface with respect to the global coordinate system; and $n = (n_x, n_y, n_z)^T$ is the outward normal of the contact surface. It has been shown (Fang and Liou, 1993) that any kind of constraint as expressed by equation (67) can be substituted by a set of point-to-surface constraints as expressed in equation (16).

During motion, two or more bodies may interpenetrate or separate, which is called a collision or a separation, respectively. If two bodies collide and remain in contact after collision, one or more point-to-surface constraints are formed. However, an impact in the system may cause bodies to separate due to the resulting impulsive motion. Two other reasons exist for a constraint breaking, when the contact point between two bodies ceases to be in compression, or when a point in contact slides off the surface.

Consider the motion of the mechanical assembly that is subjected to a set of constraints described in equation (16) prior to time t_0, at which a new/old point-to-surface constraint is detected to form/break. It is assumed that the active constraint equations are $g = (g_1, g_2, \ldots, g_m)^T$ before the event time t_0. At time t_0, a new/old constraint (g_n/g_o) is added/deleted into/from the system's constraint list. The new constraint list is then $g^* = g + g_n$ if a new contact is formed, and $g^* = g - g_o$ if an old contact is destroyed. After time t_0, the new constraint list is used in the equations of motion and the configurations, positions and velocities which may change due to collision, of the system are used as the initial conditions to integrate the equations of motion continuously. The method is very flexible in handling the topology changes of the system since the user is not required to predict any constraint change and the resulting topology.

6. CASE STUDY

In order to demonstrate the idea, a part-feeding system which consists of four main bodies: a conveyor, two feeders, and a parts are used an example. Part feeding systems are usually used before assembly or manufacturing to systematic feed the parts in certain orientation. Figure 7 shows the simulation series in three-dimensional view. The part, a rectangular block, is originally moving with the conveyor at the same velocity. The conveyor, as shown in Figure 7 is the shaded area moving to the right. The two feeders are oriented in angles with respect to the horizontal and vertical axes. After the contact between the part and feeders, the orientation of the part is changed and the part moves in a new orientation. The results also show that the final velocity of the part is the same as the conveyor due to friction.

Fig. 7a Simulation of a part in a part feeding system. (See Color Plate I.)

Fig. 7b (See Color Plate II.)

Fig. 7c (See Color Plate III.)

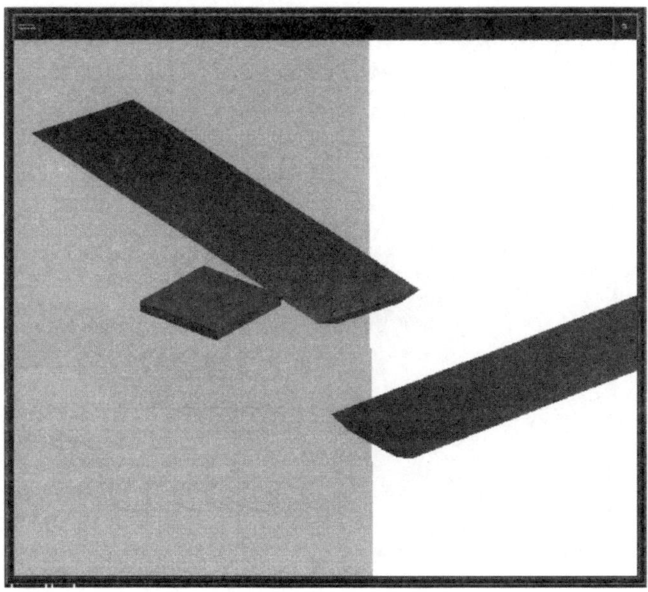

Fig. 7d (See Color Plate IV.)

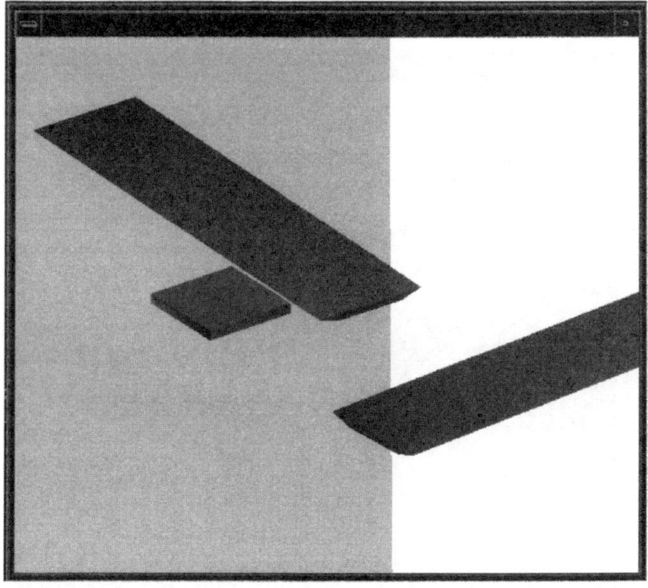

Fig. 7e (See Color Plate V.)

Fig. 7f (See Color Plate VI.)

Fig. 7g (See Color Plate VII.)

Fig. 7h (See Color Plate VIII.)

Fig. 7i (See Color Plate IX.)

Fig. 7j (See Color Plate X.)

Fig. 7k (See Color Plate XI.)

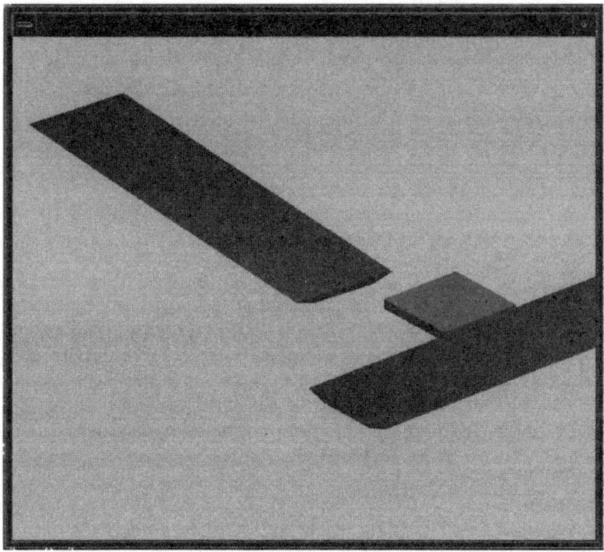

Fig. 71 (See Color Plate XII.)

7. CONCLUSION

Most of the manufacturing systems must be described by discontinuous governing equations of motion with the discontinuities caused by unforeseen kinematic changes. The constraint changes become difficult to foresee when the changes are due to the dynamics of the system. Frictional forces and stiction make the analysis even more complicated. It is impossible to predict the kinematic constraint changes before they really occur.

Using the force closure point-to-surface constraint joint model and the collision detection and response algorithms described above, the system can automatically detect the kinematic constraint changes and reformulate the equations of motion of a manufacturing system accordingly. No user interface is required. Even though only polyhedrons are considered here, any kind of body can be analyzed and simulated through the appropriate approximation. This system can be used as a virtual rapid prototyping tool for a general manufacturing system. A designer can create the geometric model of the desired manufacturing system, and the computer can automatically simulate and display the dynamic behavior of the system.

Acknowledgements

This research was supported by National Science Foundation Grant No. DDM-9210839, and the Missouri Department of Economic Development through the

MRTC grant. Their financial support is greatly appreciated. The authors are also would like to thank Richard Meyer for reviewing this manuscript.

References

J. E. Bobrow, A direct minimization approach for obtaining the distance between convex polyhedra. *The International Journal of Robotics Research*, **8**, pp. 65–76 (1989).

Y. Fang and F.W. Liou, Computer simulation of three dimensional mechanical assemblies-part 1: general formulation. *Proceedings of ASME International Computers in Engineering Conference*, pp. 579–587 (1993).

Y. Fang, Geometric Modeling and Computer Simulation of Mechanical Assemblies with Changing Topologies, Ph.D Thesis, University of Missouri-Rolla, Feb (1995).

R.L. Fox, *Optimization Methods for Engineering Design*, Addison-Wesley, Reading, MA (1971).

B.J. Gilmore and R.J. Cipra, Simulation of planar dynamic mechanical systems with changing topologies-part 1: characterization and prediction of the kinematic constraint changes. *Journal of Mechanical Design*, **113**, pp. 70–76 (1991).

B.J. Gilmore and R. J. Cipra, Simulation of planar dynamic mechanical systems with changing topologies-part 2: implementation strategy and simulation results for example dynamic systems, *Journal of Mechanical Design*, **113**, pp. 77–83 (1991).

I. Han, B.J. Gilmore and M.M. Ogot, The incorporation of arc boundaries and stick/slip friction in a rule-based simulation algorithm for dynamic mechanical systems with changing topologies. *Journal of Mechanical Design*, **15**, pp. 423–434 (1993).

E.J. Haug, *Computer Aided Kinematics and Dynamics of Mechanical Systems Volume I: Basic Methods*, Allyn and Bacon, MA (1989).

E.J. Haug, S.C. Wu and S.M. Yang, Dynamics of mechanical systems with coulomb friction, stiction, impact and constraint addition-deletion-I. *Mechanism and Machine Theory*, **21**(5), pp. 401–406 (1986).

P.E. Nikravesh, *Computer Aided Analysis of Mechanical Systems*, Prentice-Hall, Inc., Englewood Cliffs, NJ (1988).

B. Paul, Analytical dynamics of mechanisms – a computer oriented overview. *Mechanism and Machine Theory*, **10**, pp. 481–507 (1975) .

F. Reuleaux, *The Kinematics of Machinery*, MacMillan and Co., London (1876).

A.A. Shabana, *Dynamics of Multi-body Systems*, John Wiley & Sons (1989).

D.A. Smith, Reaction Forces and Impact in Generalized Two-Dimensional Mechanical Systems, Ph.D Thesis, University of Michigan, Sept (1971).

R.A. Wehage and E.J. Haug, Dynamic analysis of mechanical systems with intermittent motion. *Journal of Mechanical Design*, **104**, pp. 778–784 (1982).

J. Wittenburg, *Dynamic of System of Rigid Bodies*, B.G. Teubner. Stuttgart (1977).

Index

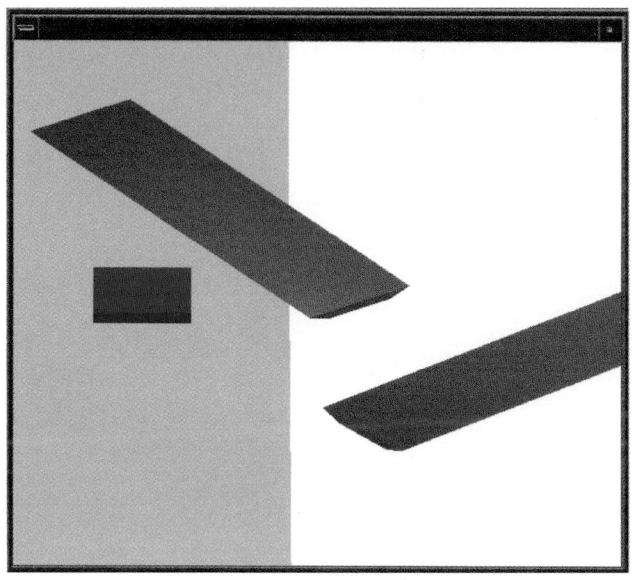

COLOUR PLATE I See Yong Fang and F.W. Liou, Figure 7a, Page 357.

COLOUR PLATE II See Yong Fang and F.W. Liou, Figure 7b, Page 358.

COLOUR PLATE III See Yong Fang and F.W. Liou, Figure 7c, Page 358.

COLOUR PLATE IV See Yong Fang and F.W. Liou, Figure 7d, Page 359.

COLOUR PLATE V See Yong Fang and F.W. Liou, Figure 7e, Page 359.

COLOUR PLATE VI See Yong Fang and F.W. Liou, Figure 7f, Page 360.

COLOUR PLATE VII See Yong Fang and F.W. Liou, Figure 7g, Page 360.

COLOUR PLATE VIII See Yong Fang and F.W. Liou, Figure 7h, Page 361.

COLOUR PLATE IX See Yong Fang and F.W. Liou, Figure 7i, Page 361.

COLOUR PLATE X See Yong Fang and F.W. Liou, Figure 7j, Page 362.

COLOUR PLATE XI See Yong Fang and F.W. Liou, Figure 7k, Page 362.

COLOUR PLATE XII See Yong Fang and F.W. Liou, Figure 7l, Page 363.